Progress in Optical Science and Photonics

Volume 38

Indexed by Scopus

The purpose of the series Progress in Optical Science and Photonics is to provide a forum to disseminate the latest research findings in various areas of Optics and its applications. The intended audience are physicists, electrical and electronic engineers, applied mathematicians, biomedical engineers, and advanced graduate students.

Balaji Ramachandran · Ragini Singh ·
Santosh Kumar
Editors

Advanced Optical Sensors

Noble Metal-Based Hybrid Composites

Springer

Editors
Balaji Ramachandran
Department of Electronics
and Communication Engineering
Centre of Excellence for Nanotechnology
Koneru Lakshmaiah Education Foundation
Guntur, Andhra Pradesh, India

Ragini Singh
Department of Biotechnology
Koneru Lakshmaiah Education Foundation
Guntur, Andhra Pradesh, India

Santosh Kumar
Department of Electronics
and Communication Engineering
Centre of Excellence for Nanotechnology
Koneru Lakshmaiah Education Foundation
Guntur, Andhra Pradesh, India

ISSN 2363-5096 ISSN 2363-510X (electronic)
Progress in Optical Science and Photonics
ISBN 978-981-95-1686-5 ISBN 978-981-95-1687-2 (eBook)
https://doi.org/10.1007/978-981-95-1687-2

This Springer imprint is published by the registered company Springer Nature Singapore Pte Ltd.
The registered company address is: 152 Beach Road, #21-01/04 Gateway East, Singapore 189721, Singapore

Preface

The field of optical sensing has witnessed a transformative evolution with the integration of nanotechnology, particularly through the utilization of noble metal-based hybrid composites. This book, *Advanced Optical Sensors: Noble Metal-Based Hybrid Composites*, aims to provide a comprehensive understanding of how noble metals such as gold, silver, and platinum—celebrated for their remarkable optical, electronic, and catalytic properties—are revolutionizing modern sensing technologies. With growing demands for highly sensitive, real-time, and miniaturized sensors in sectors like healthcare, environmental monitoring, energy, and defense, noble metal nanostructures have emerged as pivotal components. Their unique phenomena, including surface plasmon resonance (SPR), surface-enhanced Raman scattering (SERS), and exceptional electron mobility, make them ideal for next-generation sensing applications. This book adopts an interdisciplinary approach, bridging materials science, electronics, photonics, biotechnology, and machine learning to explore the design, fabrication, characterization, and practical applications of noble metal-based hybrid sensing platforms. Beginning with foundational principles, it delves into diverse types of noble metal nanostructures and their functional enhancements when combined with materials such as transition metal oxides and 2D materials. The chapters place a strong emphasis on how these hybrid structures enhance the performance of devices like plasmonic FET biosensors, nanogenerators for energy harvesting, gas sensors, photodetectors, and solar cells.

In addition, the book explores the integration of artificial intelligence and machine learning with optical sensors, enabling intelligent signal analysis, predictive modeling, and real-time optimization. Emerging technologies including self-powered sensors, metamaterials, and spintronic devices are also examined in the context of noble metal hybrid composites. Designed to serve as a valuable resource for researchers, engineers, and graduate students in the fields of nanotechnology, materials science, photonics, and electronic engineering, this book blends foundational knowledge with cutting-edge research. Through collaborative contributions from leading experts, it aspires to inspire innovation and guide future advancements in the rapidly evolving domain of optical sensor technology.

Organization of This Book

This book is structured into eleven chapters, each sequentially building from basic concepts to advanced technological applications, offering a logical and in-depth exploration of noble metal-based hybrid composites in optical sensors. Chapter "Synthesis, Properties and Significance of Noble Metals Used for Optoelectronics and Sensors" introduces the synthesis, key properties, and significance of noble metals such as gold, silver, and platinum, particularly highlighting their roles in optoelectronic and sensing devices. Chapter "Probing Noble Metal-Based Hybrid Nanostructures for SERS Opti-sensing" focuses on the design and application of noble metal-based hybrid nanostructures for Surface-Enhanced Raman Spectroscopy (SERS), emphasizing substrate reusability, synthesis strategies, and cost-efficiency. Chapter "Noble-Metal-Based Composites for High-Performance SPR Sensors" delves into the development of high-performance surface plasmon resonance (SPR) sensors using noble metal composites, showcasing their capabilities in real-time, label-free detection across various fields, including biomedical diagnostics and environmental monitoring.

Chapter "Recent Advance in Plasmonic FET Biosensors Using Noble Metal-Based Composites" explores recent advancements in plasmonic field-effect transistor (FET) biosensors, particularly the integration of noble metal nanostructures to enhance device sensitivity and selectivity. Chapter "Noble Metal-Based Hybrid Nanogenerators for Sustainable Energy and Optoelectronic Innovation" discusses noble metal-based hybrid nanogenerators and their significance in sustainable energy solutions and optoelectronic applications, particularly through triboelectric and piezoelectric mechanisms. Building upon this, Chapter "Metal Oxide-Enhanced TENG/PENG Nanogenerators for Self-Powered Optoelectronic Devices" examines the synergistic role of metal oxides in enhancing nanogenerator-based devices for self-powered optoelectronics, with potential applications in wearable and adaptive technologies.

Chapter "Detection of Nitrous Oxide Gas Using 2D Molybdenum Diselenide Monolayer" presents a detailed investigation into gas sensing using two-dimensional materials, specifically focusing on the detection of nitrous oxide (N_2O) using molybdenum diselenide ($MoSe_2$) monolayers, supported by theoretical modeling and analysis of electronic and optical changes. Chapter "Improvement of Solar Cell Performance Using Plasmonics Nanoparticles and Metamaterials" shifts the focus to solar energy, examining how plasmonic nanoparticles and metamaterials can significantly enhance solar cell efficiency by improving light absorption and carrier dynamics in both thin and thick silicon structures. Chapter "Integrating ML Approaches for Photodetectors" introduces machine learning approaches in the design and optimization of photodetectors, discussing their role in material selection, performance prediction, and device miniaturization. Finally, Chapter "Next Generation Optical Sensor Devices" encapsulates the current state and future outlook of next-generation optical sensor devices, highlighting advanced micro/nano-structured sensors, their multifaceted applications, and the technological challenges and opportunities that lie ahead. Finally, Chapter "Advanced 2D Nanomaterials for Remediation of Environmental Pollutants" explores the growing role of two-dimensional (2D) materials in addressing critical environmental challenges such as water contamination and air pollution. It highlights the unique physicochemical properties of materials like graphene, boron nitride, molybdenum disulfide, and black phosphorus, which make them ideal candidates for the removal of heavy metals, dyes, oil, and particulate matter. The chapter also reviews recent advances in air filtration and water treatment technologies, comparing the effectiveness of 2D materials with conventional methods. Challenges and future prospects for their large-scale environmental applications are also discussed.

Each chapter, authored by domain experts, incorporates recent research findings, case studies, technical illustrations, and forward-looking insights. Together, the chapters provide a cohesive narrative that captures both the scientific fundamentals and the emerging innovations in noble metal-based optical sensing technologies.

Guntur, India

Balaji Ramachandran
Ragini Singh
Santosh Kumar

Acknowledgments

The authors and editors of *Advanced Optical Sensors: Noble Metal-Based Hybrid Composites* would like to extend their heartfelt gratitude to all those who played a significant role in the successful completion of this book. This work represents the culmination of the efforts, support, and encouragement of many individuals and institutions, to whom we are deeply indebted.

First and foremost, we express our sincere thanks to Springer Nature Press for their trust in this project and for providing a global platform to share this important work with the scientific community. We are especially grateful to the editorial team for their professionalism, guidance, and commitment throughout the publication process. Their attention to detail and constructive feedback have helped ensure the highest quality of this volume.

We would also like to thank our contributing authors and colleagues whose expertise, insights, and collaborative spirit significantly enriched the content of this book. Their valuable contributions and stimulating discussions were instrumental in shaping the scope and depth of each chapter.

We would like to express our profound gratitude to Shri Koneru Satyanarayana Garu, the Honorable President of Koneru Lakshmaiah Education Foundation (Deemed to be University), Vaddeswaram, Andhra Pradesh, India, for his unwavering encouragement and institutional support during the writing of this book.

Dr. Balaji Ramachandran is also deeply thankful to his wife, Dr. Vyoma Singh, and daughter, Abhrisha Singh Balaji, for their endless support, patience, and motivation, as well as to his loving parents for their blessings.

Dr. Ragini Singh and Dr. Santosh Kumar would like to jointly express their heartfelt appreciation to their son, Ayaansh Singh, whose love, support, and understanding have been a constant source of strength throughout this journey. They are also deeply grateful to their beloved parents for their enduring love, blessings, and encouragement, which have guided them every step of the way.

This book is a reflection of the shared dedication, hard work, and passion of everyone involved. We hope that it will serve as a valuable resource for researchers, scholars, and students, and inspire further advancements in the field of optical sensing.

Balaji Ramachandran
Ragini Singh
Santosh Kumar

About This Book

Advanced Optical Sensors: Noble Metal-Based Hybrid Composites presents a comprehensive exploration of the next-generation sensor technologies enabled by noble metal nanostructures and hybrid composites. With a strong emphasis on gold, silver, and platinum-based materials, this book brings together cutting-edge research across fields such as surface-enhanced Raman scattering (SERS), surface plasmon resonance (SPR), plasmonic field-effect transistors (FETs), and sustainable nanogenerators.

Structured across ten well-curated chapters, the volume delves into the synthesis, properties, and wide-ranging applications of noble metal-based materials in optoelectronics and biosensing. It discusses innovative fabrication techniques, hybridization with materials like TMOs, TMDCs, and 2D monolayers, and the integration of advanced concepts like artificial intelligence (AI) and machine learning (ML) for predictive and adaptive sensing.

Key topics include:

- Noble metal nanostructures for high-performance SERS and SPR sensors
- Hybrid plasmonic FET biosensors for medical diagnostics
- TENG- and PENG-based nanogenerators for self-powered optoelectronic devices
- Optical gas sensing using 2D materials like $MoSe_2$ monolayers
- Solar energy enhancement through plasmonic nanoparticles and metamaterials
- Machine learning integration in photodetector design and performance optimization.

This book also presents a forward-looking perspective on next-generation optical sensors, addressing challenges such as material cost, scalability, and interdisciplinary integration. Rich in technical detail and practical insights, this work is ideal for researchers, academics, and engineers in materials science, nanotechnology, sensor development, and applied photonics.

Contents

Editors and Contributors

About the Editors

Dr. Balaji Ramachandran earned his master's in nanoscience and technology in 2016 at Anna University, India, and his Ph.D. in chemical engineering in 2022 at National Taipei University of Technology, Taiwan. He was a postdoctoral researcher at National Taiwan University, Taiwan, working on a printing technology for health care devices. He currently holds a full-time assistant professor position at Koneru Lakshmaiah Education Foundation (KLEF), India. He has published 90+ research articles in international Scopus and Web of Science-indexed journals. His research areas include SERS, electrochemical sensors, and bioinspired surfaces. He is also a reviewer for various leading journals.

Dr. Ragini Singh obtained her Ph.D. degree in the nanobiotechnology area from Ahmedabad University in 2018. She is working as an associate professor in the Department of Biotechnology at Koneru Lakshmaiah Education Foundation (KLEF), India, and has more than five years of research and teaching experience. Notably, she successfully completed the Shandong provincial fund project, China (ZR2020QC061), in 2023. Her expertise encompasses the synthesis of diverse nanomaterials and nanocomposites as well as the establishment of their application for sensing and therapy purposes. In her research journey, she has worked on the development of several optical-based sensors (colorimetric and surface plasmon resonance), lateral flow assay-based sensors, and photothermal-based sensors. Her scholarly contributions in terms of several book chapters (Royal Society of Chemistry, Elsevier, Springer) and more than 150 research papers (h-index 50) feature prominently in reputed journals like *American Chemical Society*, *Biosensors and Bioelectronics* and *IEEE Journals* focusing on various biomolecule detection, cancer cell analysis, and microbial detection. She has also published 10 patents till now. She serves as a potential reviewer for reputed journals like *IEEE Journals*, *Biosensors and Bioelectronics* and *Nanoscale*. She is a senior member of IEEE and a lifetime member of the Materials Research Society of India (MRSI) and the Indian Society of Nanomedicine.

Dr. Santosh Kumar is a professor in the Department of Electronics and Communication Engineering at Koneru Lakshmaiah Education Foundation (KLEF), India. He also serves as the Lab Head of the Centre of Excellence for Nanotechnology (CoE-NANO), Group Head of the Nanotechnology Research Group (NTRG), and Advisor—R&D at KLEF. He earned his Ph.D. from IIT (ISM) Dhanbad in 2014 and was previously with Liaocheng University, China (2018–2023). Recognized among the world's top 2% scientists by Stanford University for six consecutive years, Dr. Kumar has supervised 20 M.Tech. dissertations and nine Ph.D. candidates. He has published over 550 research articles in prestigious SCI journals, including 130 IEEE journal papers, with more than 13,000 citations and an h-index of 66. His 130 papers are recognized under various Sustainable Development Goals (SDGs) according to Scopus. His research has been featured in leading journals such as *Biosensors and Bioelectronics*, *Journal of Lightwave Technology*, *Optics Express*, *Optics Letters*, *Applied Physics Letters*, *ACS Applied Nano Materials* and *Biosensors*. His expertise spans fiber optic sensors, photonics and plasmonic devices, nano and biophotonics, waveguides, interferometers, and the Internet of Things. He has authored seven scholarly books with CRC Press (Taylor & Francis), IEEE-Wiley, and Springer Nature. Dr. Kumar was recently granted a patent for his groundbreaking optical fiber sensing technology and has published 16 patents on biosensing applications. A highly regarded expert, he has reviewed over 2800 manuscripts for top-tier SCI journals published by IEEE, Elsevier, Springer, OPTICA, SPIE, and *Nature*. His collaborative research extends across institutions in India, China, Portugal, Iraq, Kazakhstan, Brazil, France, and Italy.

Dr. Kumar is a fellow of SPIE and IETE, a life fellow of the Optical Society of India (OSI), and a senior member of IEEE, OPTICA, and SPIE. He is an OPTICA Traveling lecturer and serves as the chair of the Optica Optical Biosensor Technical Group, USA. He is also an Associate Editor for *Scientific Reports (Nature)*, *IEEE Internet of Things Journal*, and *Biomedical Optics Express*. His recent accolades include the Best Performing Associate Editor Award from *IEEE Sensors Journal, USA*, the Best IEEE JLT Paper Award from USA, and the Foreign Distinguished Professor Award from China.

Contributors

Manoj Bhatnagar Materials and Nano Engineering Research Laboratory, Department of Physics, School of Physical Sciences, DIT University, Dehradun, India

Sayantani Bhattacharya Department of ECE, College of Engineering and Technology, Faculty of Engineering and Technology, SRM Institute of Science and Technology, Kattankulathur, Chennai, Tamil Nadu, India

Surjeet Chahal Department of Physics, University Centre for Research and Development, Chandigarh University, Gharuan, Mohali, India

Naren Das Department of Electronics and Communication Engineering Department, Birla Institute of Technology, Mesra, Patna Campus, Patna, India

Debajit Deb Department of ECE, Koneru Lakshmaiah Education Foundation, Vaddeswaram, AP, India

Sukanya Ghosh Department of ECE, College of Engineering and Technology, Faculty of Engineering and Technology, SRM Institute of Science and Technology, Kattankulathur, Chennai, Tamil Nadu, India

Praveenkumar Hiremath Mechanics, Materials and Component Design, Lund University, Lund, Sweden

M. A. Ibrar Jahan Department of ECE, RNS Institute of Technology, Bengaluru, Karnataka, India

Dharmendra Kumar Department of Electronics and Communication Engineering, Guru Ghasidas Vishwavidyalaya Bilaspur, Bilaspur, Chhattisgarh, India

Parveen Kumar Materials and Nano Engineering Research Laboratory, Department of Physics, School of Physical Sciences, DIT University, Dehradun, India

Ravi Ranjan Kumar Department of Electronics and Communication Engineering, Motilal Nehru National Institute of Technology, Allahabad, Prayagraj, India

Priyatam Kumar Department of Electronics and Communication Engineering, Shri Phanishwar Nath Renu Engineering College, Araria, India

Ritesh Kumar Department of Electronics and Communication Engineering, Shri Phanishwar Nath Renu Engineering College, Araria, India

Santosh Kumar Centre of Excellence for Nanotechnology, Department of Electronics and Communication, Engineering, Koneru Lakshmaiah Education Foundation, Vaddeswaram, Andhra Pradesh, India

Shatrughna Kumar Department of Physical Electronics, Faculty of Engineering, School of Electrical Engineering, Centre for Light-Matter Interaction, Tel Aviv University, Tel Aviv, Israel;
School of Engineering and Sciences, Department of ECE, SRM University, Amaravati, Andhra Pradesh, India

Sarita Kumari Department of ECE, Amity University, Ranchi, Jharkhand, India

Smita Kumari Department of Basic and Applied Sciences, School of Engineering and Sciences, G D Goenka University, Gurugram, Haryana, India

J. Lakshmi Prasanna Department of Electronics and Communication Engineering, Koneru Lakshmaiah Education Foundation, Gajularamaram, Hyderabad, Telangana, India

Boris A. Malomed Department of Physical Electronics, Faculty of Engineering, School of Electrical Engineering, Centre for Light-Matter Interaction, Tel Aviv University, Tel Aviv, Israel

Kaliyannan Manojkumar Center for Flexible Electronics, Department of Electronics and Communication Engineering, Koneru Lakshmaiah Education Foundation, Vaddeswaram, India

Neha Mishra Department of Electronics and Communication Engineering, Madan Mohan Malaviya University of Technology, Gorakhpur, India

Mukilan Muthuramalingam Center for Flexible Electronics, Department of Electronics and Communication Engineering, Koneru Lakshmaiah Education Foundation, Vaddeswaram, India

Debarati Nath Department of ECE, College of Engineering and Technology, Faculty of Engineering and Technology, SRM Institute of Science and Technology, Kattankulathur, Chennai, Tamil Nadu, India

Bramha P. Pandey Department of Electronics and Communication Engineering, Madan Mohan Malaviya University of Technology, Gorakhpur, India

Deepak Punetha Department of Electronics and Communication Engineering, Motilal Nehru National Institute of Technology, Allahabad, Prayagraj, India

Amit Ranjan Department of Electronics and Communication Engineering, Shri Phanishwar Nath Renu Engineering College, Araria, India

Arijit Bardhan Roy Department of Electronics and Communication Engineering, SRM Institute of Science and Technology, Kattankulathur, Chengalpattu, Tamil Nadu, India

Moumita Saha Department of Botany, Shri Gnanambica Degree College, Madanapalle, Andhra Pradesh, India

P. Saleem Akram Department of Electronics and Communication Engineering, Koneru Lakshmaiah Education Foundation (Deemed to Be University), Guntur, India

Chella Santhosh Department of Electronics and Communication Engineering, Center of Excellence for Nanotechnology, Koneru Lakshmaiah Education Foundation, Vaddeswaram, Guntur, Andhra Pradesh, India

Armaan Siddiqui Department of Electronics and Communication Engineering Department, Birla Institute of Technology, Mesra, Patna Campus, Patna, India

Ragini Singh Department of Biotechnology, Koneru Lakshmaiah Education Foundation, Vaddeswaram, Andhra Pradesh, India

P. S. Srinivasa Babu Center for Flexible Electronics, Department of Electronics and Communication Engineering, Koneru Lakshmaiah Education Foundation, Vaddeswaram, India

R. Srujana Department of ECE, Jain University, Bengaluru, Karnataka, India

Shriya Subramanyam Materials and Nano Engineering Research Laboratory, Department of Physics, School of Physical Sciences, DIT University, Dehradun, India

Arunmetha Sundaramoorthy Center for Flexible Electronics, Department of Electronics and Communication Engineering, Koneru Lakshmaiah Education Foundation, Vaddeswaram, India

Venkateswaran Vivekananthan Center for Flexible Electronics, Department of Electronics and Communication Engineering, Koneru Lakshmaiah Education Foundation, Vaddeswaram, India

Archana Yadav Department of ECE, Integral University, Lucknow, Uttar Pradesh, India

Abbreviations

2D	Two-Dimensional
Ag	Silver
AI	Artificial Intelligence
ALD	Atomic Layer Deposition
Au	Gold
CMOS	Complementary Metal-Oxide-Semiconductor
CO	Carbon Monoxide
Eads	Adsorption Energy
EBL	Electron Beam Lithography
FET	Field-Effect Transistor
IoT	Internet of Things
LSPR	Localized Surface Plasmon Resonance
MAE	Magnetic Anisotropy Energy
MDS	Magnetically Dilute Semiconductor
ML	Monolayer/Machine Learning (context-dependent)
Mo	Molybdenum
$MoSe_2$	Molybdenum Diselenide
N_2O	Nitrous Oxide
NIL	Nanoimprint Lithography
NPs	Nanoparticles
PD	Photodetector
PDOS	Partial Density of States
PENG	Piezoelectric Nanogenerator
PL	Photoluminescence
PV	Photovoltaic
QT	Charge Transfer
SERS	Surface-Enhanced Raman Scattering
Si	Silicon
SPR	Surface Plasmon Resonance
TDOS	Total Density of States
TENG	Triboelectric Nanogenerator

TMDC	Transition Metal Dichalcogenide
TMO	Transition Metal Oxide
UV-Vis	Ultraviolet–Visible Spectroscopy
μB	Bohr Magneton

Synthesis, Properties and Significance of Noble Metals Used for Optoelectronics and Sensors

J. Lakshmi Prasanna and Chella Santhosh

Abstract The remarkable features of noble metal nanostructures have sparked a great deal of research attention in recent years. Gold and silver nanoparticles and nanorods are examples of noble metal nanoparticles that have special photonic, electrical, and catalytic qualities. Since they link the internal molecular circuits, they have a wide range of electrical and technological uses in the electromechanical system. Noble metal nanostructures have the unusual ability to adjust their properties based on the size, shape, and surface composition of the crystal arrangement. When produced under carefully regulated morphological conditions, noble metal nanostructures have a wide range of applications. A fascinating new field with enormous potential is the controlled production of noble metal nanostructures with particular shapes and low toxicity. Here, we examine noble metal nanostructures and their uses in a variety of industries, including photothermal therapy, clean energy, electrocatalysis, infrared biosensing, bioimaging, and more. The new area of use for noble metal nanostructures is surface-enhanced Raman scattering, or SERS. Although the application areas are very broad, we primarily concentrate on biosensing, SERS (Surface Enhanced Raman Scattering), and the most recent developments in these domains. The rapidly expanding field of noble metal nanostructure research will pave the way for the development of opto-electronics and sensors in the present era.

Keywords Nanomaterials · Noble metals · Synthesis · Characterization · Opto-electronic sensors

J. Lakshmi Prasanna (✉)
Department of Electronics and Communication Engineering, Koneru Lakshmaiah Education Foundation, Bowrampet Campus, Gajularamaram, Hyderabad, Telangana, India
e-mail: lakshmiprasannanewmail@gmail.com

C. Santhosh
Department of Electronics and Communication Engineering, Center of Excellence for Nanotechnology, Koneru Lakshmaiah Education Foundation, Vaddeswaram, Guntur, Andhra Pradesh, India

B. Ramachandran et al. (eds.), *Advanced Optical Sensors*, Progress in Optical Science and Photonics 38, https://doi.org/10.1007/978-981-95-1687-2_1

This chapter reviews the recent advancements in noble metal nanoparticles from three perspectives: (1) the synthesis of noble metal nanoparticle systems using different synthesis techniques; (2) the properties of noble metal nanoparticles; and (3) the importance and uses of such systems in opto-electronics and sensors.

1 Introduction

A material is considered a nanomaterial if its maximum dimension is 100 nm, which is also one billionth of a meter, or 10^{-9} m [1–6]. About ten H or five Si atoms are arranged in a line. The Research and Development (R&D) sector continues to grow at the fastest rate in recent decades, as seen by the numerous billions of dollars invested in this field each year [7, 8]. Nanomaterials and nanoparticles (NPs) have special properties that enable their use in a broad range of nanotechnology applications, including materials science, microelectronic and metallurgical research, medical science, chemistry, bio-network, and applied physics. Numerous investments are being made in the field of medical science, namely in the field of therapeutics, which is a combination of the terms diagnostics and therapies [9]. It is a cutting-edge method for early cancer identification and cure that combines cancer diagnosis and treatment [10, 11]. In order to accomplish this, the periodic table contains unique metals such as alkaline to alkaline metallics, rare metallics, and noble metallics employed in theragnostic applications [12]. In contrast to these metals, noble metals exhibit remarkable resistance to oxidation, especially at high temperatures, and to corrosive conditions (in acidic or basic media). Furthermore, at the nanoscale, certain metals—particularly gold and silver, display remarkable and adjustable plasmonic characteristics that have captivated researchers for the past 10 years and continue to get more interest today [13–15]. In the domains of biomedicine and biosensors, their plasmonic characteristics offer further benefits [5]. Noble metals have been used since the time of the ancient Greeks and Romans [5, 16]. Gold and silver were among the precious metals that were utilized in the production of coins as well as pots and containers for liquids. Prior to the development of antibiotics in the early 1940s, silver colloids were widely utilized in antimicrobial therapy [17]. Furthermore, gold and its compounds were utilized as remedies in ancient China. Additionally, during the Middle Ages, Europeans also utilized colloidal gold, especially in laboratories run by chemists and alchemists, to heal a variety of infectious disorders and even to prolong life [18]. Noble Laureate Dr. R. Feynman used the word "nanotechnology" in his groundbreaking speech, "There is Plenty of Room at the Bottom," at the American Physical Society meeting held in December 1959 at the California Institute of Technology in the United States [19].

1.1 Synthesis of Nobel Metal Nanoparticles

Nanoparticle preparation can be done in two basic ways: top-down and bottom-up (Fig. 1). The bottom-up approach is a fundamental method for creating metal nanoparticles by lowering their ions, and an agent like a stabilizer or surfactant typically stops the nanoparticles growth. Chemical reduction [20–22], photochemical reduction [23], electrochemical reduction [24], templating [25], and thermal procedures [26, 27] are examples of bottom-up approaches. Large-scale nanoparticle synthesis is possible using bottom-up approaches, but there is a clear drawback: the nanoparticles are typically not as uniformly sized and shaped as those made with top-down approaches. The desired nanostructures are left behind by removing material from the bulk substrate using the top-down method. Photolithography, electron beam lithography, and nanosphere lithography (NSL) are examples of popular top-down techniques [28, 29].

Noble metal nanoparticles (NPs) are synthesized using a variety of methods, each with special benefits and difficulties. These techniques are vital for modifying the physicochemical characteristics of nanoparticles, which are necessary for uses in engineering, medicine, and catalysis, among other domains. The synthesis methods can be broadly divided into three categories: chemical, physical, and biological.

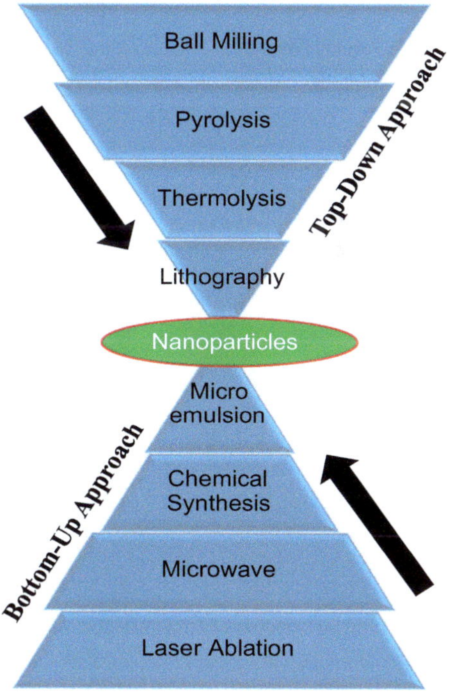

Fig. 1 Top-down and bottom-up approaches for synthesis of NPs

Each of these processes offers a unique way to produce the required properties of nanoparticles.

In general, the production of noble metal nanoparticles can be separated into two categories: chemical (Fig. 2) and physical (Fig. 3).

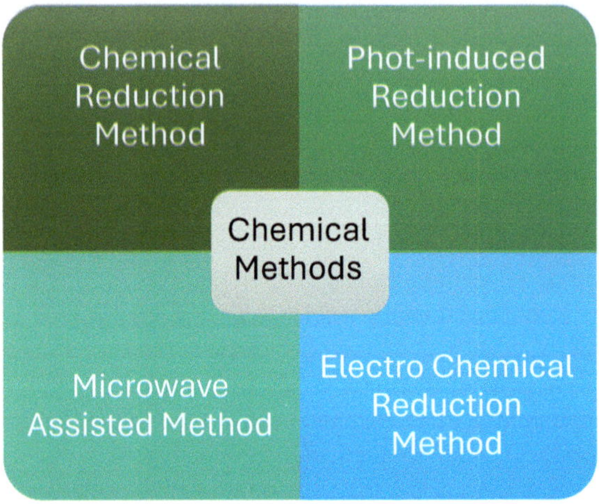

Fig. 2 Various chemical synthesis methods of nanomaterials

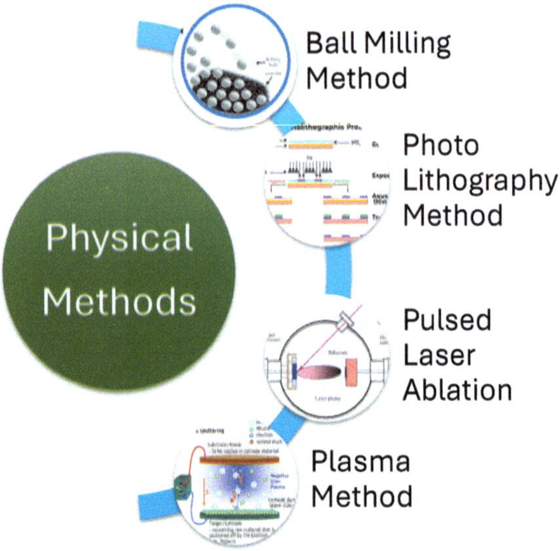

Fig. 3 Various physical synthesis methods of nanomaterials

1.2 Chemical Synthesis Techniques

Chemical Reduction: In this conventional technique, metal precursors are reduced in a solution, frequently in batch reactors. But problems like reagent stability and temperature control can occur. An alternative is provided by microreactor devices, which allow for the creation of nanoparticles with exact characteristics and improved control over reaction conditions [30].

Photochemical Synthesis: Starts the reduction of metal salts by using light. As shown with gold and silver NPs, the size and distribution of nanoparticles are strongly influenced by the concentration of metal salts. Under light circumstances, this thermodynamically advantageous technique enables regulated synthesis [31].

1.3 Physical Synthesis Techniques

Laser Ablation and Lithography: These techniques create nanoparticles by physically removing material. Although they can be expensive and contain dangerous chemicals, they are effective [32, 33].

Mechanical Milling and Sputtering: These methods offer an alternative strategy for reaching specified nanoparticle sizes and shapes since they create nanoparticles physically [33].

1.4 Biological Synthesis Techniques

Green Synthesis: Involves the synthesis of nanoparticles using biological entities such as microbes and plants. This process is sustainable and eco-friendly, yet it frequently produces less monodispersed particles and slower synthesis rates. This strategy can be made more effective by optimizing microbial cultivation and extraction methods [32].

Even though every synthesis method has advantages, problems including expense, environmental effect, and managing nanoparticle characteristics still exist. These issues might be resolved by the creation of hybrid approaches, which combine elements of several approaches to enable the synthesis of nanoparticles in a more effective and sustainable manner [34].

The future of NMNPs' biological uses has nearly been endangered by the overuse of chemicals in chemical synthesis. This led to the investigation of alternative environmentally friendly techniques that require less chemicals. Green synthetic techniques that include microorganisms, biopolymers, and plant extracts have shown promise as a replacement for chemical NP synthesis methods (Fig. 4) [35].

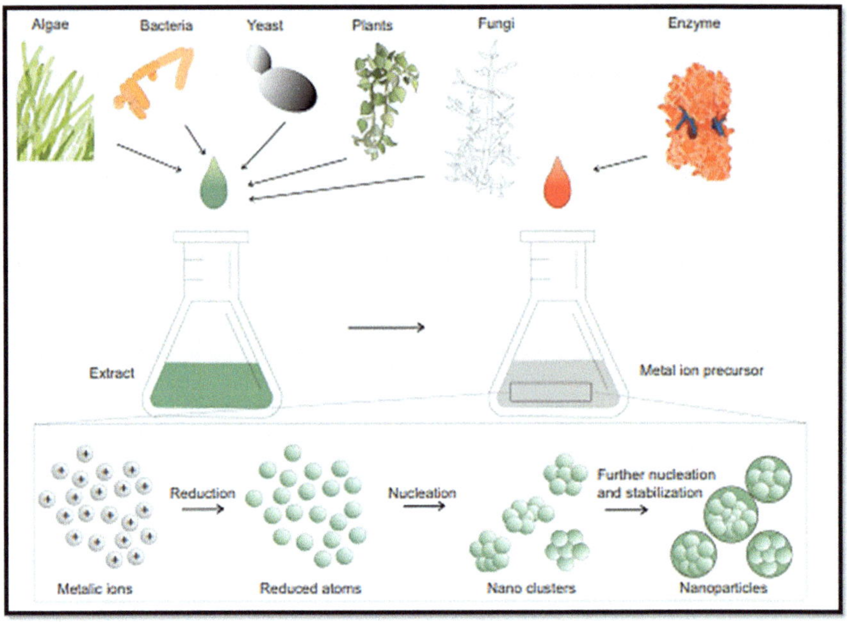

Fig. 4 Schematic representation of green synthesis methods. Adopted from the Ref. [36]

2 Bottom-Up Approach

By assembling atoms or molecules into nanoscale structures, the bottom-up method of creating noble metal nanoparticles provides exact control over the size and makeup of the final product. Because of the special qualities of nanoparticles, this technique is essential for applications in industries including electronics, medicine, and catalysis. The bottom-up method, which frequently makes use of chemical reactions and self-assembly techniques, is distinguished by its capacity to produce homogeneous and precisely defined nanoparticles.

2.1 Chemical Vapor Deposition (CVD)

Using this method, gaseous reactants are deposited onto a substrate, where they undergo chemical reactions to produce nanoparticles. Its capacity to generate homogeneous, high-purity nanoparticles makes it popular [37, 38].

2.2 *Sol–Gel Process*

This versatile technique creates nanoparticles with regulated porosity and composition by transforming a solution into a solid gel phase. Metal oxide nanoparticles can be made with this method, which is especially helpful [39, 40].

2.3 *Self-assembly*

This method depends on molecules spontaneously arranging themselves into ordered configurations due to non-covalent interactions [41] state that it is important for the creation of complicated nanostructures with particular functions, such those utilized in biomedical applications [42].

2.4 *Applications and Advantages*

Catalysis: The high surface area and active sites of noble metal nanoparticles produced via bottom-up techniques are utilized in catalysis to improve reaction speeds and selectivity [13].

Biomedical Applications: Nanoparticles are perfect for medication delivery and imaging applications because of their exact control over size and surface characteristics, especially in cancer therapies [40].

Environmental Impact and Scalability: Compared to top-down approaches, bottom-up methods are typically more scalable and ecologically benign, which makes them appropriate for industrial applications [39].

Although the bottom-up technique has many benefits, it also has drawbacks, including the requirement for exact control over reaction conditions and the possibility of nanoparticle aggregation. Furthermore, research into integrating these techniques into large-scale production is still ongoing in order to improve their suitability for use in a variety of industries.

3 Top-Down Approach

By disassembling bulk materials into nanoscale particles, the top-down method for creating noble metal nanoparticles provides a simple and frequently more scalable substitute for bottom-up techniques. This method is very useful for creating nanoparticles without the need for stabilizers or intricate chemical reactions, which makes it appropriate for uses needing little contamination and high purity.

3.1 Electrochemical Methods

One well-known top-down technique is electrochemical synthesis, which creates nanoparticles by submerging bulk metal wires in an electrolyte and applying a strong alternating voltage. This approach simplifies the synthesis process by doing away with the necessity for surfactants or reducing agents [13].

Another top-down method for producing nanoporous metal nanosheets from non-layered alloy precursors is the electrochemical dealloying technique. This process is cost-effective for large-scale manufacturing and enables high production rates [43].

3.2 Advantages of Top-Down Approaches

In contrast to bottom-up approaches, which can need numerous synthesis steps and intricate procedures, top-down techniques, like electrochemical synthesis, are typically more suited to large production rates [13].

These techniques can create nanoparticles with special structural characteristics, like nanoporous architectures, which are advantageous for energy conversion devices and catalytic applications [43].

3.3 Challenges and Considerations

Although top-down methods are simpler and more scalable, they might not provide as much control over the size and shape of the particles as bottom-up alternatives. When particular nanoparticle properties are needed for particular applications, this could be a drawback [43]. Furthermore, the characteristics of the final nanoparticles can be greatly impacted by the selection of precursor materials and synthesis parameters, including potential amplitude and electrolyte composition [13].

Bottom-up techniques, including chemical reduction in microreactor systems, provide finer control over nanoparticle size and shape than top-down techniques. Although they frequently need more intricate synthesis procedures and the use of stabilizers, these techniques are especially helpful when specialized physicochemical qualities are required [40].

3.4 Synthesis of Noble Metal Nanoparticles in Microreactor Systems

Conventional reactors have been used to create nanoparticles for a long time. The synthesis of noble metals is made possible by the advancement of microreactor

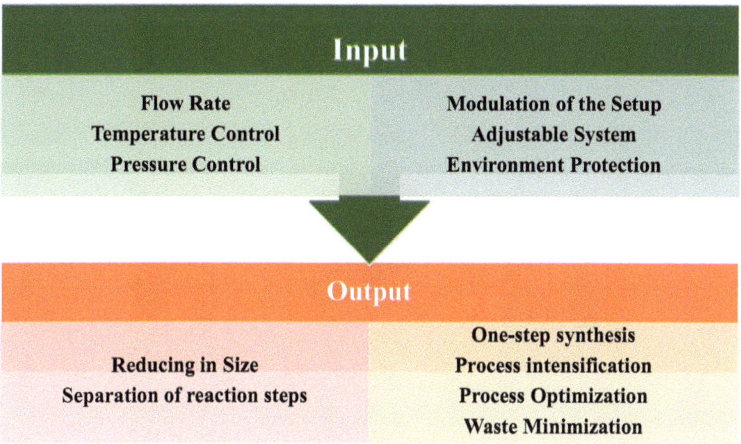

Fig. 5 The microreactor input (set parameter control, design of the system for the individual process) and output (effect on the products and process). Adopted from the Ref. [30]

devices. Nanoparticle size and form can be controlled with this method. Additionally, it offers improved process parameter control. These days, a lot of research papers explain the production of noble nanomaterials [44]. In general, there are possible advantages to using microreactors to create nanomaterials. Figure 5 provides a summary of the parameters' quantifiable inputs and outcomes.

4 Applications for Noble Metal NPs

Functional molecules or systems that detect or sense a certain analyte molecule in the surrounding environment are often defined. When the analyte species concentration in the sample changes, the output signal is produced. Thus, the sensing application of NPs is dependent on three elements: (i) the analyte; (ii) an analyte's binding sensitivity detecting component; and (iii) transducers. The transducer can be described as a device that can translate binding events into equivalent voltage or current. For any sensor to be effective, the three previously mentioned factors must be present in order to provide a minimal error with a higher signal-to-noise (S/N) ratio and a minimum analyte concentration. Nevertheless, noble metal nanoparticles are less stable in open spaces and may have challenging surface compounds, which limit their use in biosensing.

To fully satisfy sensing needs, the noble metal's functional nanoparticles may thus have outstanding photophysical properties such strong fluorescence and visible region absorption, conductivity, and redox characteristics. As seen in Fig. 6, the Ag NPs exhibit varying hues depending on their size. Additionally, the Ag NPs

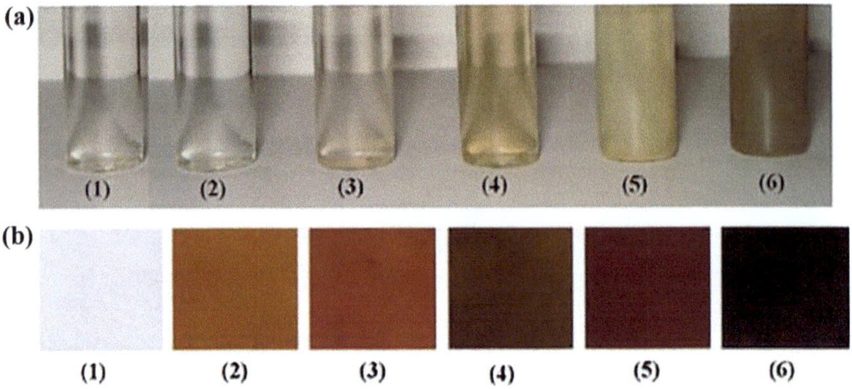

Fig. 6 Ag nanocolloids dissolved in water **a** and nanocomposite films cast upon silicate glass **b** exhibit color variation. Ag is present in 0.0, 0.2, 0.5, 1.0, 2.0, and 5.0 weight percent in samples 1–6 (with permission Ref. [30])

exhibit remarkable stability both in colloidal solutions and as nanocomposite coatings. Because of variations in particle size, the Ag colloidal NPs in Fig. 6a, b exhibit a range of colors.

The polyol technique was used to create NPs of different sizes as the Ag level increased. The excess water solvent was then evaporated, and a viscous solution was poured into a glass petri dish in the shape of a film. Figure 6b shows the color shift in the film as the silver level rises. Additionally, the size of the gold particles showed remarkable color changes, and the infrared zone's absorption wavelength was adjustable (Fig. 7). El-Sayed and his colleagues [45] observed the color shift in Au NPs because of a different aspect ratio (Fig. 7a). Furthermore, it was demonstrated that the silica NPs were tailored to around 600–11,000 nm and that the gold shell SPR on them was core (Fig. 7b).

A lot of research has recently been done on the use of noble metal nanoparticles, such as silver nanoparticles (AgNPs) and gold nanoparticles (AuNPs), in the colorimetric detection of lead ions [46, 47]. Because of their desirable qualities (such as chemical inertness, resistance to oxidation, and ease of surface functionalization), AuNPs and AgNPs show interparticle distance-dependent optical characteristics and exceptionally large extinction coefficients for the effective identification of metal ions and biomolecules [48, 49]. When AuNPs and AgNPs bind or interact with lead ions, they aggregate and undergo changes in their optical characteristics (such as etching, cross-linking, and refractive index), which are accompanied by a change in hue [50, 51]. Additionally, nanozymes based on noble metals have been employed as colorimetric signal producers. In the presence of metal ions, AuNPs and PtNPs can function as nanozymes because they can imitate the enzymatic activity of peroxidases, such as horseradish peroxidase, to cause color changes when the substrate is added [52, 53].

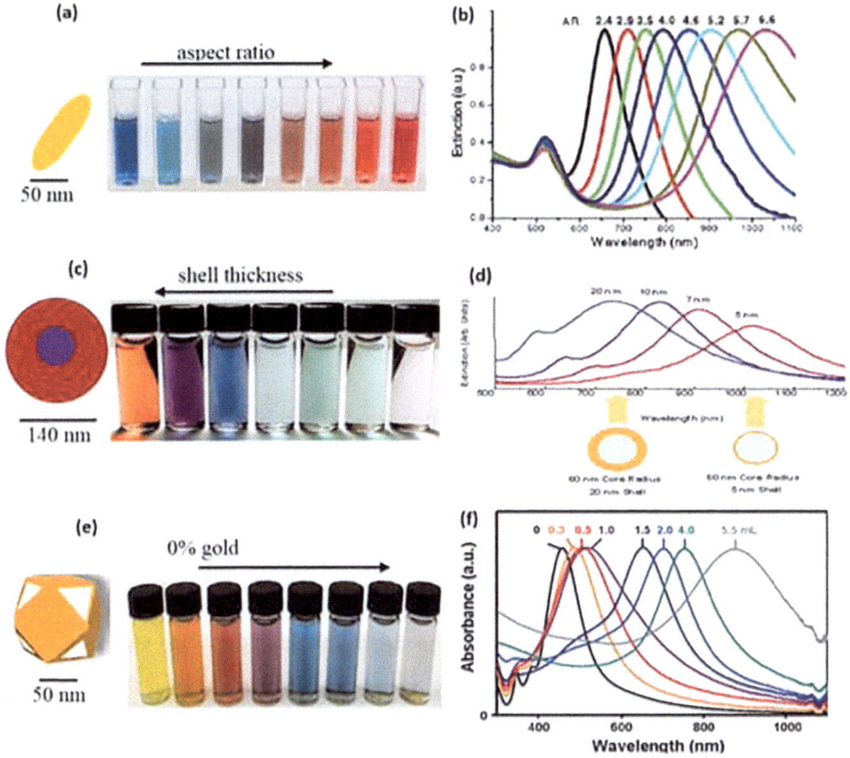

Fig. 7 Optical characteristics of Au include: **a** color change in Au nanorods with varying gold nanoparticle aspect ratios; **b** tunable SPR of gold nanoparticles with varying gold nanoparticle aspect ratios; **c** color change observed in Au nanoshells with varying shell thickness (**d**); **e** the physical appearance of Au nanocages with varying $HAuCl_4$ content in their synthesis; and **f** wavelength tunability (with permission from [45])

4.1 Gold Nanoparticles (AuNPs) as Sensors

For colorimetric detection of chemical and biological analytes, gold nanoparticles and their surface functionalization are considered next-generation sensing instruments. Over the past ten years, [54, 55] AuNPs have been widely employed in colorimetric lead ion detection [56, 57]. AuNPs have been used as colorimetric nanoprobes for the quick and instantaneous detection of Pb^{2+}, according to a recent study [58]. In the presence of additional interfering metal ions, the sensor demonstrated high selectivity. The results showed that the LOD was 18 μM and the LOQ was 53 μM.

4.2 Silver Nanoparticles (AgNPs) as Sensors

Over the past few decades, AgNPs significance as plasmonic nanoparticles has increased dramatically. Three chemical functional components are typically needed to fabricate AgNPs, a solvent, a reducing/stabilizing agent, and a silver precursor. By altering reaction parameters like temperature, pH, reaction time, and chemical precursors, AgNPs size and shape can be controlled [59, 60]. Because bare or naked AgNPs are thermodynamically unstable, storing them in aqueous solutions alters their shape. For the intended plasmonic functionality to be maintained, the produced AgNPs must be stabilized. AgNP structures may be disrupted in the absence of a stabilizer. As a result, AgNPs lose their plasmonic behavior when they dissolve or aggregate.

To give AgNPs colloidal stability, appropriate stabilizing (or capping) agents must be applied during nucleation and growth. Highly stable AgNPs can be produced using a variety of stabilizing agents, including polymers (including PVP, PEG, and silica), biomolecular ligands (including DNA and proteins), and surfactants and ligands (such citrate and CTAB) [61].

As illustrated in Fig. 8, pyridine-formaldehyde functionalized AuNPs were introduced in a recent work on the colorimetric method for lead ions [62]. The absorbance peak of AuNPs at 520 nm was diminished in the presence of Pb^{2+} ions, and a new peak appeared at about 800 nm. The color shift from red to blue was also discernible to the unaided eye. The sensor's detection limit was 1 μM with a UV–vis spectrophotometer and 4 μM with the unaided eye.

Green synthesis of AgNPs is simple and cost-effective. Because of this, AgNPs made from the leaf extract of Aconitum violaceum were used to create a basic colorimetric sensor. When Pb^{2+} ions were added, the NPs' color changed from yellow to red, and a new absorbance band at 520 nm appeared. At the same time, the absorbance band at 404 nm decreased (Fig. 9). In order to encourage NPs to aggregate and change color, it was deduced that the catechin polyphenolic groups found in leaf extracts acted in concert with Pb^{2+} ions to create a Pb^{2+} catechin complex. Within the linear detection range of 0.5–25 μM, the nanosensor displayed a limit of detection of 0.1 μM.

4.3 Bimetallic (Au/Ag) Nanoparticles as Sensors

Compared to their monometallic counterparts (AuNPs and AgNPs), bimetallic nanoparticles (BNPs) made of two distinct metal elements (Ag and Au) may have greater promise in terms of the quantum-size domain and catalytic efficiency [64–67]. A variety of monometallic NP designs can be combined to create BNPs. It is possible to tailor the metallic mixture's energy of plasmon absorption band in BNPs for their versatile use in sensing applications [68, 69]. In general, they differ from

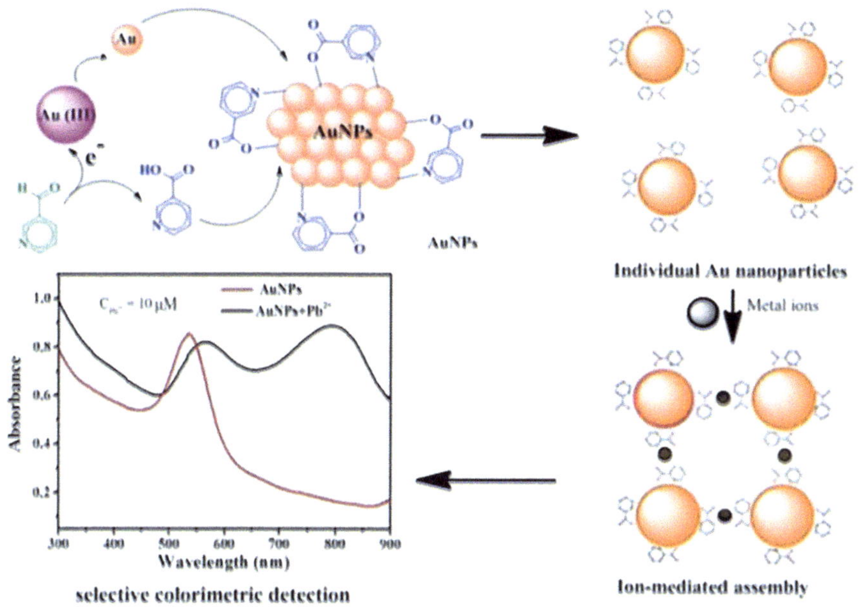

Fig. 8 The principle of detection of Pb^{2+} ions by a colorimetric method using functionalized AuNPs [62]

pure elemental particles in their optical, electrical, thermal, and catalytic characteristics. It has been observed that various BNPs composed of Au/Ag, Au/Pt, Cu/Ag, Au/Pd, and Au/Pt elements are effective colorimetric sensors for visual detection [70–73].

Since the electrons are more exposed to the metal's surface for easier delocalization, noble metals show surface plasmon, which is enhanced when the materials are in nanoforms. The presence of the surface plasmon is responsible for many photochemical reactions, light absorption and energy conversion, sensing, optoelectronic, and biomedical applications.

Because of their surface plasmon resonance in the visible spectrum, which may include surface plasmon polaritons (SPPs) and localized surface plasmon resonances (LSPRs), gold and silver nanometals have been the subject of extensive research. SPPs happen when plasmons propagate on the metal surface because of their longer wavelength than the incident light, while LSPRs occur when the incident light's wavelength and radiation mean free path are significantly higher than the nano-metal. The ensuing resonance between the input light's electric field and the oscillating electrons on the nanoparticle surfaces is what causes plasmon absorption. These interactions have some potential optical applications and can be modified for a variety of uses, particularly when a core–shell structure is present.

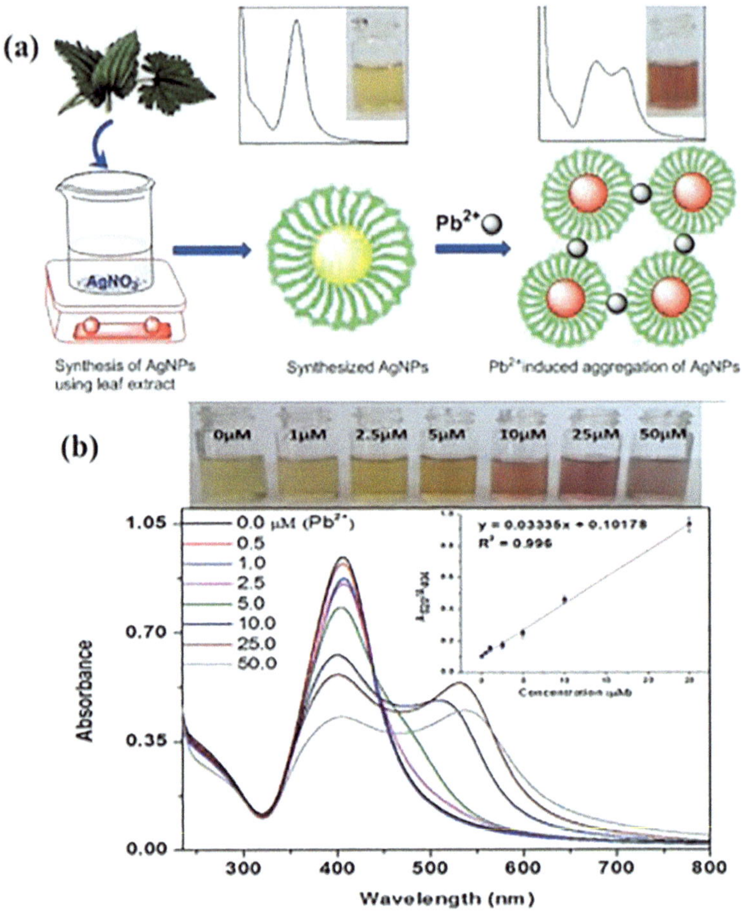

Fig. 9 Use of green generated AgNPs for Pb^{2+} ion colorimetric detection. **a** AgNPs production schematic; **b** AgNPs (diluted ten times) UV–vis spectra as a function of different Pb^{2+} concentrations (0.5–50 μM) and associated graphs [63]

4.4 Nobel Metals (Au and Ag) in Solar Cell Applications

By integrating Au silica nanoribbions (NRs) in the active layer and Au nanospheres in the buffer layer simultaneously using small molecular solar cells. Xu et al. achieved an efficiency of 8.54%, which they attributed to better hole transport in the buffer layer and wider optical absorption in the active layer. An efficiency of 7.41% was found for the addition of Au nanospheres to the buffer layer [74]. When Au@Ag was added to the buffer layer of PT7B:PC70BM as opposed to the buffer layer of PCDTBT:PC71BM, the maximum efficiency was noted [75]. The device characterization information for OSCs with AuNP(s) inserted in the buffer layer is shown in Table 1.

Table 1 Organic photovoltaic (OPV) device characterization data using buffer layer metallic AuNPs

Buffer layer (Au)	PCE (%)	V_{oc}	J_{sc}	FF (%)	References
Structure with AuNPs	8.2	0.73	16.7	68	[74]
Structure + Au	3.32	0.61	9.94	53.3	[76]
Structure + AuAgNP prism	7.1	0.75	15	64	[77]
Structure + Au	2.2	0.54	8.1	50	[78]
Structure + Au	2.5	0.56	7.1	65	[78]
Structure + AuND (LD)	3.65	0.58	11.13	57	[79]
Structure + AuNPs	5.87	0.87	10.71	63	[80]
Structure + Au@AgNCs	6.08	0.87	11.10	63	[80]
Structure + AuNP	8.14	0.73	16.64	67	[80]
Structure + Au@AgNCs	8.74	0.74	17.38	68	[80]

In contrast to a rise of 3.41% in only the active layer or 3.61% in only the buffer layer, Xie et al. demonstrated that the coalescing of AuNPs in both the P3HT: PCBM layer and the PEDOT: PSS layer further raised the PCE to 3.85% [81]. The additional NPs added to the separate layers were responsible for this additional increase. According to Spyropoulus et al. [82] the concatenation of surfactant-free AuNPs in the active layer caused the PCE to rise to 3.71%, which was attributed to a decrease in exciton quenching in the capped AuNP layer caused by recombination.

According to Spyropoulos et al. [81], smaller AuNPs cause the LPSR, but bigger AuNPs cause multiple scattering. The OPVs with AuNP(s) integrated into the active layer device characterization data are displayed in Table 2.

While both of their reference cells without AgNPs present similar efficiency of ≈ 5.20%, Jung et al. reported on the assembly of nanobumps (NBA) on the MoO_3 buffer layer concealing the AgNPs to further intensify the PCDTBT:PC70BM device's

Table 2 Data on the OPV device characterisation using metallic AuNPs in the active layer

Active layer (Au)	PCE (%)	V_{oc}	J_{sc}	FF (%)	References
Structure + Au	6.45	0.89	11.16	65.0	[75]
Structure + Au-SiO_2	4.40	0.59	12.70	58.1	[81]
Structure + Au-SiO_2NR	8.20	0.77	15.40	70.4	[82]
Structure + AuRD/Ag	4.14	0.61	10.39	65.3	[83]
Structure + AuECTO/Ag	3.76	0.60	9.94	63.0	[83]
Structure + Aucube/Ag	3.75	0.60	9.95	62.8	[83]
Structure + Autriangle/Ag	3.62	0.60	9.57	63.1	[83]
Structure + Au/Pt	5.94	0.74	14.24	56.0	[84]
Structure + Au/LiF	3.85	0.61	9.74	65.0	[85]
Structure + AuNPs/Al	3.71	0.6	9.77	63.3	[86]

performance to 6.07%. This is in contrast to the device with a flat PEDOT:PSS buffer layer, which has an efficiency of 5.63% [87]. The NBA structure's multi-reflection and high forward scattering effect were attributed to the improved performance. Additionally, the entirely covered AgNPs by MoO_3 further prevent recombination losses, isolating it from the active layer. However, because of its hygroscopic and acidic properties, the flat PEDOT:PSS layer causes instability in the device. Additionally, exposed AgNPs that come into direct contact with the active layer encourage recombination losses and exciton quenching, which worsens the device's performance. The device characterization data for OPVs with AgNPs integrated into the buffer layer is displayed in Table 3.

In C60-bis, the back type device with TNP AgNPSs was improved to a PCE of roughly 8.30%. In line with other research reports, the dual type devices produced a much higher efficiency of about 9.02%. The dual doped interfacial layer and AgNPS adjustment were credited with this increased efficiency. They further demonstrated their adaptability by using other polymer-based devices, which resulted in an efficiency of roughly 7.06% in the dual type of PCDTBT: PC71BM-based devices and 7.85% in the PIDT-PhanQ device. The device characterization information for OPVs with AgNP(s) integrated into the active layer is shown in Table 4.

Table 3 OPV device characterization data using buffer layer metallic AgNPs

Buffer layer (AgNPs)	PCE (%)	V_{oc}	J_{sc}	FF (%)	References
Structure + AgNPs	4.91	0.53	1.59	60.9	[88]
Structure + AgNWs	7.12	0.75	16.35	59	[89]
Structure + $Ag@SiO_2$	3.35	0.59	9.67	58.3	[90]
Structure + AgNP-67	8.6	0.75	16.33	70	[91]
Structure + Ag	3.4	0.67	9.3	53.9	[82]
Structure + Ag (or Au) NPs	7.52	0.71	16.80	63.0	[92]
Structure + AgNPs	7.21	0.85	12.86	66.0	[93]
Structure + Ag NPs-MoO_3 (NBA)	6.07	0.88	10.48	65.0	[87]
Structure + AgNP/PCDTBT	5.63	0.89	10.24	62.0	[87]

Table 4 Data on the OPV device characterisation using metallic AgNPs in the active layer

Active layer (AgNPs)	PCE	V_{oc}	J_{sc}	FF	References
Structure + Agnanoplates	6.6	0.87	13.19	57	[94]
Structure + AgNWs	3.91	0.66	9.32	63.9	[95]
Structure + $Au_{11}Ag_{89}$	4.73	0.63	12.21	61.5	[96]
Structure + Agblends	4.30	0.64	10.61	63.3	[97]
Structure + Agnanoprisms	4.07	0.64	9.93	64.1	[97]
Structure + $Ag@SiO_2$	3.94	0.61	9.72	66.5	[98]
Structure + Ag	9.02	0.96	14.36	63	[99]

5 Summary

Even with recent developments in this field, there are still several obstacles to overcome before their usefulness can be fully realized. For instance, consistent optical and chemical properties with repeatable synthesis techniques are essential for noble metal nanoparticles, such as AuNPs and AgNPs. Every synthesis batch should consistently retain the size and shape of the NPs. For large-scale nanoparticle synthesis, additional work is needed to develop physical techniques (like lithography) in addition to chemical synthesis. The functionalization agents' stability and denaturation characteristics should also be considered since they may affect the sensor's performance. When AuNPs are subjected to complicated water samples, matrix interference may occur, changing the sensor's functionality. The robustness of the devices and the sensor's selectivity towards the intended analyte are two major challenges when handling the various matrices and challenging conditions that arise during the on-site measurements (such as biological samples and wastewater). Increased efforts should be made to improve colorimetric sensors' analyte detection range. Furthermore, there are presently no technologies that can give ultra-sensitive detection that is visible to the naked eye. As of right now, color changes that are visible to the naked eye only produce a low resolution. The creation of new signal readout channels (such as cellphones) for target analyte monitoring can make sensor use even easier.

Creating novel nanomaterials based on size-/shape-controlled techniques and controlled growth/etching would be a creative way to advance the creation of trustworthy colorimetric probes. Plasmonic nanosensors based on growth or etching will become possible thanks to developments in material science. Noble-metal semiconductor nanocomposites have a bright future ahead of them in applications related to electronics, energy, and optics. The semiconductors or noble metals can be adjusted so that their combination to create a nanocomposite provides the material with the most effective use. The noble metals' surface plasmon effects are an excellent characteristic that helps to explain energy absorption, electron transfer, and hole formation with semiconductor materials, all of which are essential to comprehending their interaction mechanism. Additionally, it is anticipated that the use of SPR and SERS-active probes would pave the way for the development of biosensors and in situ remote sensors for environmental and biological applications.

References

1. D. Huo, M.J. Kim, Z. Lyu, Y. Shi, B.J. Wiley, Y. Xia, One-dimensional metal nanostructures: from colloidal syntheses to applications. Chem. Rev. **119**, 8972–9073 (2019)
2. J.L. West, N.J. Halas, Engineered nanomaterials for biophotonics applications: improving sensing, imaging, and therapeutics. Annu. Rev. Biomed. Eng. **5**, 285–292 (2003)
3. N. Chen, Y. Zhang, H. Liu, X. Wu, Y. Li, L. Miao, Z. Shen, A. Wu, High-performance colorimetric detection of Hg^{2+} based on triangular silver nanoprisms. ACS Sens. **5**, 521–527 (2016)

4. A. Gautam, P. Komal, Probable ideal size of Ln^{3+}-based upconversion NPs for single and multimodal imaging. Coord. Chem. Rev. **376**, 393–404 (2018)
5. E.C. Dreaden, A.M. Alkilany, X. Huang, C.J. Murphy, M.A. El-Sayed, The golden age: Gold NPs for biomedicine. Chem. Soc. Rev. **41**, 2740–2779 (2012)
6. K. Sztandera, M. Gorzkiewicz, B. Klajnert-Maculewicz, Gold NPs in cancer treatment. Mol. Pharm. **16**, 1–23 (2019)
7. R. Paull, J. Wolfe, P. Hébert, M. Sinkula, Investing in nanotechnology. Nat. Biotechnol. **21**, 1144–1147 (2003)
8. L. Jia, Global governmental investment in nanotechnologies. Curr. Nanosci. **1**, 263–266 (2005)
9. A. Gharatape, R. Salehi, Recent progress in theranostic applications of hybrid gold NPs. Eur. J. Med. Chem. **138**, 221–233 (2017)
10. Y. Xia, W. Li, C.M. Cobley, J. Chen, X. Xia, Q. Zhang, M. Yang, E.C. Cho, P.K. Brown, Gold nanocages: from synthesis to theranostic applications. Acc. Chem. Res. **44**, 914–924 (2011)
11. W. Zhu, J. Zhao, Q. Chen, Z. Liu, Nanoscale metal-organic frameworks and coordination polymers as theranostic platforms for cancer treatment. Coord. Chem. Rev. **398**, 113009–113023 (2019)
12. Y. Liu, L. Yang, Y. Shen, Hydrothermal synthesis of gold nanoplates and their structure-dependent LSPR properties. J. Mater. Res. **33**, 2671–2679 (2018)
13. A. Gautam, S. Ram, Preparation and thermomechanical properties of Ag-PVA nanocomposite films. Mater. Chem. Phys. **119**, 266–271 (2010)
14. S. Ram, P. Tripathy, H.J. Fecht, Gold NPs reinforced poly (vinyl alcohol) of self-standing optical films. J. Nanosci. Nanotechnol. **7**, 3200–3206 (2007)
15. S. Ram, A. Gautam, H.J. Fecht, J. Cai, J. Bansmann, R.J. Behm, A new allotropic structure of silver nanocrystals nucleated and grown over planar polymer molecules. Philos. Mag. Lett. **87**, 361–372 (2007)
16. M. Cametti, Z. Džolić, New frontiers in hybrid materials: noble metal NPs supramolecular gel systems. Chem. Commun. **50**, 8273–8286 (2014)
17. J. Alexander, History of the medical use of silver. Surg. Infect. **10**, 289–292 (2009)
18. L.A. Dykman, N.G. Khlebtsov, Gold NPs in biology and medicine: recent advances and prospects. Acta Nat. **3**, 34–55 (2011)
19. "Plenty of Room" Revisited. Nat. Nanotechnol. **4**, 781 (2009)
20. S.H. Chen, D.L. Carroll, Synthesis and characterization of truncated triangular silver nanoplates. Nano Lett. **2**, 1003–1007 (2002)
21. G.S. Metraux, C.A. Mirkin, Rapid thermal synthesis of silver nanoprisms with chemically tailorable thickness. Adv. Mater. **17**, 412–415 (2005)
22. T.K. Sau, C.J. Murphy, Room temperature, high-yield synthesis of multiple shapes of gold nanoparticles in aqueous solution. J. Am. Chem. Soc. **126**, 8648–8649 (2004)
23. R.C. Jin, Y.C. Cao, E.C. Hao et al., Controlling anisotropic nanoparticle growth through plasmon excitation. Nature **425**, 487–490 (2003)
24. B.S. Yin, H.Y. Ma, S.Y. Wang et al., Electrochemical synthesis of silver nanoparticles under protection of poly(N-vinylpyrrolidone). J. Phys. Chem. B **107**, 8898–8904 (2003)
25. C.A. Foss, G.L. Hornyak, J.A. Stockert et al., Template-synthesized nanoscopic gold particles: optical spectra and the effects of particle size and shape. J. Phys. Chem. **98**, 2963–2971 (1994)
26. R. Jin, S. Egusa, N.F. Scherer, Thermally-induced formation of atomic Au clusters and conversion into nanocubes. J. Am. Chem. Soc. **126**, 9900–9901 (2004)
27. Y.G. Sun, Y.N. Xia, Shape-controlled synthesis of gold and silver nanoparticles. Science **298**, 2176–2179 (2002)
28. S.L. Zhu, F. Li, C.L. Du et al., A localized surface plasmon resonance nanosensor based on rhombic Ag nanoparticle array. Sensors Act. B **134**, 193–198 (2008)
29. A.J. Haes, R.P. Van Duyne, A nanoscale optical biosensor: Sensitivity and selectivity of an approach based on the localized surface plasmon resonance spectroscopy of triangular silver nanoparticles. J. Am. Chem. Soc. **124**, 10596–10604 (2002)
30. A. Pach, A. Szot, K. Fitzner, M. Luty-Błocho, Opportunities and challenges in the synthesis of noble metal nanoparticles via the chemical route in microreactor systems. Micromachines. **15**(9), 1119 (2024). https://doi.org/10.3390/mi15091119

31. S.M. Aldebasi, H. Tar, A.S. Alnafisah, L. Beji, N. Kouki, F. Morlet-Savary, J. Lalevée, et al., Photochemical synthesis of noble metal nanoparticles: influence of metal salt concentration on size and distribution. Int. J. Molec. Sci. **24**(18), 14018 (2023)
32. S.M. Naghib, H.R. Garshasbi, Green plant extract-based synthesis of multifunctional nanoparticles and their biological activities (Bentham Science Publishers, 2023)
33. R.P. Chauhan, Processes for nanoparticle synthesis, in *Discovery, Disruption, and Future Implications of Nanomaterials*. (IGI Global, 2024), pp. 69–100
34. S. Parthasarathy, S. Behera, D.K. Das, P. Patnaik, N. Das, P. Mallick, S.K. Nayak, Nanoparticle synthesis methods, in *Discovery, Disruption, and Future Implications of Nanomaterials* (IGI Global, 2024), pp. 33–68
35. K.S. Siddiqi, A. Husen, R.A.K. Rao, A review on biosynthesis of silver nanoparticles and their biocidal properties. J. Nanobiotechnol. **16**, 14 (2018)
36. G. Habibullah, J. Viktorova, T. Ruml, Current strategies for noble metal nanoparticle synthesis. Nanoscale Res. Lett. **16**, 47 (2021). https://doi.org/10.1186/s11671-021-03480-8
37. G. Oskam, R.L. Penn, Z. Hu, P.C. Searson, Synthesis and characterization of metal oxide nanoparticles, in *Emerging Fields in Sol-Gel Science and Technology* (2023), pp. 149–156.
38. Y. Etafa, S. Elangovan, K. Legesse, M. Chewaka, Synthesis and fabrication techniques for nanomaterials, in *Exploring Nanomaterial Synthesis, Characterization, and Applications* (IGI Global, 2025), pp. 241–266
39. M.Z.U. Abidin, M. Ikram, S. Moeen, G. Nazir, M.B. Kanoun, S. Goumri-Said, A comprehensive review on the synthesis of ferrite nanomaterials via bottom-up and top-down approaches advantages, disadvantages, characterizations and computational insights. Coord. Chem. Rev. **520**, 216158 (2024)
40. J. Schneider, T. Tichter, C. Roth, Electrochemical methods, in *Flow Batteries: From Fundamentals to Applications*, vol. 1 (2023), pp. 229–262
41. S. Chatterjee, M. Bhattacharya, S. Nag, K. Dhama, C. Chakraborty, A detailed overview of SARS-CoV-2 omicron: its sub-variants, mutations and pathophysiology, clinical characteristics, immunological landscape, immune escape, and therapies. Viruses **15**(1), 167 (2023)
42. T.P. Kamble, S.R. Shingte, V.D. Chavan, D.K. Kim, S.H. Mujawar, T.D. Dongale, P.B. Patil, $CoFe_2O_4$ nanoparticle embedded carbon nanofibers: a promising non-noble metal catalyst for oxygen evolution reaction. Int. J. Hydrogen Energy **92**, 1099–1107 (2024)
43. X. Huang, M.A. El-Sayed, Gold NPs: optical properties and implementations in cancer diagnosis and photothermal therapy. J. Adv. Res. **1**, 13–28 (2010)
44. M. Rahman, E. Rebrov, Microreactors for gold nanoparticles synthesis: from Faraday to flow. Processes **2**, 466–493 (2014)
45. E. Oliveira, C. Núñez, H.M. Santos, J. Fernández-Lodeiro, A. Fernández-Lodeiro, J.L. Capelo, C. Lodeiro, Revisiting the use of gold and silver functionalized nanoparticles as colorimetric and fluorometric chemosensors for metal ions. Sens. Actuators B **212**, 297–328 (2015)
46. M. Sabela, S. Balme, M. Bechelany, J.M. Janot, K. Bisetty, A review of gold and silver nanoparticle-based colorimetric sensing assays. Adv. Eng. Mater. **19**, 1700270 (2017)
47. L. Qin, G. Zeng, C. Lai, D. Huang, P. Xu, C. Zhang, M. Cheng, X. Liu, S. Liu, B. Li, "Gold rush" in modern science: fabrication strategies and typical advanced applications of gold nanoparticles in sensing. Coord. Chem. Rev. **359**, 1–31 (2018)
48. E. Priyadarshini, N. Pradhan, Gold nanoparticles as efficient sensors in colorimetric detection of toxic metal ions: a review. Sens. Actuators B **238**, 888–902 (2017)
49. N. Ratnarathorn, O. Chailapakul, W. Dungchai, Highly sensitive colorimetric detection of lead using maleic acid functionalized gold nanoparticles. Talanta **132**, 613–618 (2015)
50. B. Chen, Z. Wang, D. Hu, Q. Ma, L. Huang, C. Xv, Z. Guo, X. Jiang, Scanometric nanomolar lead (II) detection using DNA-functionalized gold nanoparticles and silver stain enhancement. Sens. Actuators B **200**, 310–316 (2014)
51. K.N. Han, J.-S. Choi, J. Kwon, Gold nanozyme-based paper chip for colorimetric detection of mercury ions. Sci. Rep. **7**, 1–7 (2017)

52. W. Chen, S. Li, J. Wang, K. Sun, Y. Si, Metal and metaloxide nanozymes: bioenzymatic characteristics, catalytic mechanism, and eco-environmental applications. Nanoscale **11**, 15783–15793 (2019)
53. C.-C. Chang, C.-P. Chen, T.-H. Wu, C.-H. Yang, C.-W. Lin, C.-Y. Chen, Gold nanoparticle-based colorimetric strategies for chemical and biological sensing applications. Nanomaterials **9**, 861 (2019)
54. M.H. Jazayeri, T. Aghaie, A. Avan, A. Vatankhah, M.R.S. Ghaffari, Colorimetric detection based on gold nano particles (GNPs): an easy, fast, inexpensive, low-cost and short time method in detection of analytes (protein, DNA, and ion). Sens. Bio-Sens. Res. **20**, 1–8 (2018)
55. S.-K. Kim, S.-D. Kim, E.-J. Hong, M.-S. Han, Alkyl phosphate functionalized gold nanoparticles-based colorimetric probe for Pb^{2+} ions. Bull. Korean Chem. Soc. **31**, 3806–3808 (2010)
56. Y.-L. Hung, T.-M. Hsiung, Y.-Y. Chen, C.-C. Huang, A label-free colorimetric detection of lead ions by controlling the ligand shells of gold nanoparticles. Talanta **82**, 516–522 (2010)
57. R.M. Tripathi, S.H. Park, G. Kim, D.H. Kim, D. Ahn, Y.M. Kim, S.J. Chung, Metal-induced redshift of optical spectra of gold nanoparticles: an instant, sensitive, and selective visual detection of lead ions. Int. Biodeterior. Biodegrad. **1444**, 104740 (2019)
58. L. Lei, H. Song, J. Zhao, Q. Yang, Z. Chen, Preparation of gold nanoparticles using pyridine-formaldehyde as a reducing agent and its application in high sensitivity colorimetric detection of Pb^{2+}. Anal. Methods **11**, 4362–4369 (2019)
59. H. Kang, J.T. Buchman, R.S. Rodriguez, H.L. Ring, J. He, K.C. Bantz, C.L. Haynes, Stabilization of silver and gold nanoparticles: preservation and improvement of plasmonic functionalities. Chem. Rev. **119**, 664–699 (2018)
60. A. Heuer-Jungemann, N. Feliu, I. Bakaimi, M. Hamaly, A. Alkilany, I. Chakraborty, A. Masood, M.F. Casula, A. Kostopoulou, E. Oh, The role of ligands in the chemical synthesis and applications of inorganic nanoparticles. Chem. Rev. **119**, 4819–4880 (2019)
61. N.A. Khan, A. Niaz, M.I. Zaman, F.A. Khan, M. Tariq, Sensitive and selective colorimetric detection of Pb^{2+} by silver nanoparticles synthesized from Aconitum violaceum plant leaf extract. Mater. Res. Bull. **102**, 330–336 (2018)
62. L. Guo, X. Zhou, D.-H. Kim, Facile fabrication of distancetunable Au-nanorod chips for single-nanoparticle plasmonic biosensors. Biosens. Bioelectron. **26**, 2246–2251 (2011)
63. M.M. Kumari, J. Jacob, D. Philip, Green synthesis and applications of Au–Ag bimetallic nanoparticles. Spectrochim. Acta, Part A **137**, 185–192 (2015)
64. A. Zaleska-Medynska, M. Marchelek, M. Diak, E. Grabowska, Noble metal-based bimetallic nanoparticles: the effect of the structure on the optical, catalytic and photocatalytic properties. Adv. Colloid Interface Sci. **229**, 80–107 (2016)
65. G. Sharma, A. Kumar, S. Sharma, M. Naushad, R.P. Dwivedi, Z.A. ALOthman, G.T. Mola, Novel development of nanoparticles to bimetallic nanoparticles and their composites: a review. J. King Saud Univ. Sci. **31**, 257–269 (2019)
66. N. Kumari, S. Kour, G. Singh, A. Chauhan, R. Verma, R.K. Sharma, A brief review on the synthesis of bimetallic nanoparticles for biomedical and solar energy applications. AIP Conf. Proc. **2220**, 020149 (2020)
67. P. Srinoi, Y.-T. Chen, V. Vittur, M.D. Marquez, T.R. Lee, Bimetallic nanoparticles: enhanced magnetic and optical properties for emerging biological applications. Appl. Sci. **8**, 1106 (2018)
68. L. Feng, G. Gao, P. Huang, K. Wang, X. Wang, T. Luo, C. Zhang, Optical properties and catalytic activity of bimetallic gold-silver nanoparticles. Nano Biomed. Eng. **2**, 258–267 (2010)
69. A. Sannigrahi, S. Chowdhury, I. Nandi, D. Sanyal, S. Chall, K. Chattopadhyay, Development of a near infrared Au–Ag bimetallic nanocluster for ultrasensitive detection of toxic Pb^{2+} ions in vitro and inside cells. Nanoscale Adv. **1**, 3660–3669 (2019)
70. A. Arshad, H. Wang, X. Bai, R. Jiang, S. Xu, L. Wang, Colorimetric paper sensor for sensitive detection of explosive nitroaromatics based on Au@ Ag nanoparticles. Spectrochim. Acta, Part A **206**, 16–22 (2019)
71. L. Zhang, W. Hou, Q. Lu, M. Liu, C. Chen, Y. Zhang, S. Yao, Colorimetric detection of hydrogen peroxide and lactate based on the etching of the carbon based Au-Ag bimetallic nanocomposite synthesized by carbon dots as the reductant and stabilizer. Anal. Chim. Acta **947**, 23–31 (2016)

72. B. Yin, W. Zheng, M. Dong, W. Yu, Y. Chen, S.W. Joo, X. Jiang, An enzyme-mediated competitive colorimetric sensor based on Au@ Ag bimetallic nanoparticles for highly sensitive detection of disease biomarkers. Analyst **142**, 2954–2960 (2017)
73. X. Xu, Plasmonic organic solar cells with gold nanoparticles (Doctoral dissertation) (2015)
74. A. Ng, W.K. Yiu, Y. Foo, Q. Shen, A. Bejaoui, Y. Zhao, H.C. Gokkaya, A.B. Djurišić, J.A. Zapien, W.K. Chan, C. Surya, Enhanced performance of PTB7: PC71BM solar cells via different morphologies of gold nanoparticles. ACS Appl. Mater. Interfaces **6**(23), 20676–20684 (2014)
75. F.X. Xie, W.C. Choy, C.C. Wang, W.E. Sha, D.D. Fung, Improving the efficiency of polymer solar cells by incorporating gold nanoparticles into all polymer layers. Appl. Phys. Lett. **99**(15), 219 (2011)
76. F. Otieno, N.P. Shumbula, M. Airo, M. Mbuso, N. Moloto, R.M. Erasmus, A. Quandt, D. Wamwangi, Improved efficiency of organic solar cells using Au NPs incorporated into PEDOT: PSS buffer layer. AIP Adv. **7**(8), 085302 (2017)
77. D. Liu, Q. Liang, G. Li, X. Gao, W. Wang, Q. Zhan, T. Ji, Y. Hao, Y. Cui, Improved efficiency of organic photovoltaic cells by incorporation of AuAg-alloyed nanoprisms. IEEE J. Photovolt. **7**(4), 1036–1041 (2017)
78. S. Zhang, Z. Chen, L. Xiao, B. Qu, Q. Gong, Organic solar cells with 2 Thenylmercaptan/AU self-assembly film as buffer layer. Sol. Energy Mater. Sol. Cells **95**(3), 917–920 (2011)
79. J.H. Lee, J.H. Park, J.S. Kim, D.Y. Lee, K. Cho, High efficiency polymer solar cells with wet deposited plasmonic gold nanodots. Org. Electron. **10**(3), 416–420 (2009)
80. S.W. Baek, G. Park, J. Noh, C. Cho, C.H. Lee, M.K. Seo, H. Song, J.Y. Lee, Au@ Ag core–shell nanocubes for efficient plasmonic light scattering effect in low bandgap organic solar cells. ACS Nano **8**(4), 3302–3312 (2014)
81. G.D. Spyropoulos, M.M. Stylianakis, E. Stratakis, E. Kymakis, Organic bulk heterojunction photovoltaic devices with surfactant-free Au nanoparticles embedded in the active layer. Appl. Phys. Lett. **100**(21), 213904 (2012)
82. D.H. Wang, D.Y. Kim, K.W. Choi, J.H. Seo, S.H. Im, J.H. Park, O.O. Park, A.J. Heeger, Enhancement of donor–acceptor polymer bulk heterojunction solar cell power conversion efficiencies by addition of Au nanoparticles. Angew. Chem. **123**(24), 5633–5637 (2011)
83. X. Xu, A.K.K. Kyaw, B. Peng, D. Zhao, T.K. Wong, Q. Xiong, X.W. Sun, A.J. Heeger, A plasmonically enhanced polymer solar cell with gold–silica core–shell nanorods. Org. Electron. **14**(9), 2360–2368 (2013)
84. X. Xu, A.K.K. Kyaw, B. Peng, Q. Xiong, H.V. Demir, Y. Wang, T.K. Wong, X.W. Sun, Influence of gold-silica nanoparticles on the performance of small-molecule bulk heterojunction solar cells. Org. Electron. **22**, 20–28 (2015)
85. W.H. Tseng, C.Y. Chiu, S.W. Chou, H.C. Chen, M.L. Tsai, Y.C. Kuo, D.H. Lien, Y.C. Tsao, K.Y. Huang, C.T. Yeh, J.H. He, Shape-dependent light harvesting of 3D gold nanocrystals on bulk heterojunction solar cells: plasmonic or optical scattering effect? J. Phys. Chem. C **119**(14), 7554–7564 (2015)
86. J.N. de Freitas, M.A. Mamo, M. Maubane, W.A. van Otterlo, N.J. Coville, A.F. Nogueira, Nanocomposites of gold and poly (3-hexylthiophene) containing fullerene moieties: Synthesis, characterization and application in solar cells. J. Power. Sources **215**, 99–108 (2012)
87. K. Jung, H.J. Song, G. Lee, Y. Ko, K. Ahn, H. Choi, J.Y. Kim, K. Ha, J. Song, J.K. Lee, C. Lee, Plasmonic organic solar cells employing nanobump assembly via aerosol-derived nanoparticles. ACS Nano **8**(3), 2590–2601 (2014)
88. H.S. Noh, E.H. Cho, H.M. Kim, Y.D. Han, J. Joo, Organic solar cells using plasmonics of Ag nanoprisms. Org. Electron. **14**(1), 278–285 (2013)
89. B.Y. Wang, T.H. Yoo, J.W. Lim, B.I. Sang, D.S. Lim, W.K. Choi, D.K. Hwang, Y.J. Oh, Enhanced light scattering and trapping effect of Ag nanowire mesh electrode for high efficient flexible organic solar cell. Small **11**(16), 1905–1911 (2015)
90. Y. Hao, J. Song, F. Yang, Y. Hao, Q. Sun, J. Guo, Y. Cui, H. Wang, F. Zhu, Improved performance of organic solar cells by incorporating silica-coated silver nanoparticles in the buffer layer. J. Mater. Chem. C **3**(5), 1082–1090 (2015)

91. S.W. Baek, J. Noh, C.H. Lee, B. Kim, M.K. Seo, J.Y. Lee, Plasmonic forward scattering effect in organic solar cells: a powerful optical engineering method. Sci. Rep. **3**, 1726 (2013)
92. T. Jägeler-Hoheisel, F. Selzer, M. Riede, K. Leo, Direct electrical evidence of plasmonic near-field enhancement in small molecule organic solar cells. J. Phys. Chem. C **118**(28), 15128–15135 (2014)
93. P.P. Cheng, L. Zhou, J.A. Li, Y.Q. Li, S.T. Lee, J.X. Tang, Light trapping enhancement of inverted polymer solar cells with a nanostructured scattering rear electrode. Org. Electron. **14**(9), 2158–2163 (2013)
94. D.H. Wang, J.K. Kim, G.H. Lim, K.H. Park, O.O. Park, B. Lim, J.H. Park, Enhanced light harvesting in bulk heterojunction photovoltaic devices with shape controlled Ag nanomaterials: Ag nanoparticles versus Ag nanoplates. RSC Adv. **2**(18), 7268–7272 (2012)
95. C.H. Kim, S.H. Cha, S.C. Kim, M. Song, J. Lee, W.S. Shin, S.J. Moon, J.H. Bahng, N.A. Kotov, S.H. Jin, Silver nanowire embedded in P3HT: PCBM for high efficiency hybrid photovoltaic device applications. ACS Nano **5**(4), 3319–3325 (2011)
96. H.C. Chen, S.W. Chou, W.H. Tseng, I.W.P. Chen, C.C. Liu, C. Liu, C.L. Liu, C.H. Chen, C.I. Wu, P.T. Chou, Large AuAg alloy nanoparticles synthesized in organic media using a one-pot reaction: their applications for high-performance bulk heterojunction solar cells. Adv. Funct. Mater. **22**(19), 3975–3984 (2012)
97. X. Li, W.C.H. Choy, H. Lu, W.E. Sha, A.H.P. Ho, Efficiency enhancement of organic solar cells by using shape-dependent broadband plasmonic absorption in metallic nanoparticles. Adv. Funct. Mater. **23**(21), 2728–2735 (2013)
98. L.W. Jang, H. Park, S.H. Lee, A.Y. Polyakov, R. Khan, J.K. Yang, I.H. Lee, Device performance of inverted polymer solar cells with $AgSiO_2$ nanoparticles in active layer. Opt. Express **23**(7), A211–A218 (2015)
99. K. Yao, M. Salvador, C.C. Chueh, X.K. Xin, Y.X. Xu, D.W. Dequilettes, T. Hu, Y. Chen, D.S. Ginger, A.K.Y. Jen, A general route to enhance polymer solar cell performance using plasmonic nanoprisms. Adv. Energy Mater. **4**(9) (2014)

Probing Noble Metal-Based Hybrid Nanostructures for SERS Opti-sensing

Shriya Subramanyam, Manoj Bhatnagar, Surjeet Chahal, and Parveen Kumar

Abstract Surface-enhanced Raman Spectroscopy (SERS), attains its high sensitivity with a large signal amplification due to the usage of noble-metal nanoparticles as substrates. This surpasses the underachieving reactivity of Raman Spectroscopy and can be further used in the detection of various analytes and for molecular recognition in the fields of food processing, biosensing along with ecological conservation. However, due to SERS substrates being non-reusable, and extremely valuable and expensive, studies are being undertaken on feasible and cost-effective SERS substrates, consisting of noble-metal hybrid nanostructures. While ensuring reusability, with an evolved morphology and chemical profile, the hybrid nanostructures help improve SERS activity. This chapter reviews the noble-metal nanostructures, consisting of silver (Ag), and gold (Au), in the form of monolayers, nanorods, nanotubes, and nanoclusters. The hybrid structures are constructed with other noble-metal nanostructures, as well as transmission metal oxides (TMOs), transition metal dichalcogenides (TMDCs) and other nanomaterials, as substrates, while discussing their synthesis techniques, characterization techniques, along with the mechanism for SERS, followed by the substrate. It also studies the impact of dielectric environment, change in nanostructure along with method of synthesis and assembly of the substrate on SERS activity.

Keywords SERS · Optical sensing · Noble metals · Hybrid nanostructures · Ag NPs

S. Subramanyam · M. Bhatnagar · P. Kumar (✉)
Materials and Nano Engineering Research Laboratory, Department of Physics, School of Physical Sciences, DIT University, Dehradun, India
e-mail: parveenpaliwal@gmail.com

S. Chahal
Department of Physics, University Centre for Research and Development, Chandigarh University, Gharuan, Mohali, India

B. Ramachandran et al. (eds.), *Advanced Optical Sensors*, Progress in Optical Science and Photonics 38, https://doi.org/10.1007/978-981-95-1687-2_2

1 Introduction

Advancements in biomedical applications of nanomaterials have seen their usage for non-invasive imaging, as well as sensing of any hazardous materials. With the increase in usage of antimicrobials to keep fruits and vegetables safe, as well as pesticides, herbicides and fertilizers during cultivation, a need for safe sensing methods has risen in recent times [1]. Along with this, it is necessary to have the ability to detect pharmaceuticals and antibiotics, as well as pathological biomarkers [2] and pathogenic factors [3]. Presence of pollutants added to the aerosol systems and well as water bodies [4] has also been a source of heightened anxiety, which are responsible for the decay of aquatic, terrestrial and human life. Overexposure to such particles can indeed increase the number of health problems faced in a day-to-day life. While these are only some of the growing issues of today, they have indicated a strong need for development in effective and efficient detection [5].

Optical sensors are often known for their high sensitivity and selectivity for the target species, are characterized by the usage of light for the detection of molecules [6]. Research into the sensors has been increased in the past few years due to their reusability, low costs, low maintenance, as well as high biocompatibility [7]. Due to the high sensitivity of the method, SERS is often used for the in-situ detection of target molecules [8]. One study has employed a miniaturized thermal purge-and-trap colorimetric/SERS dual-model optical sensor to quantify the presence of sulphur dioxide in wine, employing the Karl Fischer reaction [9].

Nanomaterials are used in optical detectors thanks to their high surface to volume ratio, providing a large region for detection of target molecules. Changes in intensity of light is often observed using interpreters which convert the light rays into quantifiable electrical signals. Noble metals are often defined by their resistivity towards corrosion, leading to their increased usage in various industries. With the growing research into noble metal nanoparticles such as silver (Ag), and Gold (Au), it is necessary to gain a thorough understanding of their abilities. It is noteworthy that these noble metals have been observed to not only improve SERS based optical sensing methodologies, but also SPR optical sensing methods, for the detection of metal ions, proteins, DNA, and other molecules [10]. Superlattice formations of noble metals are known to increase SERS sensitivity of probes prepared, having detected ~ 10^{-8} M of crystal violet (CV) using a laser with wavelength of 633 nm [11]. Noble metals have played a vital role in the research of SERS optical sensing. Raspberry-like Ag nanoparticles have been synthesized, which have been noted to quantitatively detect the presence of colorants which adulterated black tea. Results showed 0.5 to 300 μg/mL as the range of detection for the SERS substrate [12]. Levels of human α-thrombin, a measure of blood coagulation, and vascular endothelial growth factor, VEGF-165, a measure for angiogenesis, the formation of new blood vessels, were observed using silica coated substrate, consisting of gold nanoprisms costed with the target substance. They were detected in their respective trace concentrations of 100 fM and 1000 fM, using a diode laser with 785 nm, with an overall enhancement factor of ~ 10^5 [13]. A nanostructure consisting of gold nanoparticles functionalized with 4-mercaptophenylboronic acid and c-DNA, in the form of a fibre optic probe coated with chromium using magnetron sputtering. It was used to detect mercury ions, Hg^{2+}, known to cause cardiovascular diseases. With a limit of detection of 3.94 $\times 10^{-13}$ M, the probe could optically detect Hg^{2+} in the trace concentrations of 10^{-12} to 10^{-6} M, using an excitation of wavelength 633 nm [14]. Noble metal heterostructures have shown improved detectivity for target molecules, which are elaborated in further sections.

2 SERS-Opti Sensing

Optical sensing is one of the up-and-coming technologies, due to its non-invasive molecule detection abilities. Without damaging the particles or samples in its entirety, the presence of the target molecule is efficiently identified. The most generic optical sensing equipment generally consists of three parts: a light source, detector and an interpreter. A light source is often in the form of a laser, while a detector involves the use of the chosen target molecule, which may reflect, transmit or scatter the light source in certain directions or with a certain amount of intensity.

However optical sensing techniques do not include just imaging, but also help identify materials. Raman spectroscopy is one such method. When a ray of light interacts with a single molecule, a large majority of the photons simply change their direction, without any loss of energy, that is, they undergo Rayleigh scattering, also known as elastic scattering. However, due to molecular interaction, a very small fraction of photons ends up losing their energy, described as "degraded frequency" during the phenomena's discovery by C. V. Raman in 1928 [15]. This is known as Raman scattering and has since been used for the detection of specific molecular bonds via Raman spectroscopy. Specific molecular bonds are identified through Raman shifts, which are characterized by changes in wavelength. Discovered in 1974, Surface-enhanced Raman Spectroscopy (SERS) essentially enhances the intensity of a general Raman Spectroscopy for many nanoparticles.

While Raman spectroscopy enhances the signal of the wave pertaining to a particular molecule, SERS, Surface enhanced Raman spectroscopy, is known to enhance signals by ~ 10^{10}. Using a laser source often constructed in the wavelength range of 400–800 nm, SERS is able to detect trace amounts of the target molecule. The effectiveness of the SERS substrate so synthesized is described by the enhancement factor and limit of detection. While the limit of detection is measured by using the following expression [16]:

$$Limit\ of\ Detection = \frac{3\sigma}{S}$$

having σ as the standard deviation of the intensities from the blank platform, with S as the slope of the corresponding calibration curve. Meanwhile, the enhancement factor is calculated using the following formula [17]:

$$Enhancement\ Factor = \frac{I_{NC}}{I_{TM}} = \frac{I_{pNC} - I_{bNC}}{I_{pTM} - I_{bTM}}$$

where, I_{NC} is the peak intensity of the nanocomposite when used as a SERS substrate, and I_{TM} is the peak intensity of the target molecule. I_p indicates the peak intensity, while I_b describes the base intensity of the particles. But enhancement factor can be mathematically expressed in multiple ways, such as [18]:

$$Enhancement\ Factor = \frac{I_{SERS}/N_{Surf}}{I_{RS}/N_{vol}}$$

having I_{SERS} as the SERS intensity while I_{RS} is the Raman intensity. N_{Surf} is the number of molecules adsorbed on the surface of the SERS probe, while $N_{vol} = c_{RS}V$, the number of molecules in the scattered volume during Raman Spectroscopy. It is also interesting to note that enhancement factor in Raman Spectroscopy (say, E) is related to that due to SERS as $|E|^4$ [19].

SERS allows for the magnification of intensity of applied electromagnetic field, which are commonly known as "hots spots". Hots spots are often found between nanogaps or spacings, in any deformities in the lattice, as well as on any edges, or rough surfaces, making their instances dependent on the morphologies of the SERS probes prepared. This is attributed to molecular vibrations which cause localized surface plasmon resonance, in turn amplifying the intensity of applied rays [20].

3 Applications in Food-Safety

With the growing instances of usage of materials unfit for human consumption during the preparation, storage and preservation of food products, methods for detection of agrochemicals have also proved their importance, as presented in Table 1. The thiram pesticide was detected by a nanocomposite, AgNPs@4-ABT, consisting of 4-aminothiophenol (4-ABT) functionalized by silver nanoparticles (AgNP), as shown in Fig. 1a. The SERS optical sensor was built on a 0.3 cm × 0.3 cm piece of pre-treated cotton fabric ultrasonically immersed in a solution consisting of the nanocomposite, making a solid SERS substrate. Studies conducted on 2 g of apples, cabbage, and tomatoes found 1×10^{-8} M to 5×10^{-5} M of the pesticide detected, with a detection limit of 2.56×10^{-12} M, with a laser of wavelength 532 nm. This was attributed due to the creation of the dual hots spot construction which counters inhibition of hot spot formation by thiram, leading to the high enhancement factor [21]. Figure 1b presents another silver-based capillary-assisted SERS optical sensor was used to detect the presence of thiram on the surface of apples. The sensor consisted of a paper like substrate where silver nanoparticles were grown on laboratory filter paper of size 1 cm^2 which were dried in a nitrogen environment, after washing in water and ethanol. A real food sample was prepared by adding 10 µL of the standard thiram solution followed by the absorption of added methanol, followed by the addition of 50 µL of ultra-pure water. The SERS measurement look place using the real food sample prepared. It was done in an excitation of wavelength 785 nm and 0.005 ng/cm^2 was found to be the limit of detection for the pesticide, with the sampling time being less than one minute, while the substrate can be stored for forty days [22]. A paper-based metal organic framework, Zeolitic Imidazolate Framework-8 (ZIF-8), integrated with artificial antibodies (AAB), with Au@Ag coreshell nanostructures, and added to Ag electroplated on ZIF-8, making the ZIF-8@Ag/AAB/Au@Ag SERS substrate, as seen in Fig. 1c. The material was used in the optical sensing operation to detect the presence of carbyl, an insecticide which is harmful for the surrounding ecosystem. Using a laser with energy corresponding to 532 nm, the limit of detection for carbyl was found to be 5.72 × 10^{-3} μg/mL. The substrate proved effective in real samples, consisting of apples, cabbage, cucumber, pears and tomatoes, with high recovery ranging from 93.47% to 105.51% [23].

Table 1 Summarizing heterostructures used for detection of agrochemicals

Sr. No.	SERS substrate	Target molecule	Usage of target	Limit of detection	Wavelength of laser (nm)	References
1	AgNPs@4-ABT	Thiram	Pesticide	2.56 × 10^{-12} M	532	[21]
2	Ag coated filter paper	Thiram	Pesticide	0.005 ng/cm^2	785	[22]
3	Ag NP/GP	Thiram	Pesticide	7.20 × 10^{-7} M	785	[25]
4	ZIF-8@Ag/AAB/Au@Ag	Carbyl	Insecticide	5.72 × 10^{-3} μg/mL	532	[23]
5	AgNCs-GO	Permethrin	Insecticide	10^{-5} M	488	[26]
6	Au NPs & ZIF-8	Parathion-methyl	Insecticide	10^{-10} M	633	[27]
7	Au NPs & ZIF-8	Thiabendazole	Fungicide	10^{-10} M	633	[27]
8	pNIPAM@AuNR	Malachite green	Fungicide	1.58 × 10^{-9} M	633	[28]
9	AgNP@PA	Malachite green	Fungicide	10^{-7} M	532	[29]
10	SA-LB/Ag film	Paraquat	Herbicide	10^{-8} M	532	[24]
11	a-AuAg@Ti	Melamine	Food additive	10^{-7} M	785	[30]
12	MIL-101/Au-ETTH-NR	Chloramphenicol	Aquaculture antibiotic	1.5 × 10^{-9} M	652	[31]
13	MIL-101/Au-ETTH-NR	Chloramphenicol	Aquaculture antibiotic	0.01 × 10^{-9} M	785	[31]
14	Ag@r-TNR	Chloramphenicol	Aquaculture antibiotic	10^{-10} M	532	[32]

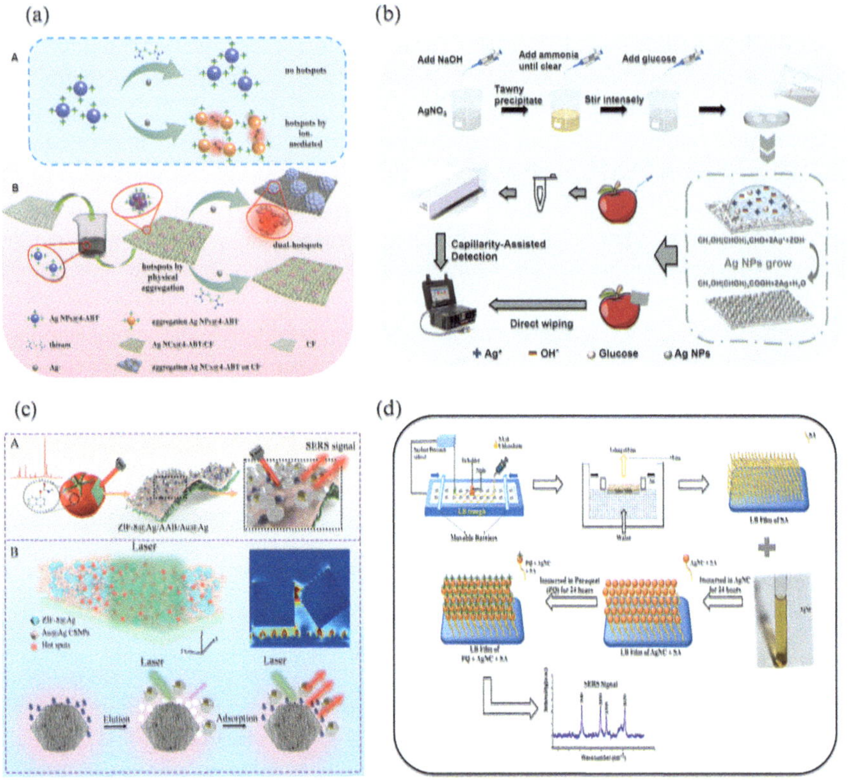

Fig. 1 Schematics of preparation of SERS sensors, as well as their application to detect various toxic chemicals: **a** AgNPs@4-ABT SERS substrate constructed on cotton fabric to optically detect thiram, (A) presence hotspots via ion-mediation, (B) schematic for hotspot creation during substrate preparation [21]; **b** silver nanoparticles/filter paper based, capillary assisted SERS optical sensor to detect thiram on surface of apples [22]; **c** ZIF-8@Ag/AAB/Au@Ag optical SERS substrate for quantitative carbyl detection, (A) the schematic showing detection on surface of tomatoes, (B) schematic for interaction between laser and substrate for carbyl detection [23]; **d** detection of Paraquat using SA-LB/Ag optical SERS sensor [24]

A nano-hybrid, pNIPAM@AuNR, composed of silver nanorods (AuNR) templated on poly N-isopropyl acrylamide (pNIPAM), was prepared by the layer-by-layer method, found to have petal-like hexagonal packing. The SERS measurement was carried using three laser sources of wavelengths 532 nm, 633 nm and 785 nm. The SERS substrate proved to have a detection limit of 1.58×10^{-9} M, equivalent to 0.73 ng/g, in the detection range 1 ng/g to 100 ng/g, along with a qualification limit of 5.21×10^{-9} M for the fungicide malachite green, often found in fish fillets. The study noted that the detection ability of the substrate was within the minimum required performance limit, as per regulations set by the European Commission. The SERS substrate so prepared was used on tilapia fillets, with ~ 57.86 ppb detected with ~ 116% of recovery [28]. 'Paraquat', chemically known as 1,1′-dimethyl-4,4'dipyridylium, is an herbicide used for weed control but also well known to cause Parkinson's disease. The herbicide's detection using a SERS substrate, SA-LB/Ag is shown using the schematic in Fig. 1d. It consisted of a bilayer Langmuir–Blodgett film, prepared using stearic acid (SA-LB), which was infused with Ag nano-colloidal system (Ag-NC), which was synthesized using Creighton's recipe. The SA-LB/Ag film composition was constructed by immsersing the SA-LB film in the Ag-NC. Under a laser with a wavelength of 532 nm, 10^{-8} M, that is, 15 ppb was found to be the substrate's limit of detection, it also proved effective in detecting the herbicide in animal systems, as per the studies done with zebrafish embryo in agricultural water found near the University of Kalyani in the city of Kalyani in India. 24 embryos were nurtured in the E3 medium, and 50 ppb was found to be the maximum non-lethal amount for the herbicide while concentrations at 150 ppb had a 50% mortality rate, with SA-LB/Ag substrate having high reproducibility while detecting the herbicide [24].

One study made a SERS probe consisting of an anisotropic gold-silver nanostructure deposited on a titanium structure, a-AuAg@Ti, to detect the presence of melamine in milk, which can falsely indicate high presence of protein but is a major health hazard. The a-AuAg nanostructure was drop-casted on 1 cm^2 sheets of titanium, having a thickness of 0.05 cm. The SERS optical sensing was carried out under an excitation of 785 nm. A low limit of detection of 0.1 μM was observed when SERS measurement was carried out, and signal reproducibility was indicated by the high enhancement factor of 1.7×10^8 [30]. A dual mode colorimetric/SERS optical sensor, MIL-101/Au-ETTH-NR consisting of elongated tetrahexahedral gold nanorods (Au-ETTH-NR), loaded on the MIL-101 metal organic framework. This was used to detect the presence of chloramphenicol, an antibacterial antibiotic used in aquaculture, but can pose danger for human life. The probe had the low limit of detection of 0.01×10^{-9} M, with detection conducted at 785 nm, while at an excitation of 652 nm the detection limit was 1.5×10^{-9} M [31].

4 Industrial Dye Detection

It is necessary to combat the unjustified disposal of textile dyes into fresh water sources, it is also crucial to study and create systematic and sustainable methods for their detection, as summarized in Table 2. A composite consisting of nanotriangle Au with Cu_2O and the $Ti_3C_2T_x$ MXene of thickness 6 nm, MXene/Au/Cu_2O, as shown in Fig. 2a. It was prepared using hydrothermal and chemical etching methods, followed by sonication, for the detection of methylene blue (MB) dye. Under an excitation wavelength of 785 nm of 5 mW power and accumulation time of 5 sec, an analytical enhancement factor of 2.37×10^9 was observed at a minimal order of ~ 10^{-12} M, as shown in Fig. 2b. Such observations were attributed to electron transfer from MB to the surface of the three nano components of the composite described [33]. MB molecules were also detected by heterostructures consisting of Au and Ag nanoparticles on a green synthesized graphene sheet, to make AuNPs/

G and AgNPs/G, respectively. After which the prepared substrate was sonicated and vacuum filtered. The heterostructures were synthesized by an initial agitation of the corresponding precursor salt in the green synthesized graphene extract, followed by vacuum filtration of the resultant colloidal solution. The corresponding enhancement factors of the AuNPs/G and AgNPs/G heterostructures were 17.02 and 5.96, under a laser of wavelength 532 nm, with 5 collections, each of 30 s [17]. Another study found a minimal limit of detection being 25 nM and 100 nM of MB was observed, as seen in the SERS spectra in Fig. 2c when silver nanospheres (Ag NPs) decorated with polydopamine (PDA), to make the AgNPs@PDA heterostructure, shown in Fig. 2d. These nanoparticles were employed as SERS substrates in ~ 600 nm excitation, in pure water and tap water, respectively. The PDA used had a thickness of 2 mm, and was coated using the colloidal solution of the prepared Ag NPs [34].

CV dye was detected using a gold (Au) rough shell heterostructure, Au NBP@Au dimers. It consisted of the deposition of Au-on-Au nanobipyramids (NBP), seen in Fig. 2e, which were synthesized using a seed mediated growth method, via the mediation using mercaptobenzimidazole-5-carboxylic acid (MBIA), creating a rough surface on the NBP due to use of lower amounts of MBIA, as shown in Fig. 2f. An excitation wavelength of 633 nm was used during the dye detection process. An enhancement factor during SERS opti-detection was 1.06×10^6, an improvement with respect to the original Au NBP, which had an enhancement factor of 3.5×10^5, for a solution with CV concentration of 1 mM [35]. Rhodamine 6G (R6G) particles were detected by a Cu_2O/Ag nanocomposite prepared via an *in-situ* reduction process. When it was used as a SERS substrate, with a laser of wavelength 532 nm, ~ 10^{-9} M of the dye could be detected, through a range of 10^{-5} M to 10^{-9} M, as shown in Fig. 2g [36]. A SERS chip, Au/CCN NWs, using carbon nitride nanowalls (CCN NWs) decorated with gold nanoparticles (Au) was employed for identifying the presence of benzidine in laundry wastewater, which is known to be used in manufacture of azo dyes and for its carcinogenic nature. Using a laser of wavelength 633 nm, power 5 mW and excitation time of 3 s, benzidine was detected in a linear range of 1–600 μg L^{-1}, with an enhancement factor of 1.76×10^6 [42].

A heterojunction, AgNPs@Co_3O_4NRs, between Ag nanoparticles (Ag NPs) and Co_3O_4 nanorods (NRs), detected the presence of MB dye and R6G. As the SERS activity was conducted using an excitation of wavelength 633 nm with a 10 s accumulation time, it was observed that the localized surface plasmon resonance (LSPR) bands of the nanomaterials were instigated. The dyes were identified to have a limit of detection at the trace concentration of 10^{-10} M, with enhancement factors of 1.12 $\times 10^6$ and 2.55×10^6, for the two respective dyes [16]. Another nanostructure, Au HN/NS composite consisting of Au nanospheres (Au NSs) embedded on Au hexagonal nanoplates (Au HNs) was synthesized using triple gap-induced plasmon coupling. The material was then used as a SERS substrate to detect CV, MB, R6G, and rhodamine B (RhB). With an enhancement factor of 1.51×10^{12}, the limit of detection was minimal at 10^{-12} M, wherein the SERS activity tests were being conducted using lasers with wavelengths of 532 nm and 785 nm. Results found at 785 nm were consistent with those at 532 nm, showcasing the enhanced efficiency of the prepared nanostructure [37]. Localized surface plasmon resonance (LSPR) was attributed to

Table 2 Summarizing heterostructures used for industrial dye detection

Sr. No.	SERS substrate	Target molecule	Limit of detection	Wavelength of laser (nm)	References
1	Mxene/Au/Cu_2O	MB	~ 10^{-12} M	785	[33]
2	AuNPs/G	MB		633	[17]
3	AgNPs/G	MB		633	[17]
4	AgNPs@PDA	MB in pure water	25 nM	~ 600	[34]
5	AgNPs@PDA	MB in tap water	100 nM	~ 600	[34]
6	AgNPs@Co_3O_4NRs	MB	10^{-10} M	633	[16]
7	Au HN/NS	MB	10^{-12} M	532	[37]
8	Au-MoO_2	Methyl blue	10^{-9} M	532	[38]
9	Ag@Fe_3O_4	MB	0.1 μM	532	[39]
10	AgNP@PA	MB	10^{-9} M	532	[29]
11	Au NPs & ZIF-8	MB	10^{-11} M	633	[27]
12	Au NBP@Au dimers	CV	1 mM	633	[35]
13	Au HN/NS	CV	10^{-12} M	532	[37]
14	Au-MoO_2	CV	10^{-9} M	532	[38]
15	Ag@Fe_3O_4	CV	0.1 μM	532	[39]
16	Ag@r-TNR	CV	10^{-12} M	532	[32]
17	Cu_2O/Ag	R6G	~ 10^{-9} M	532	[36]
18	AgNPs@Co_3O_4NRs	R6G	10^{-10} M	633	[16]
19	Au HN/NS	R6G	10^{-12} M	532	[37]
20	Au-MoO_2	R6G	10^{-9} M	532	[38]
	CarbonF/Ag	R6G	10^{-6} M	532	[40]
21	Ag NP/GP	R6G	8.73×10^{-9} M	785	[25]
22	AgNCs-GO	R6G	8.5×10^{-9} M	488	[26]
23	AuNP@Si	R6G	8.8×10^{-5} M	532	[41]
24	AuNP@Al	R6G	1.5×10^{-9} M	532	[41]
25	AuNP@Cu	R6G	1.6×10^{-9} M	532	[41]
26	Au NPs & ZIF-8	R6G	10^{-11} M	633	[27]
27	Au/CCN NWs	benzidine in laundry wastewater	1 μg L^{-1}	633	[42]
28	Au HN/NS	RhB	10^{-12} M	532	[37]

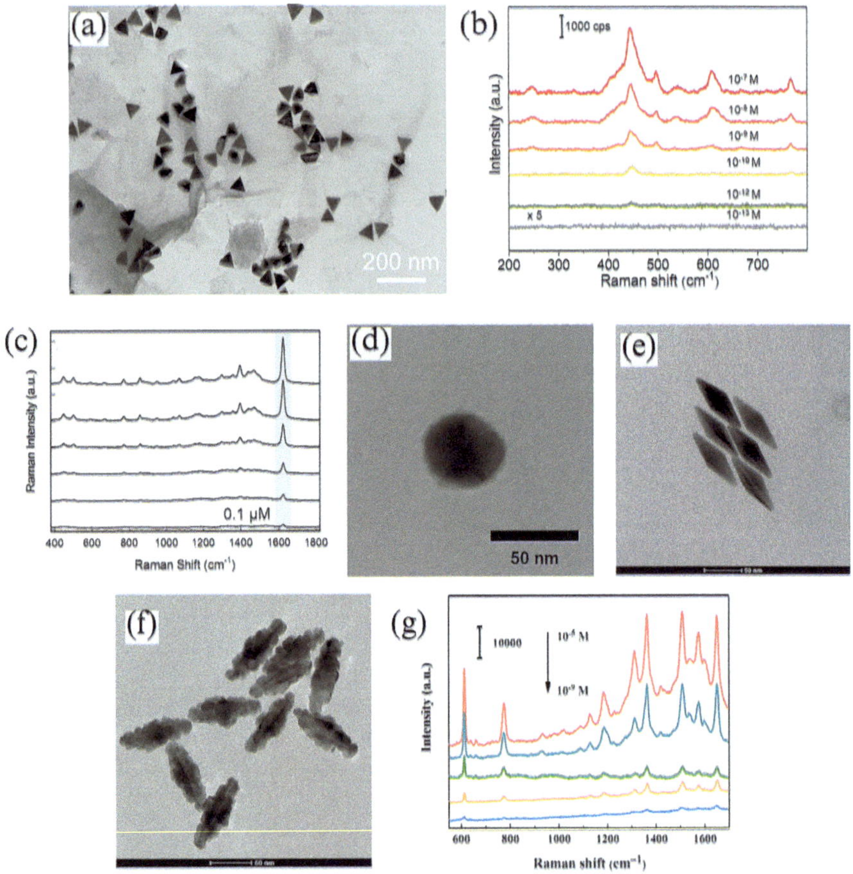

Fig. 2 **a** TEM of nanotriangular MXene/Au/Cu_2O SERS substrate to detect MB dye, upto 10^{-12} M concentration of the dye, as shown in **b** [33]; **c** detection of 100 nM (= 0.1 μM) of MB dye using AgNPs@PDA, **d** shown in TEM image of the substrate [34]; **e** TEM image of Au-on-Au nanobipyramids, **f** TEM image of enhanced Au-on-Au nanobipyramids by MBIA [35]; **g** detection of various concentrations of R6G dye using Cu_2O/Ag optical SERS substrate [36]

the enhancement of SERS by a factor of 4.9×10^6, when a Au-MoO_2 composite was used as the substrate. A low concentration of 10^{-9} M of R6G, methyl blue and CV were used as targets. The SERS optical sensing process used a laser of wavelength 532 nm. Additionally, it was noted that at a higher wavelength of 808 nm, the composite reached a temperature of 529 °C, eliminating the dye residue completely, without compromising the reliability and sensitivity of the SERS substrate [38]. MB and CV were also detected using a SERS substrate consisting of a heterostructure consisting of magnetic core–shell nano-particles Ag@Fe_3O_4, silver core–shell nanoparticles with iron (III) oxide, Fe_3O_4, added as a spacer. The SERS activity was measured using a laser with excitation of 532 nm. The substrate could detect trace concentrations of the two dyes in the range of 0.1–10 μM, with the enhancement factor for photoluminescence increasing by a factor of 24.8 [39].

5 Biomedical Applications

With the growing interest in non-invasive detection and treatment techniques, SERS substrates have intensively proved their applicability in the field, as shown in Table 3. A three-dimensional carbon framework on silver (Ag) molecules was made using a carbonizing method, followed by magnetron sputtering of Ag molecules onto the heterostructure, to create CarbonF/Ag nanoparticles. These were further used in the detection of uric acid in synthetic urine, which was followed by clinical samples. SERS measurement took place while using a laser of wavelength 532 nm, with an accumulation time of 10 s and 1 mW intensity. When R6G was used as a Raman detector, 10^{-12} M was the limit of detection found, along with an enhancement factor of 3.26×10^{9}. Using CarbonF/Ag as a SERS substrate enhanced uric acid detection, by detecting the molecule with concentrations as low as ~ 10^{-6} M, as shown in Fig. 3a [40]. Substrate performance of an Ag-WO_3/TiO_2 crystal heterostructure was studied using 4-mercaptobenzoic acid (4-MBA), as a raman detector. At an excitation of 532 nm, 10^{-13} M of 4-MBA could be detected using the heterostructure, which was attributed to the synergy of plasmic, charge transfer and slow photon effects [43]. A silver loaded graphitic silver nitrate nanocomposite (Ag@g-C_3N_4) were found to have detected 10^{-14} M of the ciprofloxacin drug, represented in Fig. 3b, often used for the treatment of any potential anti-bacterial acitivities. This was associated with the charge transfer between the two constituents, which agreed with the corresponding DFT calculations. SERS based optical detection of the drug was carried out using a He–Ne laser with a wavelength of 633 nm, having an intensity of 17 mW [44]. Another study explored the detection of methotrexate in human blood, an immunosuppressant that is often used for treating leukemia but being toxic in larger concentrations, while employing a silver nanoparticle-based SERS probe, consisting of truncated triangular silver nanoparticles on quartz, which were prepared by using silver nanospheres as shown in Fig. 3c. SERS measurements were carried using a laser with a wavelength of 532 nm with 2 mW laser power. The probe involved deposition of truncated triangular silver nanoparticles on optically transparent quartz glass. While being able detect 10^{-9} M of the immunosuppressant on synthesized surfaces, concentrations of methotrexate in the range of 10^{-6} M were detected in a clinical study consisting of 24 participants. It also observed a reduction of cost for patients, which went from 30 USD for traditional testing methods, to 7 USD, by using this novel optical SERS based detection method [45]. A high enhancement factor of 2.75×10^{10} with an extremely minimal detection limit of 1.86×10^{-12} M was observed while optically detecting nitrofurantorin, as seen in Fig. 3d, which is used for treating urinary tract infections, but is toxic is larger quantities, using SERS while employing a heterostructure, Ag-NSs/BiOBr-MFs, shown in Fig. 3e, composed of silver nanospheres (Ag-NSs) loaded on bismuth oxybromide microflowers (BiOBr-MFs), using a laser of excitation wavelength of 532 nm [46].

A microfluidic heterostructure, Au@PS PhC, consisting of photonic crystals (PhC) semi-coated by colloidal Au polystyrene (Au@PS) nanospheres was employed, as a biosensor using SERS activity for melanoma diagnosis. SERS activity was measured using a laser of wavelength 785 nm. With an enhancement factor of 2.8×10^9, the B16M10 melanoma cells were used as a model for detection, with the very low limit of detection of the biomarker, 0.82 pg/mL. The melanoma cells could be detected in the concentration range of 10^{-11} g/mL to 10^{-6} g/mL [47]. A signature component of the mitochondrial apoptotic pathway, caspase-9 is a vital member of the cascade activation reaction. A nanoprobe substrate, LEHD-AuNSs in the form of protruding Au nanostars (AuNSs), functionalized using polypeptides, was used to detect the presence of activated caspase-9 proteins. A fluorescence-based SERS optical sensing process was used, conducted using an excitation of wavelength 494 nm, where the limit of detection of the protein was found to be 3.49 ng/mL, which was attributed to localized surface plasmon resonance (LSPR) [48]. A functionalized magnetic nanostructure, Fe_3O_4@Au, constructed using a poly-ethyl-enimine (PEI)-mediated seed growth method, which was employed for the detection of the protein carbohydrate antigen 19-9 (CA19-9), which is the biomarker for colorectal cancer.

Table 3 Summarizing heterostructures used for biomedical applications

Sr. No.	SERS substrate	Target molecule	Usage of target	Limit of detection	Wavelength of laser (nm)	References
1	CarbonF/Ag	Uric acid	Present in excreted urine	10^{-6} M	532	[40]
2	Ag-WO_3/TiO_2	4-MBA	Raman detector	10^{-13} M	532	[43]
3	Ag@g-C_3N_4	Ciprofloxacin drug	Treating anti-bacterial activity	10^{-14} M	633	[44]
4	Silver nanoparticles on quartz	Methotrexate	Treating leukimia	10^{-6} M	633	[45]
5	Ag-NSs/ BiOBr-MFs	Nitrofurantorin	Treating urinary tract infections	1.86×10^{-12} M	532	[46]
6	Au@PS PhC	B16M10	Biomarker for melanoma	0.82 pg/ mL	785	[47]
7	LEHD-AuNSs	Caspase-9	Indicator or cell apoptosis	3.49 ng/ mL	494	[48]
8	Magnetic Fe_3O_4@Au	CA19-9	Biomarker for colorectal cancer	0.27 U m L^{-1}	785	[49]
9	AuNP@Cu	Micro- and nanoplastics	hazardous for living beings	3.36 µg/ mL	532	[41]

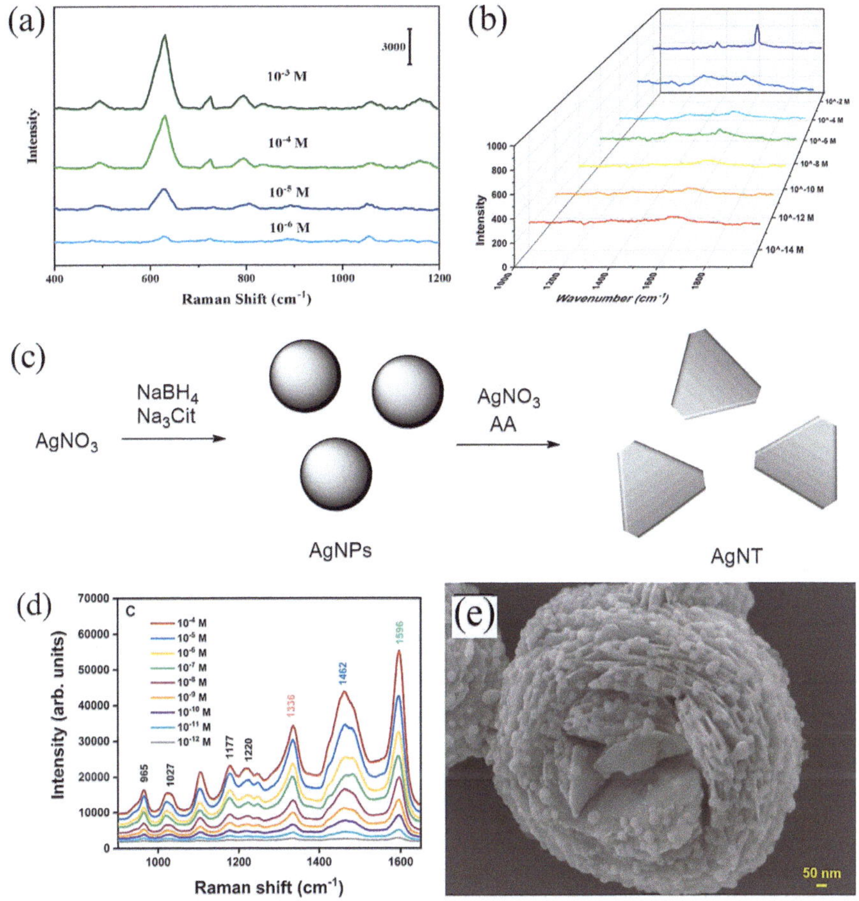

Fig. 3 **a** Detection of uric acid in a clinical sample of urine by an optical CarbonF/Ag SERS substrate [40]; **b** detection of ciproflaxin drug in clinical samples by Ag@g-C_3N_4 as a substrate [44]; **c** process of synthesis of truncated triangular silver nanoparticles for detection of methotrexate in clinical samples of human blood [45]; **d** SERS analysis to analyse the least concentration of nitrofurantorin using Ag-NSs/BiOBr-MFs, **e** FESEM of Ag-NSs/BiOBr-MFs [46]

As the SERS measurement was being carried out in a laser with 50 mW intensity, 785 nm wavelength and a 10 s exposure time, an enhancement factor of 2.76×10^7, with a detection limit of 0.27 U m L^{-1} [49].

A layer of gold nanoparticles was magnetically sputtered on an opal-structured silk fibroin, followed by layers Ag nanospheres to prepare a wearable optical SERS probe. This helped create uniformly distributed hots spots to enhance the sensitivity of the non-invasive biosensor. It was used to measure the concentration of sweat cortisol in the range of 0.1–1000 nmol/L, as well as to analyze the pH of sweat produced in the range 5.5–7.0 [50].

6 Explosive Devices

A plasma coupled Ag nanosphere internalized nanocavity using sputtering, with plasma enhanced atomic layer deposition of Al_2O_3 was used for the SERS optical detection of trinitroluene (TNT), which is often used in explosive devices. Before conducting the sensing activity, the substrate was dipped in 10^{-7} M of 4-aminothiophenol (4-ATP), allowing the 4-ATP molecules to be adsorbed on it. With a laser source with 532 nm wavelength, ~ 10^{-15} M of TNT was detected, while the substrate had an enhancement factor of 3.5×10^8. It should also be noted that the same substrate could detect 10^{-9} M of R6G [51]. A drop casting method was used to decorate silver nanoparticles on silica, SiO_2, which was used to detect the presence of different explosive materials using an optical SERS measuring system. Using a laser of excitation corresponding to 457 nm, 10^{-4} M of 1-nitroaphthalene, 10^{-7} M of 5-nitroisoquinoline, 10^{-5} M of 4-nitrophenol, and 10^{-6} M of picric acid were found to be the limits of detection for these substances [52].

7 Multiple Applications Using the Same Substrate

Another study found silver nanoparticles to be assembled on graphene paper (GP), to form Ag NP/GP. It was used to detect R6G and thiram with the respective limit of detections at 8.73×10^{-9} M and 7.20×10^{-7} M, respectively, with a maximum enhancement factor of 9.94×10^5. The pesticide was effectively detected at 0.24 ppm, which is sufficiently low below the maximum residue limit set by the U. S. Environmental Protection Agency, that is, 7 ppm. It is to be noted that the SERS measurement used a Raman spectrometer employing laser with a wavelength of 785 nm, working at an intensity of 45 mW and collection time of 3 s, conducted at room temperature. [25]. Magnetron sputtering was used to prepare a flexible SERS substrate AgNP@PA, consisting of Ag nanoparticles (AgNP) on polyamide fabrics (PA). The SERS optical sensing activity had taken place by using lasers of 532 nm and 785 nm, with an intensity of 4.5 mW and irradiation time of 10 s, accumulated over 2 turns. With an enhancement factor of 2.8×10^6, the substrate was able to detect 10^{-9} M of MB and 10^{-7} M of malachite green found on the surface of Pacific sauries, which performance attributed to localized surface plasmon resonance (LSPR) [29]. Another insecticide, permethrin, was detected by employing a fiber taper waist SERS substrate, AgNCs-GO, consisting of gold nanocoubes (AuNCs), with graphene oxide (GO). The SERS measurement took place using a micro-Raman spectrometer with a laser of wavelength 488 nm, with a power intensity of 1mW, with a 10 sec integration time, that being done only once. While the substrate could detect 8.5×10^{-9} M of R6G, it was able to show in-situ detection for permethrin of ~ 10^{-5} M in lake water. Due to the largely minimized size of the detector in comparison to microorganisms, there is no intrusion in the detection process due to the presence of aquatic life [26]. Presence of pollutants was detected in water bodies using a SERS substrate consisting of silver nanoparticles grown upon rutile titanium oxide nanoarrays (r-TNR), Ag@r-TNR, using a laser of wavelength 532 nm. Trace concentrations of the CV dye, 10^{-12} M, and the chloramphenicol aquaculture antibiotic, 10^{-10} M, were reported to have been detected with the respective enhancement factors of 4.9×10^9 and 5.6×10^8, respectively. The substrate showed high reusability with a recovery rate of 93% [32].

Another study involved the deposition of Au nanoparticles (AuNP) over metal sheets, namely, silicon (Si), aluminum (Al) and copper (Cu), to get AuNP@Si, AuNP@Al, and AuNP@Cu, respectively. With limits of detection being as low as 8.8×10^{-5} M, 1.5×10^{-9} M, and 1.6×10^{-9} M, for the three respective substrates, they had the corresponding enhancement factors of 6.19×10^2, 5.74×10^4, and 1.47×10^5, while detecting the R6G dye. It must also be noted that 1 μg/mL of 100 nm polysterene beads were also sensed by the AuNP@Cu SERS substrate using a laser of 532 nm. The substrate had a limit of detection of 3.36 μg/mL for the polystyrene nanobeads, and had also detected significant amounts of micro- and nanoplastics from samples of saliva [41]. Another SERS sensor, Au NPs & ZIF-8, consisting of Au nanoparticles (Au NPs) over zeolitic imidazolate framework-8 (ZIF-8) photonic crystals, was used to identify the presence of organic dyes, a fungicide and an insecticide. Using a monochromatic excitation of 633 nm, 10^{-10} M of thiabendazole and parathion-methyl were detected using the sensor, which pinpointed the scant concentration at 10^{-11} M for R6G and MB dyes. The sensor was observed to have an enhancement factor of 1.439×10^{11} [27].

8 Summary

This book chapter reviews the recent advancements in SERS probes prepared using noble metal-based hybrid nano-heterostructure over recent times. It highlights the current progress and work done for the detection of materials which will be hazardous or toxic to human beings in higher concentrations or if left unchecked. And with the growing use of such materials in the mundane life, it is equally necessary to develop simple, effective and affordable methods to detect them as well. Upon describing the advantages of SERS for opti-detecting over Raman spectroscopy, it covers applications across various fields including agrochemical detection and food safety, which include herbicides, pesticides, fungicides, additives and aquatic antibiotics, which are meant to be used as medicines for nursing aquatic life back to health. Optical SERS substrates to be used in biomedical applications were also studied, exploring clinical studies, as well as industrial dyes, such as R6G, CV and MB in water sources. This text reflects on the various heterostructures prepared for their use in optical detection using SERS while analyzing the wavelength in which the target molecule was detected, as well as their respective limits of detection, while describing the concentrations in which the detector works optimally.

References

1. X. Peng, Q. chen, X. Li, L. Yang, Y. Yuan, C. Zuo, Z. Zhou, Z. Bai, A semiconductor SERS sensor of corrosion-resistant PPy/GO composite film by electrochemical growth for detecting crystal violet residues in fresh fish tissue. Talanta **281** (2025). https://doi.org/10.1016/j.talanta.2024.126906
2. M.T. Campos, L.S. Pires, F.D. Magalhaes, M.J. Oliveira, A. Pinto, Self-assembled inorganic nanomaterials for biomedical applications. Nanoscale (2025). https://doi.org/10.1039/D4NR04537H
3. M. Pourmadadi, I. Ahmadi, P. Taromi, A. Rahdar, S. Fathikarkan, Z. Bradley, A. Ganguly, N. Bhalla, Cancer biosensing using plasmonic metal doped graphene-based materials. ECS Sens. Plus (2025). https://doi.org/10.1149/2754-2726/ada4bb
4. A. Srivastava, A.K. Srivastava, P.K. Dwivedi, S.K. Jha, Metal-free NITROGEN-DOPED reduced graphene oxide (N-rGO) SERS platform for detecting Rhodamine B. Sens. Act. A Phys. **383**, 116263 (2025). https://doi.org/10.1016/j.sna.2025.116263
5. I.M. Onyemaobi, Y. Xie, J. Zhang, L. Xu, L. Xiang, J. Lin, A. Wu, Nanomaterials and clinical SERS technology: broad applications in disease diagnosis. J. Mater. Chem. B (2025). https://doi.org/10.1039/D4TB02525C
6. P. Sangeetha, N. Ayyanar, G. Prabhakar, S. Rajaram, Study review of optical biosensors based on 2d materials. Plasmonics (2025). https://doi.org/10.1007/s11468-024-02646-1
7. P. Malara, A. Crescitelli, V. Di Meo, A. Giorgini, S. Avino, E. Esposito, A. Ricciardi, A. Cusano, I. Rendina, P. De Natale, G. Gagliardi, Resonant enhancement of plasmonic nanostructured fiber optic sensors. Sens. Act. B Chem. **273**, 1587–1592 (2018). https://doi.org/10.1016/j.snb.2018.07.030
8. X.X. Han, R.S. Rodriguez, C.L. Haynes, Y. Ozaki, B. Zhao, Surface-enhanced Raman spectroscopy. Nat. Rev. Methods Primers **1** (2021). https://doi.org/10.1038/s43586-021-00083-6
9. Y. Wu, M. Wang, Y. Gu, K. Ge, Miniaturized thermal purge-and-trap device combined with self-calibration colorimetric/SERS dual-model optical sensors for highly rapid and selective detection of sulfur dioxide in wine. Food Chem. X **25** (2025). https://doi.org/10.1016/j.fochx.2024.102091
10. P.C. Chen, P. Roy, L.Y. Chen, R. Ravindranath, H.T. Chang, Gold and silver nanomaterial-based optical sensing systems. Part. Part. Syst. Charact. **31**, 917–942 (2014). https://doi.org/10.1002/ppsc.201400043
11. C. Yao, W. Yan, R. Dong, S. Dou, L. Yang, Superlattice assembly strategy of small noble metal nanoparticles for surface-enhanced Raman scattering. Commun. Mater. **5** (2024). https://doi.org/10.1038/s43246-024-00506-3
12. Q. Cui, J. Shen, H. Jia, T. Li, S. Cao, S. Dong, Y. Wei, L. Zou, Y. Chen, Y. Wang, J. Ning, L. Li, Simultaneous detection of mixed colorants adulterated in black tea based on various morphological SERS sensors. Food Res. Int. **199** (2025). https://doi.org/10.1016/j.foodres.2024.115364
13. P. Anantha, P. Raj, P. Zheng, S. Tanwar, I. Barman, Gold nanoprism enhanced SERS aptasensor for simultaneous detection of thrombin and VEGF. Sens. Act. B Chem. **423** (2025). https://doi.org/10.1016/j.snb.2024.136811
14. L. Li, J. Zhao, M. Jin, S. Wu, F. Liu, Y. Huang, N. Feng, Y. Liu, Dual-mode SPR/SERS optical fiber sensor for ultra-trace mercury ions detection. Opt. Express **33**, 2247 (2025). https://doi.org/10.1364/OE.547601
15. C.V. Raman, K.S. Krishnan, A new type of secondary radiation. Nature **121**, 501–502 (1928)
16. R. Tamizhselvi, N.R. Barveen, C.W. Zeng, T.Y. Liu, C.L. Liu, A.A. Napoleon, Y.W. Cheng, AgNP-enriched Co_3O_4 nanorods as an SERS substrate for the simultaneous detection of methylene blue and rhodamine 6G. J. Alloys Compd. **1011** (2025). https://doi.org/10.1016/j.jallcom.2024.178388

17. R.D. Ávila-Avilés, I.G. Becerril-Juárez, A.R. Vilchis-Nestor, In situ green synthesis of graphene-based material nanocomposites decorated with silver and gold nanoparticles for improved SERS sensing. Opt. Mater. (Amst.) **160** (2025). https://doi.org/10.1016/j.optmat.2025.116730
18. E.C. Le Ru, E. Blackie, M. Meyer, P.G. Etchegoint, Surface enhanced Raman scattering enhancement factors: a comprehensive study. J. Phys. Chem. C **111**, 13794–13803 (2007). https://doi.org/10.1021/jp0687908
19. E.C. Le Ru, P.G. Etchegoin, Rigorous justification of the |E|4 enhancement factor in surface enhanced Raman spectroscopy. Chem. Phys. Lett. **423**, 63–66 (2006). https://doi.org/10.1016/j.cplett.2006.03.042
20. S.L. Kleinman, R.R. Frontiera, A.I. Henry, J.A. Dieringer, R.P. Van Duyne, Creating, characterizing, and controlling chemistry with SERS hot spots. Phys. Chem. Chem. Phys. **15**, 21–36 (2013). https://doi.org/10.1039/c2cp42598j
21. Y. Chen, H. Wang, J. Zhou, L. Zhang, Z. Xing, Q. Zhang, L. Xia, Dual hot-spot SERS test strip for picomole level analysis of thiram. Food Control **170** (2025). https://doi.org/10.1016/j.foodcont.2024.111035
22. J. Cheng, S. Zhuang, Z. Luo, S. Chen, Y. Shen, J. Ye, Z. Yu, T. Zhao, Y. Xie, Simple and rapid capillarity-assisted ultra-trace detection of thiram on apple surface with a silver nanoparticles/filter paper SERS substrate. J. Hazard Mater. **485** (2025). https://doi.org/10.1016/j.jhazmat.2024.136945
23. Y. Pan, P. Zhao, W. Cai, H. Fang, X. Wang, J. Zhao, P. Zhu, H. Yang, S. Ge, J. Yu, MOF@Ag/AAB/Au@Ag composite matrix full-dimensional divergence effect-based SERS paper sensor for rapid carbaryl quantification. Food Chem. **472** (2025). https://doi.org/10.1016/j.foodchem.2025.142885
24. A. Pal, P. Maity, M. Azaharuddin, S. Chakrabarty, S. Nandi, A. Das, S. Ghosh, U. Sett, P. Bandopadhyay, S. Nandy, J. Chowdhury, T. Basu, Silver nano-colloid particles embedded on Langmuir-Blodgett film matrix of stearic acid serving as a SERS active sensor for detecting the herbicide 'Paraquat'. Spectrochim. Acta. A Mol. Biomol. Spectrosc. **329** (2025). https://doi.org/10.1016/j.saa.2024.125514
25. J. Wang, Y. Feng, H. Zhang, L. Han, J. Xia, G. Wang, Uniformly aligned Ag NPs/graphene paper for enhanced SERS detection of pesticide residue. Spectrochim Acta A Mol. Biomol. Spectrosc. **329** (2025). https://doi.org/10.1016/j.saa.2024.125535
26. P. Tao, Y. Luo, K. Zheng, J. Cao, L. Zhu, X. Wang, W. Zhang, S. Dai, Q. Zou, P. Zhang, FTW SERS probes with Ag NCs-GO composite structure excited by evanescent wave for in situ detection of permethrin. Anal. Chim. Acta **1341** (2025). https://doi.org/10.1016/j.aca.2025.343646
27. W. Zheng, D. Teng, I. Kalsoom, G. Murtaza, Z. Meng, L. Qiu, Au-modified metal-organic framework photonic heterostructure as a SERS sensor for enrichment and detection of trace organic dyes and pesticides. Sens. Act. B Chem. **422** (2025). https://doi.org/10.1016/j.snb.2024.136684
28. B. Hu, D.W. Sun, H. Pu, Q. Wei, A dynamically optical and highly stable pNIPAM @ Au NRs nanohybrid substrate for sensitive SERS detection of malachite green in fish fillet. Talanta **218** (2020). https://doi.org/10.1016/j.talanta.2020.121188
29. S. Zhang, J. Ao, L. Ding, Y. Chen, J. Xu, Z. Sun, Y. Li, L. Wu, X. Liu, S. Jiang, Flexible, robust and reusable Ag nanoparticle-coated polyamide SERS substrates for rapid identification of toxic substances on curved surfaces. Microchem. J. **209**, 112899 (2025). https://doi.org/10.1016/j.microc.2025.112899
30. M. Shameer, K. Vijai Anand, J.B.M Parambath, S. Columbus, H. Alawadhi, Direct detection of melamine in milk via surface-enhanced Raman scattering using gold-silver anisotropic nanostructures. Spectrochim. Acta. A Mol. Biomol. Spectrosc. **327** (2025). https://doi.org/10.1016/j.saa.2024.125412
31. H. He, Z. Yang, L. Peng, F. Jiang, Y. Xu, Y. Yang, Y. Wang, Q. Chen, Q. Chen, T. Jiao, Colorimetric/SERS dual-mode detection for chloramphenicol through MIL-101/Au ETHH NR based catalysis and structural degradation. Food Biosci. 106032 (2025). https://doi.org/10.1016/j.fbio.2025.106032

32. T.N.Q. Trang, N.T.G. Bao, V.T.H. Thu, Photoassisted decoration of Ag nanoparticles onto TiO_2 nanorod arrays for continuously reusable ultrasensitive SERS sensors. Mater. Sci. Semicond. Process **186** (2025). https://doi.org/10.1016/j.mssp.2024.109057
33. Y.X. Zhao, Z.X. Zheng, L.S. Zhang, J.R. Feng, L. Ma, S.J. Ding, Optimized electromagnetic enhancement and charge transfer in MXene/Au/Cu_2O hybrids for achieving efficient SERS. Phys. Chem. Chem. Phys. (2023). https://doi.org/10.1039/d3cp00887h
34. S. Schiavi, A. Taglietti, A. Magni, P. Galinetto, B. Albini, Increasing SERS performance of silver nanoparticles with nanometric coating of polydopamine: a novel approach for methylene blue detection. Appl. Surf. Sci. **687** (2025). https://doi.org/10.1016/j.apsusc.2024.162223
35. H. Cai, G. Weng, J. Li, G. Wu, J. Zhu, J. Zhao, Controlling the growth mode of Au depositing on Au nanobipyramids via ligand coverage for SERS enhancement, Spectrochim. Acta A Mol. Biomol. Spectrosc. **325** (2025). https://doi.org/10.1016/j.saa.2024.125082
36. S. Qin, Z. Li, Y. Zhang, J. Pan, J. Cao, C. Li, J. Wang, Synthesis and optimization strategy for enhancing SERS performance of Cu_2O nanoparticles. Opt. Mater. (Amst.) **159** (2025). https://doi.org/10.1016/j.optmat.2024.116548
37. Z.X. Zheng, Z.L. Zheng, W.X. Xia, X.B. Chen, L. Ma, S.J. Ding, Triple gap-induced plasmon coupling in Au nanoplate/nanosphere bilayer with strong and uniform electromagnetic hotspots for sensitive and stable SERS detection. Sens. Act. B Chem. **429** (2025). https://doi.org/10.1016/j.snb.2025.137325
38. W. Li, J. Liu, R. Wu, A. Qian, S. Li, C. Wei, Ultrasensitive and light erasable surface-enhanced Raman scattering substrates based on Au–MoO_2 heterostructures. Talanta **287**, 127669 (2025). https://doi.org/10.1016/j.talanta.2025.127669
39. S. Guo, S. Wang, L. Wang, T. Jiao, M. Wang, Sensitive metal-enhanced photoluminescence and Raman scattering magnetic core–shell nano-particles with tunable size for the detection of methylene blue. Opt. Laser Technol. **181** (2025). https://doi.org/10.1016/j.optlastec.2024.112061
40. Y. Dong, J. Ma, J. Bian, Q. Zou, T. Jiang, C. Zhao, J. Zhang, X. Bian, C. Gu, Reliable detection of uric acid in urine by SERS technique on three-dimension carbon framework decorated with Ag nanoparticles. J. Mol. Liq. **423**, 127020 (2025). https://doi.org/10.1016/j.molliq.2025.127020
41. U. Pant, J. Tate, X. Liu, N. Birse, C. Elliott, C. Cao, From automated Raman to cost-effective nanoparticle-on-film (NPoF) SERS spectroscopy: a combined approach for assessing micro- and nanoplastics released into the oral cavity from chewing gum. J. Hazard Mater. **486** (2025). https://doi.org/10.1016/j.jhazmat.2024.136978
42. S. Zhang, J. Pei, Y. Zhao, X. Yu, L. Yang, Gold nanoparticles decorated crystalline carbon nitride nano-walls as a SERS chip for rapid and sensitive detection of benzidine. Talanta **283** (2025). https://doi.org/10.1016/j.talanta.2024.127057
43. M.A. Apostolaki, E. Sakellis, S. Gardelis, V. Likodimos, Interplay of plasmonic and charge transfer effects for ultrasensitive Ag-WO3/TiO_2 photonic crystal SERS sensors. Mater Adv (2024). https://doi.org/10.1039/d4ma00995a
44. A.K. Bello, M.T. Abdullahi, M.N. Tahir, A.A. Al-Saadi, SERS activity of silver nanoparticles and silver-modified 2D graphitic carbon nitride towards ciprofloxacin drug. Spectrochim. Acta A Mol. Biomol. Spectrosc. **326** (2025). https://doi.org/10.1016/j.saa.2024.125237
45. E.A. Demishkevich, S.A. Stefanova, A.Y. Zyubin, V.V. Rafalskiy, A.S. Zozulya, D.O. Evtifeev, A.A. Kundalevich, A.A. Tatarinova, A.A. Anoshin, I.I. Lyatun, I.G. Samusev, Sers-based methodology for nanomolar methotrexate concentration detection for clinics. Spectrochim. Acta A Mol. Biomol. Spectrosc. **331**, 125801 (2025). https://doi.org/10.1016/j.saa.2025.125801
46. T. Kokulnathan, K.M. Arun Kumar, T.J. Wang, A. Joseph Anthuvan, C.L. Chang, Unlocking the potential of bismuth oxybromide microflowers supported silver nanosphere as a SERS platform for detection of nitrofurantoin. Appl. Surf. Sci. **686** (2025). https://doi.org/10.1016/j.apsusc.2024.162145
47. W. Wang, W. Mao, H. Sun, F. Hou, W. Wang, W. Liu, Z. Shi, G. Lin, M. Wang, G. Fang, Y.Y. Cheng, C. Xu, Microfluidic SERS biosensor based on Au-semicoated photonic crystals for

melanoma diagnosis. Biosens. Bioelectron. **271** (2025). https://doi.org/10.1016/j.bios.2024.116983

48. X. Ma, Y. Pan, X. Lin, Z. Wang, Polypeptide functionalized gold nanoprobes for SERS-fluorescence detection and imaging of caspase-9 during apoptosis. Anal. Chim. Acta. **1335** (2025). https://doi.org/10.1016/j.aca.2024.343458
49. M. Xu, G. Chen, Y. Huang, H. Song, Z. Wu, F. Jiang, L. Fu, C. Bi, X. Cao, W. Wei, A simple SERS sensor based on antibody-modified Fe_3O_4@Au MNPs for the detection of CA19-9 in CRC patients. Anal. Methods (2024). https://doi.org/10.1039/d4ay01382d
50. M. Hu, K. Zhu, J. Wei, Z. Xu, K. Yang, L. Wu, S. Zong, Z. Wang, Wearable microfluidic SERS patch based on silk fibroin for the non-invasive monitoring of sweat cortisol and pH. Sens. Act. B Chem. **427** (2025). https://doi.org/10.1016/j.snb.2024.137152
51. C. Song, H. Huo, W. Shi, J. Xu, X. Lei, B. Gu, Z. Han, W. Zhang, Surface plasmon-coupled Ag nanosphere-internalized nanocavity arrays for ultrasensitive SERS sensing. Sens. Act. B Chem. **426** (2025). https://doi.org/10.1016/j.snb.2024.137069
52. N. Mazur, V. Dzhagan, O. Kapush, O. Isaieva, P. Demydov, V. Lytvyn, V. Chegel, O. Kukla, V. Yukhymchuk, SERS of nitro group compounds for sensing of explosives. RSC Adv. **15**, 252–260 (2025). https://doi.org/10.1039/d4ra07309f

Noble-Metal-Based Composites for High-Performance SPR Sensors

Archana Yadav, M. A. Ibrar Jahan, Sarita Kumari, R. Srujana, Boris A. Malomed, and Shatrughna Kumar

Abstract Surface plasmon resonance (SPR) is a very captivating sensing technology because it can find molecules without labels and analyze their interactions in real time. This chapter explores the fundamental principles of SPR, its components, and its applications in various fields. This chapter also explores the noble metal-based composites, particularly incorporation of Gold and Silver have attracted extensive attention of scientific research due to its exceptional abilities to improve the performance parameter of surface plasmon resonance (SPR) sensor. These noble materials possess unique photoelectric properties, good stability and low biotoxicity, which makes SPR sensors highly sensitive and selective detection for the various applications such as biosensing, chemical sensing, environmental sensing etc. Various noble metals, such as gold, silver, platinum, and palladium, are discussed, highlighting their unique optical and electrical properties, stability, and biocompatibility. Furthermore, the chapter explores using composite materials to further enhance the sensing parameters. During the process, different problems are discussed, such as fabrication challenges and cost efficiency. Future issues are also discussed, such as how to combine artificial intelligence (AI) and the internet of things to make advanced sensing possible. The chapter ends with a discussion of how SPR sensing could be

A. Yadav (✉)
Department of ECE, Integral University, Lucknow, Uttar Pradesh, India
e-mail: archanayadav.ay@gmail.com

M. A. Ibrar Jahan
Department of ECE, RNS Institute of Technology, Bengaluru, Karnataka, India

S. Kumari
Department of ECE, Amity University, Ranchi, Jharkhand, India

R. Srujana
Department of ECE, Jain University, Bengaluru, Karnataka, India

B. A. Malomed · S. Kumar
Department of Physical Electronics, Faculty of Engineering, School of Electrical Engineering, Centre for Light-Matter Interaction, Tel Aviv University, Tel Aviv, Israel

S. Kumar
School of Engineering and Sciences, Department of ECE, SRM University, Amaravati, Andhra Pradesh 522502, India

B. Ramachandran et al. (eds.), *Advanced Optical Sensors*, Progress in Optical Science and Photonics 38, https://doi.org/10.1007/978-981-95-1687-2_3

used for personalized medicine, smart agriculture, and security purposes. This highlights the need for people from different fields to work together to get around current problems and encourage new ideas.

Keywords Surface plasmon resonance · Noble metals · Nanomaterials · Sensitivity · Plasmonic biosensors

1 Introduction

An extremely sensitive and potent optical method for studying molecular interactions in real time without the use of labels or markers is surface plasmon resonance, or SPR. Optical surface plasmon resonance (SPR) biosensors are the most advanced and complex optical label-free biosensor technology, with a wide range of uses in biotechnology, drug screening, medical diagnostics, environmental protection, and food safety and security [1]. It is used to measure the changes in the refractive index (RI) near the surface of a sensor chip, usually because of the binding of molecules such as proteins, nucleic acids, or small molecules to a surface that is immobilized with a target molecule. Surface plasmon waves, which are collective oscillations of electrons at a metal–dielectric interface (often gold or silver) when light strikes the surface at a particular angle, are the basis for SPR.

When a biomolecule (analyte) binds to a molecule (ligand) that is immobilized on the sensor surface, the local RI near the surface changes, leading to a shift in the angle at which the plasmon resonance occurs. This shift is proportional to the amount of bound analyte and can be measured in real-time. It is used extensively in biosensors and chemical sensors [2–5] (Fig. 1).

Surface plasmon resonance happens when light energy is converted into the surface plasmon. (A) When molecules bind, the refractive index near the surface changes, generating a shift in the angle (1 → 2) that is proportional to the mass of bound material [6].

The Components of an SPR System includes—

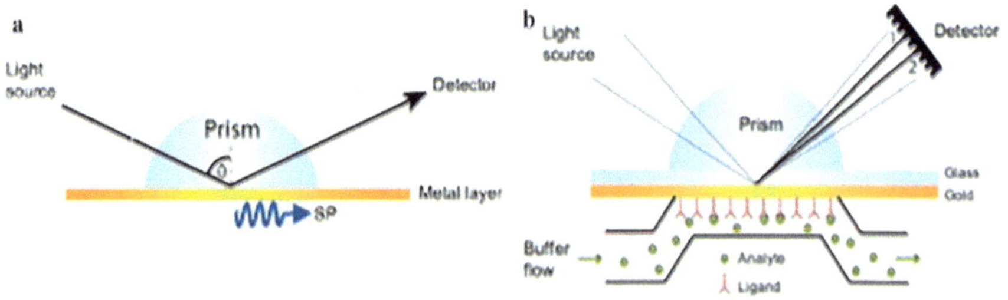

Fig. 1 An illustration of the Kretschmann-Raether setup for SP optical excitation is as follows

- **Sensor Chip**: The surface where molecules are immobilized. This is typically a thin layer of metal film, usually silver or gold, coated over a glass or plastic surface.
- **Light Source**: A monochromatic light source, typically a laser, directs light at the sensor chip at a range of angles.
- **Prism**: A coupling prism directs the light toward the sensor chip at the required angle.
- **Detector**: The detector measures the intensity of reflected light as a function of angle. The resonance angle shift can be observed here.
- **Flow Cell**: A chamber that allows the injection of liquid containing the analyte (test solution) over the sensor surface.

SPR is widely used in drug discovery and basic research to study protein–protein interactions for antibody-antigen binding, can be used to monitor DNA/RNA hybridization and interactions, Small Molecule Screening, Environmental Monitoring etc. The benefits of SPR encompass label-free detection, real-time monitoring, high sensitivity, and the provision of quantitative data that can be utilised to ascertain interaction parameters.

1.1 Noble Metals

Noble metals play a critical role in the various functions of SPR sensors. These sensors are extensively utilized in biosensing, chemical detection, and material science because of their ability to identify the changes in the refractive index (RI) near a sensor surface. Noble metals like platinum, silver, and gold are integral to SPR technology because of their unique electrical and optical properties. A detailed discussion is presented in the following sections:

1.2 Noble Metals in Surface Plasmon Resonance Phenomenon

Noble metals are responsible for the generation of surface plasmon polaritons (SPPs). SPPs are collective oscillations of electrons at the interface of a metal and a dielectric. The idea behind SPR sensing is that these oscillations are very sensitive to changes in the refractive index close to the metal surface. Gold and silver, especially, can support surface plasmon resonances at visible and near-IR wavelengths, making them ideal for SPR applications. Noble metals like gold and silver exhibit excellent electrical conductivity. This property is fundamental because the oscillating electrons needed for plasmon resonance require good conduction to keep the resonance going and make the device more sensitive. Noble metals are chemically stable and resistant to oxidation, which makes them durable in real-world sensing environments, where exposure to various chemicals and biological materials occurs. Noble metals are stable, which makes it easy to attach biomolecules or chemical probes to them. This lets them interact in specific ways to help with detection. Noble metals, especially gold and silver, can be easily fabricated into nanoparticles or thin films, which significantly enhance the sensitivity of SPR sensors. These structures can increase the surface area for interactions, improving the sensor's ability to detect smaller changes in the refractive index.

1.3 Role of Composite Materials in Enhancing SPR Performance

Surface Plasmon Resonance (SPR) is a powerful way to look at how molecules interact with each other. Composite materials make SPR work much better. SPR typically relies on metal films (usually gold or silver) as the surface where surface plasmons (oscillations of free electrons) are excited by incident light. The properties of these materials directly influence the sensitivity and resolution of SPR measurements. Composites, which are made by putting together two or more different parts, have special benefits in SPR applications that make them work better in many ways.

In summary, surface plasmon resonance is an essential tool for understanding molecular interactions in real time, especially in biosensing, drug development, and diagnostics. Its ability to provide kinetic and affinity data without labels makes it a versatile and widely used technique across many fields. Noble metals are indispensable in SPR sensors due to their ability to support surface plasmon resonances, high conductivity, chemical stability, and ease of surface functionalization. Their unique optical properties, particularly when nanostructured, enhance the sensitivity and versatility of SPR-based sensors in a wide range of scientific and industrial applications. The composite materials contribute significantly to improving SPR performance through enhancements in sensitivity, signal quality, versatility, and longevity. By optimizing the structural, optical, and chemical properties of the sensor surface, composites enable SPR to be a more effective tool for a wider range of applications, from biomolecular interactions to environmental monitoring. As new composite materials are developed, further advancements in SPR technology are expected, driving innovation in the fields of diagnostics, biotechnology, and material science [7, 8].

2 Noble Metals and Their SPR Characteristics

The noble metals are a separate group within the transition metals. They are very resistant to oxidation, corrosion, and chemical reactions, even at very high temperatures. These metals are positioned primarily in Groups 8–11 and Periods 5–6 of the periodic table and are often referred to as "noble" due to their inherent chemical stability, resistance to tarnishing, and degradation over time [9]. The key members in group 8–11 include gold (Au), platinum (Pt), silver (Ag), palladium (Pd), rhodium (Rh), iridium (Ir), and osmium (Os). Gold is particularly valued for its rarity, aesthetic appeal, and outstanding corrosion resistance. Platinum, with its remarkable thermal stability and resistance to chemical attack, is integral to catalytic processes, industrial applications, and the production of fine jewellery. While silver provides lower corrosion resistance compared to gold and platinum, it remains a noble metal due to its oxidation resistance, despite its tendency to tarnish over time. Palladium, a platinum group metal (PGM), is commonly employed in automotive catalytic converters, hydrogen storage, and electronic components [10, 11].

Also, noble metals have special surface plasmon resonance (SPR) properties that make light-matter interactions stronger when they interact with electromagnetic fields. This behaviour is essential in various technological fields, such as biosensing, plasmonic, and surface-enhanced Raman spectroscopy (SERS). Gold and silver nanoparticles, in particular, are widely used in biosensing for label-free molecular detection by tracking shifts in SPR wavelengths, which enables highly sensitive detection. These metals are very important for making plasmonic devices. SPR lets you control light very precisely in uses like photonic circuits and collecting solar energy. Raman scattering is greatly increased by SPR in SERS, which makes chemical detection and molecular analysis more sensitive. Platinum and palladium also have unique catalytic properties that are affected by their SPR features. This makes them very important for heterogeneous catalysis in chemical processes used in industry. Because they are stable, have SPR effects, and work well as catalysts, noble metals are very important for making progress in many different areas of technology [12].

2.1 SPR Characteristics of Noble Metals

Gold (Au): Gold is good for SPR-based biosensors because it doesn't oxidize or corrode easily. This means that molecular interactions can be detected accurately and sensitively, and the surface can be changed to fit biomolecules. This ensures high sensitivity and long-term sensor stability. Among the first metals used to treat diseases, gold was used for rheumatoid arthritis. Gold compounds, such as Au (I) thiolates, were used to treat it. These compounds were later turned into a drug called auronofin, which is a gold complex with triethylphosphene. For a longer period, no other gold-based drug except auranofin was used to treat rheumatoid arthritis or any other disease. But as time passed, it was observed that drugs like auranofin showed anticancer activities that were similar to cisplatin, a common chemotherapy drug. As a result, researchers started creating gold compounds that could treat cancer but were found to be too toxic for the heart in the human body [13]. Over the years, Au (III) compounds were made to treat cancer, but they eventually failed because they had the same side effects as cisplatin, including drug resistance, high levels of toxicity, and not being able to target cancer cells well. This slowed down the development of Au (III) compounds for any disease treatment. Various gold-based compounds, such as Au(I) and Au (III), have been developed for the treatment of diseases like malaria, tuberculosis, and HIV. However, research has shown that it's unclear whether these compounds directly inhibit parasite growth, suggesting that other mechanisms may be involved. While Au (I) phosphine compounds and N-heterocyclic gold complexes were believed to be effective against malaria, they also caused harm to other parts of the body. Similarly, Au (III) compounds aimed at treating HIV have not been completely effective. Therefore, further investigation is required to fully understand the potential of these gold-based therapies [14].

Silver (Ag): Light interacting with the surface of silver creates oscillations that make it sensitive to changes in the refractive index near the surface. This is useful for finding biomolecules that are bound to it. Silver is better than gold at plasmonic properties at some light wavelengths. It also has a stronger SPR signal and can pick up on small changes in the environment near its surface more easily. But with time, it was observed that when exposed to air, silver can tarnish and oxidize, leading to reduced performance in SPR biosensors. Silver oxide and other compounds form on its surface due to oxidation, which affects the immobilization of biomolecules [9]. A silver metal plate in SPR was primarily used for the medical treatment due to its anti-inflammatory and antibacterial properties but has no role to play in the human body except being toxic for the lower organisms, like bacteria. A majority of the bacteria die, and those that survive are neutralized by silver in their tiny cell walls. In the early '80s, silver nitrate was not only used to treat warts and itchiness conditions in the human body but also to cure blindness caused by infections in newborns. Due to its ability to fight bacteria, various silver compounds are developed for the treatment of numerous diseases. Silver compounds like silver sulfadiazine were used to treat burn injuries. Silver saccharinate complexes were found much more effective than silver nitrate for fighting common bacteria like Staphylococcus aureus and E. coli. Likewise, for the treatment of cancer and cystic fibrosis, which is a chronic infection in human lungs, silver nanoparticles have shown potential. Its ability to resist bacteria, cancer treatment, and low toxicity in humans makes it a valuable tool in the future of medicine.

Platinum (Pt): Cisplatin was the first metal-based chemotherapy drug used to treat several diseases, mostly cancer. The fact that it was toxic and drug-resistant raised the question of whether it could cure cancer. The major drawback of cisplatin was its ability to bind DNA in both healthy and cancerous cells, which leads to severe side effects like anemia, hair loss, and kidney failure. Also, a small portion of the drug reaches the DNA target, and the remaining starts interacting with other proteins in the body. Noble metals like gold and silver were unable to address these issues, and hence, Pt (IV) compounds like iproplatin and satraplatin were developed [10]. These compounds are less toxic and can be taken orally. With the development of these compounds, the development of cisplatin needed to be stopped. Alternative platinum compounds like picoplatin and trans-Pt (II) complexes were developed to do so. To damage the cancer cells more accurately, polynuclear platinum complexes were developed that bind the DNA in a completely different way and cause more damage to cancer cells. As years passed, several Pt-based drugs were developed that could not only damage the disease cells more accurately but also provide an alternative to cisplatin. Drugs like the trans-Pt (II)-chloride complex with 2-(2-hydroxyethyl) pyridine show higher cancer activity than cisplatin. Some complexes, such as azolato-bridged dinuclear Pt (II), kill more pancreatic cancer cells than cisplatin because they are stronger against some tumor cells. The trinuclear Pt (II) complex is more damaging to DNA than cisplatin because it can kill cells in tumors that are not sensitive to it. When platinum drugs are used to treat hormone-sensitive tumors like breast, ovarian, and prostate cancer, hormones are added to make them more selective and effective. Platinum-based compounds are being explored not only for cancer treatment but also for their antifungal properties. These complexes have shown effectiveness against various bacterial strains and parasites. The development of new platinum drugs has not only enhanced treatment effectiveness and selectivity but also enhanced safety for cancer while also investigating their potential in other health conditions [13].

Palladium (Pd): Palladium (Pd) can be considered a potential alternative to platinum (Pt). Pd and Pt have a lot in common, but they are also very different in important ways. For example, Pd is more likely to exchange ligands, which leads to rapid hydrolysis and deactivation. Pd (II) complexes have shown promise in treating a wide range of cancers, such as leukemia, breast, cervical, and ovarian cancer. Some of these complexes are polyamine-based polynuclear complexes, heterocycles based on nitrogen, and ligands containing sulfur. They kill cancer cells more effectively than cisplatin. Pd (II) complexes like Staphylococcus aureus and E. coli also show antibacterial and antifungal properties against various pathogens. Traditionally, rifampicin was used for the treatment of tuberculosis, but platinum-based compounds have now emerged as significant drugs for its treatment. Pd (II) complexes have not only shown potential in the treatment of cancer but also in antimicrobial and radiation applications. A broad comparison of all these noble metals in terms of their characteristics is shown in Table 1 [12].

2.2 *Applications of Noble Metals Based SPR*

The unique SPR properties of noble metals have led to their widespread use in various fields:

Biosensing and Medical diagnosis: Noble metals are used to find biomolecular interactions close to the surface. These interactions include protein–protein binding, antibody-antigen binding, and DNA hybridization. Due to the binding events taking place near the surface of the metal, changes in the refractive index are detected by the SPR, which makes it an essential tool in various diseases diagnosis and detection. The SPR wavelength gradually shifts due to the change in the refractive index around nanoparticles surrounded in the environment. The ability of the SPR to detect small changes in the refractive index makes it a

Table 1 Comparison of various characteristics of Au, Ag, Pt, and Pd

Characteristics	Gold (Au)	Silver (Ag)	Platinum (Pt)	Palladium (Pd)
Plasmonic properties	Supports plasmon resonance in visible and near-infrared range; stable response	Supports plasmon resonance in UV and visible spectrum; stronger plasmonic resonance	Weak surface plasmon resonance response	Moderate surface plasmon response
Oxidation resistance	Highly resistant to oxidation	Prone to oxidation, requires oxidation coating	Oxidizes faster than gold	Oxidizes relatively easily
Biocompatibility	Excellent	Moderate	High	High
Sensitivity	High	Stable	Low	Moderate
Cost	High	Lower than gold	Very high	High

useful tool for the detection of low-concentration analytes. The SPR's sensitivity is greatly affected by the shape and type of metal used. Gold and silver are the two common metals used, with silver having a smaller dielectric constant and offering sharper SPR bends. The shape of the noble metal plays a major role in the detection of various diseases. Ag nanotriangles (triangular nanoparticles), arranged in arrays, provide higher SPR sensitivity, which is useful in the detection of changes in the environment with UV–visible spectroscopy. For the diagnosis of diseases like Alzheimer's, these nanoparticles are functionalized with molecules that focus on the specific target and help in curing the disease. Thus, by selecting proper nanoparticles with the appropriate shape and size, SPR sensitivity can be enhanced, which makes it an effective tool for the detection of even very small quantities of biological and chemical substances [12–14].

Photothermal Therapy: Photothermal therapy is a cancer treatment technique that involves the process of absorbing light and converting it to heat. Noble metals such as gold and silver absorb the light in the near-infrared range (NIR); their electrons start to oscillate collectively, leading to surface plasmon resonance. The energy radiated through this resonance is then converted into heat. The converted heat is then focused on the target to destroy the tumour cells or any other abnormal tissue in the body. Thus, noble metals are highly effective in photothermal therapy because of their ability to absorb light at a specific wavelength and then convert this light energy into heat energy. The process of converting light energy into heat energy is known as plasmon absorption, which occurs due to the excitation of the electrons in the nanoparticles to the surrounding medium as heat. As a result, nanoparticles with a high cross-sectional area and low scattering are considered best for photothermal therapy. Compared to other nanoparticles, gold (Au) has a much higher absorption coefficient and a lower laser intensity [13]. This means that it can target harmful tissues without hurting healthy ones. For the treatment of deeper tumours, the shape of the nanoparticles plays a crucial role. Gold nanorods are thought to be better for treating these kinds of tumours than spherical nanoparticles because the light from the spherical nanoparticles doesn't get absorbed as well by the tissues. This makes the SPR shift to the near-infrared region, which lets the light go deeper into the body. So, these nanoparticles can be tuned to work with immune targeting strategies to help treat deeper tumours. These strategies target specific cancer cells, which makes the therapy very accurate. Gold nanocage nanostructures tuned with SPR properties and silica core gold shell particles are mostly used in photothermal therapy to kill cancer cells.

2.3 *Challenges and Future Directions*

Noble metals like gold (Au) and silver (Ag) have drawn attention due to their unique optical properties from both experimental and theoretical perspectives. These nanoparticles have great potential in biomedical applications when combined with nanostructures that contain biomolecules. This is because they can absorb light, scatter it, and improve spectroscopy at the surface. This makes them useful for biosensing and imaging. With all the developments going on, the next major step in this field will involve the challenge of nanoparticles and real-life clinical application integration. A clear understanding of how these nanoparticles interact with living organisms without damaging the useful tissues has to be taken. An additional challenge will be optimizing optical and spectroscopic signals to better treat any disease in the background of biological activities. With all the research and development, the noble metals used in SPR biosensing will experience significant and necessary growth shortly, especially in diagnostic and medical applications, which are currently a global concern [11].

SPR nanostructures made from noble metals like gold and silver are gaining attention due to their optical properties. Advanced growth in biomedical applications, such as biosensing and imaging, can be seen because they are combined with biomolecules. However, there are still significant challenges that have to be overcome before fully integrating nanoparticles with clinical applications. Still, plasmonic nanoparticles will become a useful tool in biosensing in the future if chemists, nanoparticle researchers, and researchers from other fields work together in a strong and advanced way [12].

3 Noble-Metal-Based Composites: Composition and Design

In composite nanoparticles, it is possible to integrate many functions into a single unit, and it is even possible to generate some qualities that cannot be acquired by just mixing various components. This is a property that is both distinctive and appealing. Studies on the optical characteristics of nanoparticles made of noble metals may be traced back to the year 1857, when Michael Faraday developed a variety of gold colloids with a range of colours. As of right now, we are aware that the various colours are brought about by surface plasmon resonance (SPR). A metallic nanoparticle of suitable dimensions interacts with light photons, confining a resonance photon within its perimeter, so producing this phenomenon. These one-of-a-kind optical features have sparked the development of potentially useful uses for nanoparticles made of noble metals. Various ideas and methodologies have been proposed to replicate the mutual excitation of electrons in metallic particles. These include the T-matrix, boundary element method (BEM), finite differences in the time domain (FDTD) and the discrete dipole approximation (DDA) theory proposed by Mie [15]. These theories and methods have been proposed in order to increase the level of comprehension regarding the plasmonic properties. Nanoparticles of silver and gold are examples of noble metal nanoparticles. that have been examined the most for the purpose of exploring the optical characteristics of metallic composite nanoparticles. This is due to the fact that their surface plasmon resonance (SPR) peaks are situated in the visible range. It is well knowledge that the form, content, and architecture of nanoparticles have a significant impact on the magnetic resonance (SPR) peak of these particles. The total plasmonic capacity of composite nanoparticles depends on the ratio of every component, so one should address this aspect. For instance, Kim and his colleagues have utilised the dark-field spectromicroscopy technique in order to explore the LSPR features of individual AgAuAg nanorods. The scattered spectra of hetero nanorods show longitudinal resonance wavelengths which are insensitive to Au and Ag relative composition. This suggests that the arrangement of the nanostructures themselves as well as their general shape define the plasmonic features of the hetero-metallic nanostructures. Chen et al. conducted research on the plasmonic characteristics of bimetallic nanoparticles with a star shape that were composed of gold and silver [16] (Fig. 2).

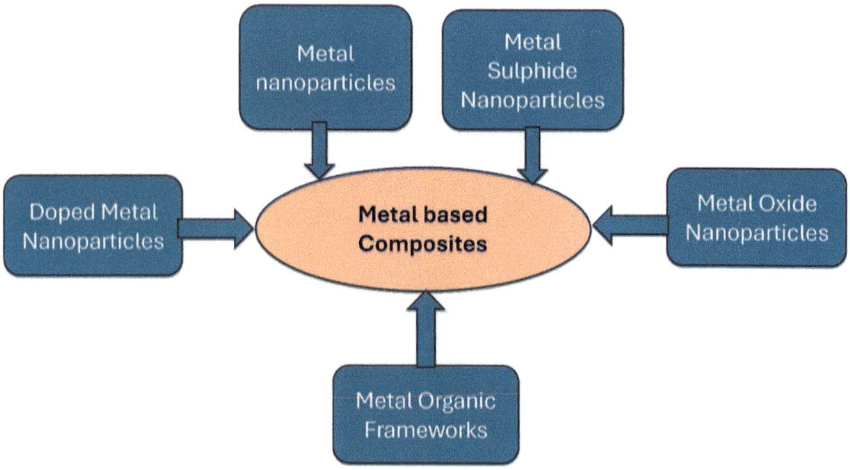

Fig. 2 Classification of metal based composites

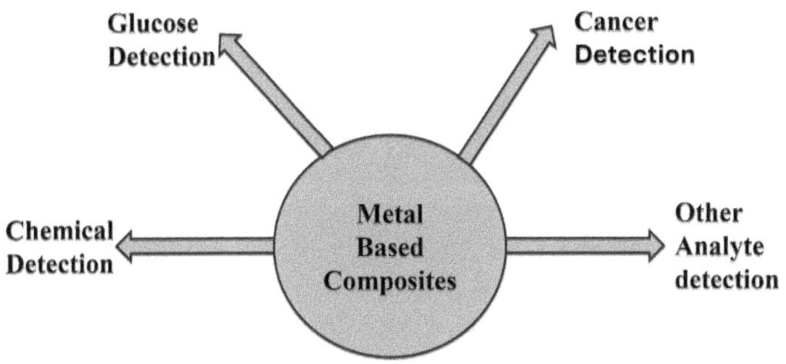

Fig. 3 Metal composites for detection of analytes

The composite nanoparticles' SPR wavelength was discovered to be quite reliant on the gold to silver ratio. Yang and colleagues studied the plasmonic properties of Ag-Au hollow nanocubes by means of the galvanic reaction between $HAuCl_4$ and silver nanocubes, Yang and associates observed at the plasmonic properties of Au–Ag void nanocubes. Their results showed that changing the gold source concentration in the combination allowed one to change the SPR wavelength of the Au–Ag void nanocubes. It was recently discovered that the SPR peak can be tuned by designing the architecture of composite nanomaterials, such as core–shell architectures, three-dimensional architectures, and other similar structures. The bonding effect between the silver and gold nanoparticles would be one illustration of this, which may dramatically boost the plasmonic characteristics of the material. These qualities can be modified by the compositions and structures of the material. When it comes to core–shell nanostructures, it has been demonstrated that the outer shell is the factor that determines the general sort of optical response (Fig. 3).

The SPR wavelength undergoes a pronounced red shift as a result of the presence of a gold core within a silver shell. In a system that is rich in gold, a single outer layer of silver can create a resonance that is similar to that of silver. Donoval and his colleagues have performed calculations to determine the influence that the silver shell has on the general plasmonic characteristics of Ag@Au core–shell nanostructures. For the purpose of gaining a better understanding of the coupling impact that the shell and the core have on one another, this was done. In accordance with the findings of theoretical research, an increase in the silver ratio within the shell is associated with a blue shift in the SPR wavelength. Based on the findings of their investigation into the effects of the Au–Ag alloy shell on the plasmonic characteristics, Maye's team came to the conclusion that plasmonic characteristics of the Au@Au–Ag composite nanoparticles could be controlled by modifying the alloy's processing temperature, thickness and composition thickness of the alloy. This conclusion was reached as a result of the findings of their investigation. Furthermore, Singh and Soni have demonstrated the effects of shell on optical characteristics by means of multilayer shell-based Al_2O_3@Al@Au@Ag core–shell nanostructures in which the composition of Au in nanostructures could be methodically tuned by means of multilayer shells. SPR wavelength could thus be systematically tuned by means of shell. It has been established that the augmentation of the electromagnetic field is greatly improved as a result of the effective decoupling from the substrate. This finding suggests that the plasmonic characteristics may also be influenced by the interactions that occur between composite nanoparticles and the substrate. [15, 17, 18]. Extensive investigation on the plasmonic characteristics of composite nanoparticles based on noble metals has been conducted on several noble metal-based composite nanoparticles including Ag–Pt, Au–Pd, Au–Ni, Au–Cu, Ag–Pd, Ag–Cu, Rh–Pd Cu–Pd and Fe–Pd, these nanoparticles might motivate some interesting uses of composite nanoparticles based on noble metals.

4 Performance Enhancement Using Noble-Metal Composites

One of the leading Bio-Sensing technologies in the world for clinical analytics and diagnostics is Surface Plasmon Resonance or SPR. As SPR offers fast, real-time, accurate, label-free sensing of organic and biological composites, it has been increasingly implemented in various capacities [19]. The classic SPR device is constructed by plating a Noble metal which is likely to be Gold or Silver onto the triangular Prism. The working of the SPR is based on "Total Internal Reflection [TIR]". This method is reliant on the generation of evanescent/localized/non-radiative SPR waves on the metal surfaces [20, 21]. For detection and analysis of bimolecular kinetics, the noble metals such as Gold, Silver and platinum have always played an integral role as they offer faster and accurate detection capability, higher absorption, better stability and greater sensitivity to the environmental changes and super conductivity etc. [22, 23]. Despite having extraordinary optical and physical properties, the Noble metals (NMs) have challenges when it comes to their implementation in optical biosensors, challenges such as: Device Fabrication, stability, Limit of detection (LOD), selectivity and replicability etc. Though NMs are crucial in SPR biosensing, mitigating their drawbacks like stability, selectivity, and sensitivity needs radical surface modifications, hybrid material amalgamation, and advanced designs in order to implement their real-world applications. This chapter addresses the aforementioned concerns and how Noble composites are implemented in modern age SPR biosensors.

4.1 Fundamentals of Noble Metal Composites

The Noble metal composites are synthesized by combining noble metals such as Gold, Silver, platinum or palladium with other elements or compounds to form hybrid nano-structures with favourable optical, chemical and physical properties to address real time challenges. The Noble Metal Composites are a refinement of the desired properties of Noble Metals while mitigating the undesirable limitations. For example, they have the inherent properties such as Plasmonic behaviour, Biocompatibility whereas drawbacks like limited functionality is mitigated and cost too can be optimized by incorporating specific low-cost materials. [24].

4.2 Different Kinds of Noble Metal Composites

Noble Metal composites can be classified on the basis of constituents of the composites. They can be classified into the four following categories (Fig. 4):

1. **Core–Shell Nanostructures**

   These are the composites where the noble metals form the core or the Shell is a dielectric material, for example, Au-SiO_2 or Ag-Polystyrene. These Core shell NCs have major benefits such as less toxicity, elevated dispersion of core particles, thermal and chemical stability and improved adsorption properties which plays a pivotal role when it comes to SPR biosensors [25, 26].
2. **Hybrid Nanoparticles**

   Noble Metals can be integrated with magnetic materials, quantum dots, biomolecules to introduce multi-functionality [27].
3. **Bi-metallic Nanostructures**

   When you layer two different metals, such as silver- gold or gold- platinum, their optical and catalytic properties can be integrated to create Bi-metallic nanostructures [28].
4. **Nanocomposites**

   Combination of polymers, carbon-based materials or ceramics with Noble Metals in the form of Nanocomposites leads to improved thermal, mechanical or optical properties [29].

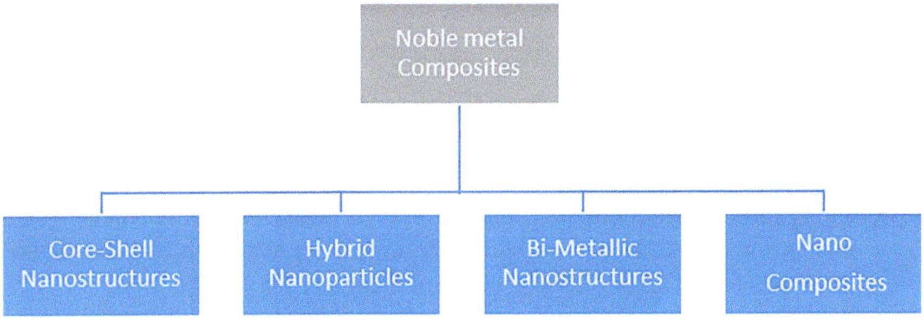

Fig. 4 Classification of global metal composites

4.2.1 Essential Components in Noble Metal Composites

1. **Noble Metals**: Due to their Outstanding optical, chemical and catalytical properties, noble metals like gold (Au), Silver (Ag), Platinum (Pt), Palladium (Pd) are chosen.
2. **Material Accoutrements**: Material additives to Noble metal composites may contain:
   - Dielectrics: Noble Metal composites may contain dielectrics like silica (SiO_2) or Titanium Dioxide (TiO_2) or any other oxides.
   - Polymers: Noble Metal composites may also be comprised of polymers like polyaniline or PEG or any other biocompatible or conducting polymer.
   - Carbon based Materials: Carbon- based materials like graphene, graphene oxides too can be implemented in noble metal composites.
   - Miscellaneous Metals: Noble Metal composites may also include miscellaneous materials such as alloys or bi-metallic amalgamations such as gold-silver or gold-copper.

4.3 Benefits of Adapting Noble Metal Composites into SPR Biosensors

In consideration of SPR biosensors, in terms of sensitivity, selectivity and stability, Noble Metal Composites have marked advantages over untainted noble metals.

The following is a list of performance parameters that are improved by the incorporation of Noble metal composites over noble metals that remain untainted.

1. **Enhanced Sensitivity**
   - Dampening in SPR biosensors refers to the loss of energy or decay in oscillations. Dampening can be caused by multitude of factors such like material properties, surface, adhesion, surrounding media etc., Dampening can lead to several breakdowns in the sensor's performance like reduced sensitivity and dull resonance peak.
   - If noble metals are combined or hybridized with a dielectric material such as silica or titanium dioxide, the loss or decay in oscillations can be assuaged. A decrease in dampening losses leads to sharper resonance and better detection.
   - Using core–shell nobke metal composites offer better field confinement, which means the electromagnetic field is concentrated which means the device possesses better sensitivity and perform at a more impressive level.
   - Additionally, adapting a non-noble metal additive material would extend the surface area for analyte adsorption which would also in turn, increase sensitivity [30].
2. **Better Stability**

- The noble metal surface of biosensors is susceptible to corrosion or degradation due to environmental factors such as oxidation
- This can impact the device performance.
- Using polymer composites or insulating the noble metal from the environment with use of additive materials as illustrated in the gold–silica or silver-polymer structures would protect the noble metals especially silver from oxidation and environmental erosion.
- Using graphene or graphene oxides in conjunction with noble metals will directly improve the thermal stability of the device.
- Particle aggregation or the 'clumping' of nano particles can be alleviated with the implementation of layered or core–shell structures to improve long term stability [31].

3. **Ameliorated Selectivity**
   - Using Noble metal composites, say graphene oxides or polymers, they can introduce surface variations to the device.
   - Using biocompatible elements, for example, silica or certain proteins can positively impact the interaction with biomolecules without impacting the device response.
4. **Augmented Range of the Spectrum**
   - Usage of noble metal composites instead of untainted noble metals can give designers better control over size, shape, composition of nanostructures and this means the device performance can be optimized across a wide range of wavelengths.
   - For biomedical applications, the device needs to perform at near infra-red regions. This can be achieved by adopting gold silver alloys, which show hybrid plasmonic resonance and improve detection, allowing the device to function in near infra-red region.
5. **Sagacious**
   - As devices using precious materials as noble metals need to be resource conscious, using noble metal composites would make any device more resource conscious and judicious.
   - Using carefully designed and engineered devices, produced using noble metal composites is a way to introduce devices that perform superior to others in many fields of science.

5 Applications in Biosensing

Biosensors can be defined as a "Compact analytical device or unit incorporating a biological or biologically derived sensitive recognition element integrated or associated with a physio-chemical transducer".

These are some of the applications of biosensors in today's world.

5.1 Glucose Monitoring

Biosensor is made up of three main parts sensitive recognition element, signal transducer element and data evaluation component. The molecular recognition elements include receptors, enzymes, antibodies, nucleic acids, microorganisms and lectins. Glucose measurements is based on interaction with any of these enzymes hexokinase, glucose oxidase (GOx) or glucose-1-dehydrogenase (GDH). These enzymes are used in the catalysis of oxidation of β-D-glucose by molecular oxygen to produce gluconic acid and hydrogen peroxide. The hydrogen peroxide generated is oxidized at a catalytic anode, typically made of platinum. This oxidation releases electron, creating an electrical current. This electrical current is propotional to the number of glucose molecules present in blood. The generated electrical signal is processed by biosensors. Three general strategies are used for glucose sensing; measurement of oxygen consumption, measurement of hydrogen peroxide produced by the enzyme reaction or by using a diffusible or immobilized mediator to transfer the electrons form the GOx to the electrode [32–34]

5.2 Pathogen Detection

The detection of pathogenic bacteria is key the prevention and identification of problems related to health and safety. In the areas of food industry, where failure to detect an infection may lead to terrible consequences. Traditional method may take 5–7 days to give result.

The development of new biosensors have made the process timing rapid to detect any pathogen in certain areas where it is required. Still some of the developments are being done to make it efficient and reduce run-time error [35].

5.3 Cancer Cell Detection

Cancer is the deadliest disease in the world. It takes life of many people per year. The preliminary detection of cancerous cell is very helpful to increase the rate of patient. for decades scientist, have focused their efforts in developing biosensors. Biosensors, in addition to being employed in other practical scenarios can essentially function as cost effective and highly efficient devices for this purpose. Traditional methods are expensive and time taking. Biomarker based cancer diagnostics is the most potential having method for early detection [36–38].

5.4 Agriculture

- Soil Monitoring: Biosensors can detect and measure soil nutrients, pH levels, and contaminants, along with helping farmers to optimize fertilization and irrigation [39].
- Crop Health: They can monitor plant health by detecting pathogens, pests, and diseases early on, allowing for timely intervention [40].
- Quality Control: Biosensors are used to assess the quality of agricultural products by detecting contaminants, toxins, and ensuring compliance with safety standards.
- Environmental Monitoring: They help in monitoring environmental conditions, such as air and water quality, to ensure sustainable farming practices.

5.5 Veterinary [41]

- Animal Health: Biosensors can detect diseases and infections in animals by monitoring biomarkers in blood or other bodily fluids.
- Drug Monitoring: They are used to monitor the levels of veterinary drugs in animals to ensure proper dosage and avoid overdosing.
- Reproduction Management: Biosensors can track reproductive hormones and cycles in animals, aiding in breeding programs.
- Food Safety: In the food industry, biosensors ensure that animal products are free from harmful pathogens and contaminants before they reach consumers [9].

6 Challenges and Limitations

Practical applications of noble metal composites in SPR biosensors, does improve its performance, they also come with obstacles and challenges of their own. The following is a list summarizing the most common limitations and challenges encountered.

6.1. Process to fabricate composites is complicated.
6.2. Concerns regarding stability.
6.3. Adapting to biological changes.
6.4. Restrictions on selectivity.
6.5. Barter of optical properties.
6.6. Inefficiency in cost and resource management.
6.7. Challenges in integrating with existing platforms.

6.1 Process to Fabricate Composites is Complicated

- Homogeneity in Nanostructures:
  - During the fabrication process of noble metal composites, it is a very arduous task to maintain consistent size, shape and structure of the nanostructures.
  - If any structure is inconsistent then the resulting sensor can have very unreliable have very unreliable performance.
- Scalability Concerns:
  - It can be a herculean and an expensive task to preserve the reproducibility of the noble metal composites and fabricate them in larger quantities [42].

6.2 Concerns About Stability

- Environmental Sensitivity:
  - One of the biggest impacts on sensor reliability happens due to environmental processes such as oxidation and degradation.
  - Noble metal composites like silver are particularly sensitive to environmental changes.
- Mechanical Stability:
  - Most biosensors need a suitable environment to work under, especially those designed using noble metal composites [43].
  - Under extreme conditions or even unsuitable conditions of temperature, fluctuations, pH variations, the composites under such conditions have a chance of undergoing biochemical instability.

6.3 Adapting to Biological Changes

- Biofouling of the surface:
  - The SPR biosensors are often used to analyze complex biological samples such as blood, saliva etc.
  - Such exposure can cause bacteria build-up on the analyte surface as even adsorption of certain proteins which may influence the refractive index of the media which can reduce specificity.
- Toxicity Concernment: In biomedical application, certain noble metal composites, especially those fabricated with non-biodegradable polymers and oxides raise concerns regarding possible toxicity [44].

6.4 Restrictions in Selectivity

- Non-Specific kind of Interactions:
  - Often surface functionalization is not adequate to prevent non-specific interactions in noble metal composites.
  - Noble metal composites that work with complex media such as urine or serum are particularly susceptible to these kinds of interactions.
- Limitations of Functionalization: In material interfaces, it can be a cumbersome task to even partially improbable to achieve steady and effective target analyte specific functionalization of composite functions [45].

6.5 The Barter of Optical Properties

- Effects of Damping: The quid pro quo arrangement of optical properties is an inherent struggle where the trad may not be worth it. For example, designing noble metal components with the use of dielectric material or polymers make a structurally sound device but it can lead to dampened plasmonic resonance in noble metals which in turn, reduces sensitivity.
- Wavelength Shifts: The margin of error in design, definition and control of material compositions is non-existent because, composites may be instrumental in fine tuning plasmonic properties, it may come at the cost of spectral range or other performance parameters [46].

6.6 Inefficiency in Cost and Resource Management

- High Costs: Noble metals like gold, silver, platinum has always been cost intensive due to their rarity. However, there is a chance that the integration of noble metal composites may not necessarily convert to relative cost reduction.
- Synthesis Costs: The biggest issue in producing noble metal composites is that they require advanced techniques and even special resources like machinery making it arduous to mass produce and the cost of synthesizing one unit alone might be too high making the design unviable from business point of view [47].

6.7 Challenges in Integrating with Existing Platforms

- Platform compatibility:

Existing SPR sensor designs would require extensive modifications to successfully integrate with noble metal composites which would increase the required development time and cost as well

- Interpretating signals:

  When analysing multiple analytes, composites with multifaceted plasmonic behaviors can create challenges in accurately interpreting the SPR signals [47].

7 Future Perspectives

It has become pertinent in recent years; one of the most effective methods for the real-time, label-free detection of chemicals, biomolecules, and environmental contaminants is Surface Plasmon Resonance (SPR) sensing technology. This technique has been used extensively in research to improve the three most effective sensor parameters: sensitivity, selectivity, and portability. When combined with artificial intelligence (AI) and the Internet of Things (IoT), new and emerging noble metal composites are anticipated to have enormous potential to further revolutionize this technology. In particular, the Surface Plasmon Resonance Technology for Intelligent Sensing for a wide range of applications is anticipated to be utilized. For the future generation of intelligent sensing technology, this combination of technologies has the potential to be portable and miniaturized, particularly when utilizing the SPR phenomena. It is possible to summarise a thorough analysis of the prospects and trends for the future in this.

- Utilising Advanced Noble Metal Composites to Improve SPR Performance
- AI-Powered Intelligent Sensing Based on SPR
- Smart SPR Sensing Networks Powered by IoT
- New and Potential Uses for SPR Sensors
- Challenges and Opportunities for Additional Study.

7.1 Utilizing Advanced Noble Metal Composites to Improve SPR Performance

Future studies will concentrate on metal composites with hybrid nanocomposites, such as next-generation materials based on platinum-graphene hybrids and gold-silver alloys. Another type of Nobel material is based on nanocomposites, which may be combined with functional materials like MXenes, quantum dots, etc. to create my Nobel material. This helps to allow multifunctional SPRS sensors.

7.2 AI-Powered Intelligent Sensing Based on SPR

The inclusion of AL algorithms in SPRS sensors makes them ideal for real-time monitoring, as they enable the autonomous detection, classification, and quantification of analytes without the need for human involvement. The SPR sensing system will also have AI models that can be explained. These models will help people understand how decisions are made, which will ensure trust and transparency in important areas like healthcare and monitoring the environment.

7.3 Smart SPR Sensing Networks Powered by IoT

IoT-enabled SPR sensors need to get better at connecting and sending data in order for smart cities and IoT ecosystems to work well with the next generation of communication networks, especially 5G and beyond. For large-scale monitoring applications, including water quality assessment, infrastructure health monitoring, and air quality monitoring, a distributed sensing network—a network of miniature sensors—can be set up.

7.4 Trends in SPR Minimisation and Portability

Wearable technology and SPR sensing technologies have the wonderful potential to operate together as wearable SPR sensors. It may be included into patches or smartwatches to provide ongoing health monitoring and early illness identification. Additionally, inexpensive metal composite incorporated SPR sensors may be disposed of. The usage of single-use sensors for a range of applications, such as environmental sampling and quick diagnostic testing, would be greatly advanced by this. The need for regular battery replacement will be eliminated with the development of new energy technologies such as solar, thermal, and kinetic energy.

7.5 New and Potential Uses for SPR Sensors

By offering real-time information on biomarkers, medication effectiveness, and patient health, AI-integrated SPR sensors will make personalized treatment possible. However, IoT-enabled SPR sensors may be utilized in smart agriculture to track pesticide levels, crop health, and soil quality, allowing for precision farming and sustainable agricultural methods. Additionally, it may be used to many security and defense-related tasks, such as real-time biological threat detection, chemical warfare agent detection, and explosives detection. Additionally, its tiny sensor can be used in space missions to track environmental conditions and find evidence of extraterrestrial life.

7.6 Challenges and Opportunities for Additional Study

The goal is to develop affordable, scalable production processes for noble metal composites and tiny SPR devices. Multidisciplinary material scientists, AI experts, Internet of Things engineers, and sensor developers should work together to promote innovation developing global standards for the design, calibration, and data interpretation of SPR sensors that integrate AI and IoT. We are addressing ethical concerns about data privacy, security, and the use of AI in sensitive applications. In industries including manufacturing, transportation, and healthcare, SPR sensors will be crucial to human–machine collaboration, enabling safer and more intelligent human-autonomous system interactions.

Taking everything into account, it can be concluded that the combination of new noble metal composites, AI, and IoT will revolutionize SPR sensing by making it more accessible, intelligent, and portable. These developments will open up new avenues for industrial automation, environmental monitoring, healthcare, and other fields.

8 Summary

SPR has grown into a strong method for detecting molecular interactions in real time and without labels. It has many uses in biomedical environmental monitoring. Noble metals, such as gold, silver, platinum, and palladium, are widely used in SPR due to their superior plasmonic properties and increased chemical stability. When composite materials were added, they made the SPR sensing performance even better. They also made the detection more accurate and stable. Despite a few challenges, such as fabrication feasibility, environmental sensitivity, and cost-related issues, widespread adoption remains possible. Future advancement can be expected with the integration of artificial intelligence and Internet of Things-enabled smart sensing devices. These innovations possess the potential to drive progress throughout a wide range of scientific and industrial applications.

References

1. X. Guo, Surface plasmon resonance-based biosensor technique: a review. J. Biophotonics **5**(7), 483–501 (2012)

2. B.A. Prabowo, A. Purwidyantri, K.C. Liu, Surface plasmon resonance optical sensor: a review on light source technology. Biosensors **8**(3), 80 (2018)
3. Y. Tang, X. Zeng, J. Liang, Surface plasmon resonance: an introduction to a surface spectroscopy technique. J. Chem. Educ. **87**(7), 742–746 (2010)
4. A. Olaru, C. Bala, N. Jaffrezic-Renault, H.Y. Aboul-Enein, Surface plasmon resonance (SPR) biosensors in pharmaceutical analysis. Crit. Rev. Anal. Chem. **45**(2), 97–105 (2015)
5. D. Capelli, V. Scognamiglio, R. Montanari, Surface plasmon resonance technology: recent advances, applications and experimental cases. TrAC, Trends Anal. Chem. **163**, 117079 (2023)
6. G.I. Janith, H.S. Herath, N. Hendeniya, D. Attygalle, D.A.S. Amarasinghe, V. Logeeshan, Y.S. Wijayasinghe, et al., Advances in surface plasmon resonance biosensors for medical diagnostics: an overview of recent developments and techniques. J. Pharm. Biomed. Anal. Open, 100019 (2023)
7. N. Ravindran, S. Kumar, C.A. Mamathi, S.N. Thirunavookarasu, C.K. Sunil, Recent advances in surface plasmon resonance (SPR) biosensors for food analysis: a review. Crit. Rev. Food Sci. Nutr. **63**(8), 1055–1077 (2023)
8. A. Kumari, A. Yadav, O.P. Singh, P. Sharan, A review of surface plasmon resonance (SPR) technology in biosensing: innovations, applications and future trends. J. Opt., 1–9 (2024)
9. Y.A. Ali, A. Hilal, P. Tabassum, A. Akil, O. Mohammad, M.I. Ismail Iqbal, A. Qari Huda, U. Khalid, M.I.M. Nasir, Recent advances in metal decorated nanomaterials and their various biological applications: a review. Front. Chem. **8** (2020)
10. M. Ahamed, H.A. Alhadlaq, M.A. Khan, P. Karuppiah, N.A. Al-Dhabi, Synthesis, characterization, and antimicrobial activity of copper oxide nanoparticles. J. Nanomater. **2014**, 17–21 (2014)
11. F.S. Arakawa, Q.L. Shimabuku-Biadola, M. Fernandes Silva, R. Bergamasco, Development of a new vacuum impregnation method at room atmosphere to produce silver–copper oxide nanoparticles on activated carbon for antibacterial applications. Environ. Technol. **1**, 1–12 (2019)
12. S. Ebrahimi, Developing perfect Plasmonic absorber as an optical surface based on pyramidal element with Nobel metal and composite. Optik **287**, 171086 (2023)
13. M. Sayed, Y. Jiaguo, G. Liu, M. Jaroniec, N.-N.-B. Photocatalysts, Chem. Rev. **122**(11), 10484–10537 (2022)
14. P.K. Jain, X. Huang, I.H. El-Sayed, M.A. El-Sayed, Noble metals on the nanoscale: optical and photothermal properties and some applications in imaging. Sens. Biol. Med. (2008)
15. Y. Xu, L. Chen, X. Wang, W. Yao, Q. Zhang, Recent advances in noble metal based composite nanocatalysts: colloidal synthesis, properties, and catalytic applications. Nanoscale **7**(24), 10559-83 (2015)
16. Y. Yang, Q. Zhang, Z.W. Fu, D. Qin, Transformation of Ag nanocubes into Ag–Au hollow nanostructures with enriched Ag contents to improve SERS activity and chemical stability. ACS Appl. Mater. Interfaces **6**(5), 3750–3757 (2014)
17. R. Singh, R.K. Soni, Plasmonics properties of trimetallic Al@ Al_2O_3@ Ag@ Au and Al@ Al_2O_3@ AuAg nanostructures. Appl. Phys. A **116**, 955–967 (2014)
18. R. Jiang, F. Qin, Q. Ruan, J. Wang, C. Jin, Ultrasensitive plasmonic response of bimetallic Au/Pd nanostructures to hydrogen. Adv. Func. Mater. **24**(46), 7328–7337 (2014)
19. H.H. Nguyen, J. Park, S. Kang, M. Kim, Surface plasmon resonance: a versatile technique for biosensor applications. Sensors (Basel). **15**(5), 10481–10510 (2015). https://doi.org/10.3390/s150510481. (PMID: 25951336; PMCID: PMC4481982)
20. A. Otto, Excitation of nonradiative surface plasma waves in silver by the method of frustrated total reflection. Z. Physik **216**, 398–410 (1968)
21. E. Kretschmann, H. Raether, Notizen: radiative decay of non-radiative surface plasmons excited by light. Zeitschrift für Naturforschung A **23**(12), 2135–2136 (1968)
22. W. Zhu, R. Esteban, A.G. Borisov, J.J. Baumberg, P. Nordlander, H.J. Lezec et al., Quantum mechanical effects in plasmonic structures with subnanometre gaps. Nat. Commun. **7**, 11495 (2016). https://doi.org/10.1038/ncomms11495

23. J.H. Choi, J.H. Lee, J. Son, J.W. Choi, Noble metal-assisted surface plasmon resonance immunosensors. Sensors (Basel). **20**(4), 1003 (2020). https://doi.org/10.3390/s20041003. (PMID: 32069896; PMCID: PMC7071002)
24. A. Pach, A. Szot, K. Fitzner, M. Luty-Błocho, Opportunities and challenges in the synthesis of noble metal nanoparticles via the chemical route in microreactor systems. Micromachines **15**(9), 1119 (2024)
25. G.F. Huseien, Potential applications of core-shell nanoparticles in construction industry revisited. Applied Nano **4**(2), 75–114 (2023)
26. K. Chatterjee, S. Sarkar, K. Jagajjanani Rao, S. Paria, Core/shell nanoparticles in biomedical applications. Adv. Coll. Interface Sci. **209**, 8–39 (2014). ISSN 0001-8686
27. N.K. Lotey, S. Pednekar, R. Chaughule, Hybrid nanoparticles in biomedical applications, in *Nanomaterials for Cancer Detection Using Imaging Techniques and Their Clinical Applications*, ed. by R.S. Chaughule, D.P. Patkar, R.V. Ramanujan (Springer, Cham, 2022)
28. H. Lin, Y. Liu, J. Deng, L. Jing, Z. Wang, L. Wei, Z. Wei, Z. Hou, J. Tao, H. Dai, Bimetallic nanoparticles: advances in fundamental investigations and catalytic applications. Environ. Sci.: Adv. **4**, 33–56 (2024). https://doi.org/10.1039/D4VA00241E
29. S. Deen, Nano-composites and their applications: a review. Charact. Appl. Nanomater. **2** (2019). https://doi.org/10.24294/can.v2i1.875
30. D. Xiang, et al., Surface plasmon resonance biosensor for exosome detection based on aptamer recognition to glypican-1. J. Nanobiotechnol. **19**(1), 1–12 (2021)
31. J. Homola, Surface plasmon resonance sensors for detection of chemical and biological species. Chem. Rev. **108**(2), 462–493 (2008)
32. E.-H. Yoo, S.-Y. Lee, Glucose biosensors: an overview of use in clinical practice. Sensors **10**(5), 4558–4576 (2010)
33. C. Sabu, T.K. Henna, V.R. Raphey, K.P. Nivitha, K. Pramod, Advanced biosensors for glucose and insulin. Biosens. Bioelectron. **141**, 111201 (2019)
34. C.A. Sunilbhai, M.S. Alam, K.K. Sadasivuni, J.R. Ansari, SPR assisted diabetes detection, in *Advanced Bioscience and Biosystems for Detection and Management of Diabetes* (Springer International Publishing, Cham, 2022), pp. 91–131.
35. O.I. Guliy, O.A. Karavaeva, A.V. Smirnov, S.A. Eremin, V.D. Bunin, Optical sensors for bacterial detection. Sensors **23**(23), 9391 (2023)
36. S. Quazi, Anti-cancer activity of human gastrointestinal bacteria. Med. Oncol. **39**(12), 220 (2022)
37. J. Loo, C. Yang, H.L. Tsang, P.M. Lau, K.-T. Yong, H.P. Ho, S.K. Kong, An Aptamer Bio-barCode (ABC) assay using SPR, RNase H, and probes with RNA and gold-nanorods for anti-cancer drug screening. Analyst **142**(19), 3579–3587 (2017)
38. Y. Guo, C. Yuan, T. Huang, Z. Cheng, Integrating UHPLC-Q-TOF-MS/MS, network pharmacology, bioinformatics and experimental validation to uncover the anti-cancer mechanisms of TiaoPi AnChang decoction in colorectal cancer. J. Ethnopharmacol. **334**, 118576 (2024)
39. L.S. Goh, N. Kumekawa, K. Watanabe, N. Shinomiya, Hetero-core spliced optical fiber SPR sensor system for soil gravity water monitoring in agricultural environments. Comput. Electron. Agric. **101**, 110–117 (2014)
40. N. Ravindran, S. Kumar, C.A. Mamathi, S. Nirmal Thirunavookarasu, C.K. Sunil, Recent advances in Surface Plasmon Resonance (SPR) biosensors for food analysis: a review. Crit. Rev. Food Sci. Nutr. **63**(8), 1055–1077 (2023)
41. F. Battaglia, V. Baldoneschi, V. Meucci, L. Intorre, M. Minunni, S. Scarano, Detection of canine and equine procalcitonin for sepsis diagnosis in veterinary clinic by the development of novel MIP-based SPR biosensors. Talanta **230**, 122347 (2021)
42. F. Otero, E. Magner, Biosensors—recent advances and future challenges in electrode materials. Sensors **20**(12), 3561 (2020)
43. T. Huang, M. Bobyr, A review of delamination damage of composite materials. J. Compos. Sci. **7**(11), 468 (2023)
44. A. Tkachenko, P. Virych, V. Myasoyedov, V. Prokopiuk, A. Onishchenko, D. Butov, Y. Kuziv, O. Yeshchenko, S. Zhong, G. Nie, N. Kutsevol, Cytotoxicity of hybrid noble metal-polymer composites. BioMed Res. Int. **1487024**, 12 (2022_

45. D. Costa, P. Pereira-Silva, P. Sousa, V. Pinto, J. Borges, F. Vaz, G. Minas, P. Sampaio, Critical issues on the surface functionalization of plasmonic Au-Ag/TiO_2 thin films with thiolated oligonucleotide-based biorecognition elements. Biosensors **14**(4), 159 (2024)
46. S. Mostufa, B. Rezaei, S. Ciannella, P. Yari, J. Gómez-Pastora, R. He, K. Wu, Advancements and perspectives in optical biosensors. ACS Omega **9**(23), 24181–24202 (2024)
47. H. Duan, T. Wang, Z. Su, H. Pang, C. Chen, Recent progress and challenges in plasmonic nanomaterials. Nanotechnol. Rev. **11**(1), 846–873 (2022)

Recent Advance in Plasmonic FET Biosensors Using Noble Metal-Based Composites

Naren Das, Armaan Siddiqui, and P. Saleem Akram

Abstract Plasmonic biosensors, which combine the optical intensification of plasmonic nanostructures with the electrical sensitivity of FETs, are a state-of-the-art development in biosensing. By introducing noble metals such as silver (Ag) or gold (Au), biomolecules can be detected with extreme sensitivity using a range of techniques, such as electrochemical transduction, Surface Enhanced Raman Scattering (SERS), and localized surface plasmon resonance (LSPR). These procedures' real-time detection has uses in environmental monitoring, food safety, and medical diagnostics, among other fields. This chapter comprehensively reviews the state-of-the-art in plasmonic FET biosensors, focusing on their unique properties, innovative fabrication methods, such as material vapor (PVD) or atomic layer (ALD) deposition procedures, imprinting on nanostructures (NIL), and electron beam lithography (EBL), and transformative applications in medical diagnostics. Priorities for the upcoming years include low-cost manufacturing, AI-based biosensors, and micro-miniaturization to facilitate the transfer from the research stage to commercialization for environmental and medical precision monitoring.

Keyword Biosensors · Plasmonic FET · SPR · LSPR · Nanofabrication techniques · Biosensing applications

1 Introduction

Over the years, there has been great progress in technological development which is driven by the ever-rising demand for the speedy, precise and affordable detection of biomolecules. Traditional biosensors have primarily relied on electrochemical,

N. Das (✉) · A. Siddiqui
Department of Electronics and Communication Engineering Department, Birla Institute of Technology, Mesra, Patna Campus, Patna, India
e-mail: narentitun@gmail.com

P. Saleem Akram
Department of Electronics and Communication Engineering, Koneru Lakshmaiah Education Foundation (Deemed to Be University), Guntur, India

B. Ramachandran et al. (eds.), *Advanced Optical Sensors*, Progress in Optical Science and Photonics 38, https://doi.org/10.1007/978-981-95-1687-2_4

optical, and piezoelectric methods, each offering varying degrees of sensitivity and specificity. However, the development of semiconductor-based biosensors, particularly field-effect transistors (FETs), has revolutionized biosensing by enabling real-time detection with enhanced sensitivity. These developments have had most profound effects in medical diagnostic, environmental, and food safety applications, where earlier identification of pathogens or biomarkers can have a gigantic effect on public health [1]. One of the most profound developments in biosensor technology has been the introduction of plasmonic nanostructures that employ surface plasmon resonance (SPR) to enhance biosensor performance.

The hybridization of FETs with plasmonic devices has developed high-sensitivity hybrid biosensors with real-time detection. Synergism offers accurate detection at ultra-low concentrations, an extremely promising tool for early diagnosis of diseases and point-of-care testing. The potential for plasmonic-FET hybridization to amplify weak biomolecular signals has drawn major interest from research groups and industries, and a large number of studies on their potential for diverse biosensing applications have been undertaken [2, 3]. In spite of the above developments, there is still a challenge in achieving strong, reproducible, and scalable biosensing platforms. Issues like non-specific binding, signal drift, and complexity in fabrication need to be overcome in order to take plasmonic FET biosensors to their full potential. The use of noble metal composites, e.g., gold (Au) and silver (Ag), has been found to significantly improve signal selectivity and stability. In addition, present nanofabrication technology and integration strategies are likely to drive plasmonic FET biosensors toward large-scale commercialization and application-ready status [4]. Field-effect transistors are increasingly being utilized as a tool of enormous potential in biosensing since they have the potential to detect charge-based molecular-level interactions.

The underlying operating mechanism of FET biosensors is the gate voltage-induced modulation of the charge carrier concentration in a semiconductor channel. The reaction between biological analytes such as nucleic acids, proteins, or microbes and the sensor surface results in measurable local electric field alteration with subsequent electrical response. Label-free detection ability of FET biosensors renders them extremely interesting towards low-cost and fast diagnosis [5].

FET biosensors possess numerous advantages like the ability to be highly sensitive, miniaturized, and even CMOS technology-compatible for large area integration. Their ability to provide real-time detection without chemical labels makes them more appropriate in clinical and environmental research. FET biosensors can also be fabricated using various semiconductor materials like silicon nanowires, graphene, and TMDs to further enhance performance. These materials are supplemented with improved electronic transport properties, which improve device stability and detection levels. However, even with their better performance in many areas, traditional FET biosensors suffer from signal attenuation, non-specific adsorption, and narrow range of detection. Integration of plasmonic nanostructures has been a possible solution to these drawbacks by increasing molecular interaction and charge transfer efficiency. Integration of noble metals with FET devices offers ultra-sensitive and selective biosensors that introduce next-generation diagnostic tools with improved performance [4].

Surface plasmon resonance (SPR) is a natural optical phenomenon where free electrons in noble metal nanoscale structures move in coordination upon irradiation with incident electromagnetic radiation. This creates a very strong intensified local electric field, which enhances molecule interactions at the sensor surface by an impressive scale. Application of SPR to biosensing has revolutionized label-free detection approaches to track biomolecular interactions in real time at high sensitivity [6]. Among the greatest advantages of SPR-based biosensors is that they have the ability to detect extremely low concentrations of target analytes. Taking advantage of the unique plasmonic nature of noble metals such as gold and silver, the sensors are able to sense as low as from femtomolar to attomolar. In addition, SPR biosensors generate controlled optical signals based on nanostructure dimension, morphology, and composition variation where the amplification of signal is more controlled. These properties make SPR become a tool which is very precious for medicine detection in medical diagnosis, drug research, and sensing in environmental application [7].

While effective, conventional SPR biosensors are bulky with intricate optical design, and miniaturization and integration into handheld systems are difficult. The problem is overcome with SPR in FET-based biosensors using charge-amplified measurement coupled with plasmonic signal amplification. The hybrid approach enables compact, ultrasensitive biosensing devices that can detect molecular interactions with record-high sensitivity [8].

Plasmonic nanostructure-based FET biosensors is a biosensing technology paradigm. The combination of the charge-sensitivity of the FET and the plasmonic nanomaterial optical amplification property results in improved sensing ability. Various strategies towards the integration of noble metal nanostructures with semiconductor channels have been explored in the past decade, such as direct nanoparticle decoration, hybrid graphene-plasmonic architectures, and localized SPR coupling methods [5].

Hybrid plasmonic-FET biosensors were extremely sensitive and specific and are consequently best utilized to detect ultra-low concentrations of biomolecules. Enhancement of the local electric field of noble metal nanostructures enables better interaction with analytes and increased signal transduction. Furthermore, improvements in nanofabrication technologies such as electron beam lithography and self-assembly processes have enabled plasmonic properties to be controlled to further maximize sensor performance. The improvements have opened up new possibilities in biosensing for personalized medicine, screening for infectious disease, and monitoring the environment [9].

While extremely promising, plasmonic FET-based biosensors are marred by a plethora of stability, reproducibility, and large-area fabrication challenges. Material processing steps, surface processing steps, and high-end integration steps will play pivotal roles in combating such challenges in a bid to bring such sensors from the lab to the market. Advances in artificial intelligence and internet of things (IoT) integration will propel further innovation in the processing of data towards real-time autonomous biosensing solutions [10].

2 Fabrication Techniques for Plasmonic FET Biosensors

Plasmonic FET biosensors are a highly promising category of biosensing devices that combine the electrical sensitivity of FETs with the plasmonic enhancement of plasmonic nanostructures. These biosensors enable highly sensitive, label-free detection of biomolecules and are considered valuable for applications in medical diagnostics, environmental monitoring, and food safety [4]. Their fabrication requires advanced nanofabrication techniques to achieve optimal electrical and plasmonic performance. This paper explores key fabrication methods used in designing plasmonic FET biosensors, including lithography, material deposition, etching, and integration approaches.

2.1 Lithographic Techniques

2.1.1 Electron Beam Lithography

Electron Beam Lithography (EBL) is a highly effective method for developing plasmonic nanostructures in FET biosensors. It offers exceptional resolution, enabling the fabrication of nanostructures with features as small as sub-10 nm. The process begins by applying a sensitive electron-beam resist onto a substrate, commonly silicon or glass. A focused electron beam selectively exposes specific areas of the resist, forming a pattern that is then developed. This patterned resist serves as a template for material deposition. Materials like gold or silver are deposited after patterning on the substrate. The metals are utilized as they have the capability of supporting surface plasmon resonances, which are essential in the operation of plasmonic FET biosensors. Deposition may be carried out in numerous different ways including thermal evaporation and sputtering. Integration of the plasmonic nanostructures in the FET biosensors allows the biological molecules to be detected through the change in the local refractive index. Upon attachment of the target molecule on the surface of the plasmonic nanostructure, a change in the plasmon resonance takes place, and this is detectable to assess the amount of the analyte. The detection by use of the optical means is very sensitive with quick response rates, and therefore the approach is adaptable for real-time biosensing applications. EBL-based fabrication techniques have certain ideal characteristics when plasmonic FET biosensors are to be fabricated. EBL has the accuracy to create intricate nanostructures with the desired optical properties. The process is versatile enough to create the dimensions and shapes of the nanostructures to meet the requirements of a particular application. The capacity of the process to create sophisticated arrays of nanostructures increases the performance and sensitivity of the biosensors.

Improved EBL technology over the last decade has also increased nanoscale fabrication ability. Advances in electron-sensitive resists with higher resolution and sensitivity have made it possible to fabricate even more intricate and smaller nanostructures [11] as depicted in Fig. 1. There have also been improvements in electron beam sources and control systems to enhance the patterning process regarding speed and precision. Such progress has made possible the fabrication of plasmonic nanostructures of unprecedented functionality and precision. Plasmonic nanostructures have shown phenomenal promise in diverse applications of biosensing. Among the key merits of plasmonic FET biosensors is their ability to identify low levels of analyte with high sensitivity. This finding is especially relevant in medical diagnosis, where identification of biomarkers in the nascent stage leads to improved outcomes for patients. Plasmonic FET biosensors have been employed successfully for the detection of various biological molecules, from proteins and nucleic acids to small molecules. The capacity to detect such molecules in real-time and in a highly specific manner makes plasmonic FET biosensors an important tool in healthcare and biomedical research [12]. Apart from their application in medical diagnostics, plasmonic FET biosensors have also been employed for environmental monitoring. Identification of pollutants and contaminants in water and air is significant for the maintenance of public health and safety. Plasmonic FET biosensors have been applied to the identification of heavy metals, pesticides, and other toxins in environmental samples. The elevated sensitivity and rapid response rates of plasmonic FET biosensors make them a good choice for monitoring environmental pollution in real time.

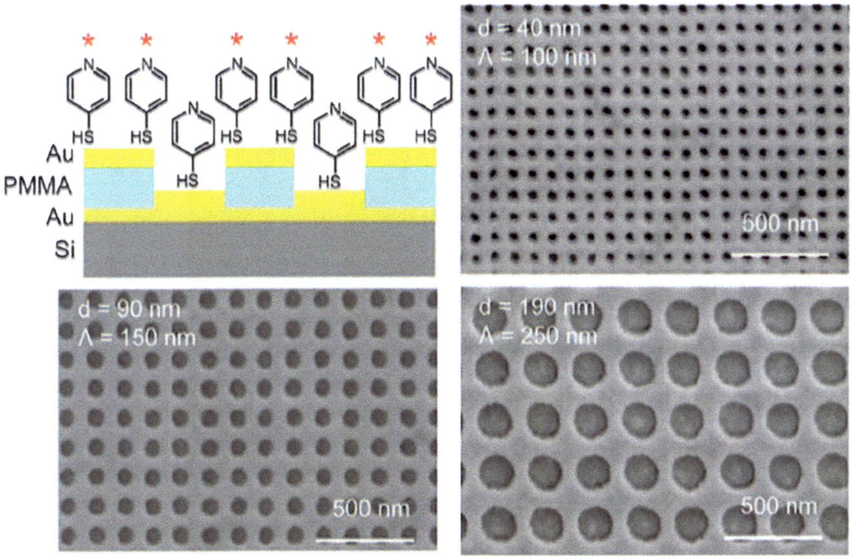

Fig. 1 Schematic of side view of gold thin film with nanoholes and a physically separated layer of isolated gold disks at the bottom of the well. Reprinted with permission from Ref. [11]

Another ongoing field of research is the customization of plasmonic nanostructures to specific uses. By varying the dimensions and configurations of the nanostructures, researchers can tailor their optical functionality for specific sensing applications. As an example, the use of nanostructures with pointed corners and edges could be utilized to enhance the sensitivity of plasmonic FET biosensors by increasing the local electromagnetic field. Besides, the use of nanostructures with periodic arrays can also improve the plasmonic FET biosensor performance via optimized coupling efficiency of surface plasmon resonances [13].

Another promising avenue for the enhancement of FET biosensor performance is the integration of plasmonic nanostructures with other nanomaterials. For example, the integration of hybrid nanostructures comprising plasmonic material and semiconductor or magnetic material has the capability to enable detection of more than one analyte with high sensitivity and selectivity. Multifunctional plasmonic nanostructures are highly promising for the advancement of the field of biosensing and expanding the area of application of FET biosensors. [14].

While there are several advantages of EBL-based fabrication techniques, issues still exist to be addressed. One of them is the scalability of the fabrication process. While EBL is high in resolution and accuracy, it is a relatively slow and expensive technique compared to other nanoscale fabrication processes [15]. Development of more effective and cheaper EBL systems is vital for common usage of plasmonic FET biosensors in many areas. Stability and longevity of plasmonic nanostructures are yet another challenge. Performance of plasmonic FET biosensors can be impacted by different variables such as oxidation, corrosion, and mechanical pressure. Synthesis of durable and stable plasmonic nanostructures is important in ensuring long-term performance and reliability of FET biosensors. Researchers are now exploring various routes, such as the use of protective coatings and the design of self-healing nanostructures, to surmount these problems.

2.1.2 Photolithography

Photolithography is a fundamental method applied in the fabrication of plasmonic nanostructures, and specifically in the fabrication of FET biosensors. Photolithography is unique with the capability to build nanostructures with high control of size, shape, and spatial distribution, which is required in enabling the function of plasmonic FET biosensors. We are going to discuss in this essay the complexity of photolithography and how photolithography applies to the making of plasmonic FET biosensors, the process itself, the material, and what benefits it bears. Photolithography is one process that duplicates patterns from a photomask onto a substrate. It goes through a range of steps that start from preparation of the substrate to deposition of the material. The major photolithography processes are substrate cleaning, resist coating, exposure, development, and pattern transfer. The first process is substrate cleaning, and it is silicon or glass. This removes any residue on the substrate that may interfere with the following processes. A light-sensitive photoresist is then coated on the substrate. The resist may be positive or negative, depending on whether exposed or unexposed regions are to be dissolved during development. A photomask in the required pattern is then placed on top of the resist-coated substrate. Ultraviolet (UV) light is shone through the mask onto the resist. The exposed regions of the resist to the UV light chemically change. The exposed substrate is then developed, usually with a chemical developer, to remove the chemically altered regions of the resist. This step creates a template for material deposition. After development, the patterned resist acts as a mask for material deposition.. Gold or silver, which have plasmonic properties, are deposited on the substrate. It can be done by methods like thermal evaporation or sputtering.

Photolithography is offering a number of advantages to plasmonic FET biosensor fabrication. For instance, photolithography is extremely accurate in the patterning and therefore the likelihood of intricate nanostructures with proper geometries as shown in Fig. 2 [16]. Photolithography has also been a highly flexible process that makes intricate arrays of nanostructures simple to design and fabricate and ready for specific biosensing applications [17]. Third, photolithography is a scalable method and thus it is suitable for large-scale production of plasmonic FET biosensors [18]. Finally, the process ensures reproducibility in the production of nanostructures, which is essential for consistent sensor operation.

The recent advancement of photolithography has also enhanced its application towards the synthesis of plasmonic FET biosensors. More recent advancements such as nanoimprint lithography and extreme ultraviolet (EUV) lithography offer still greater precision and control in nanostructure fabrication processes [18]. Nanoimprint lithography is obtained by embossing a mold with nano-scale features in a resist-coated substrate. The resist is then exposed and cured, and the mold is removed to leave a patterned resist, which can now be used for material deposition [17]. EUV lithography uses shorter wavelength light (13.5 nm) than conventional UV lithography (193 nm). This will allow for even smaller features and push the nanofabrication boundaries further [16]. Plasmonic FET biosensors have found applications

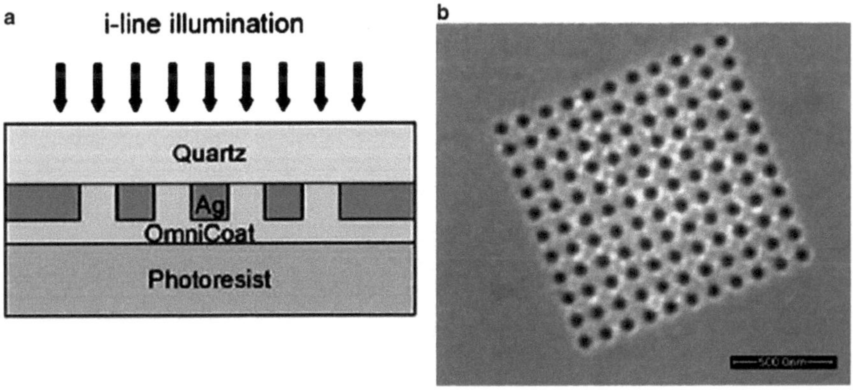

Fig. 2 Schematic of lithography setup designed by Srituravanich et al., **b** The silver mask with hole array used in this experiment. Reprinted with permission from Ref. [16]

in numerous fields, from medical diagnostics to environmental monitoring and food safety. Because they can detect low levels of analytes, they are considered to be effective tools for the early diagnosis of diseases as well as environmental pollutant tracking [18].

2.1.3 Nanoimprint Lithography

Nanoimprint Lithography (NIL) is a cutting-edge nanofabrication technique that creates manometer scale patterns with high resolution and cost-effectiveness. This method is particularly important in semiconductor manufacturing, optics, and biosensors devices. When plasmonic Field-Effect Transistor (FET) biosensors are to be made, NIL may be employed for making intricate nanostructures that enhance the sensitivity and performance of the devices.

A common NIL method is Thermoplastic Nanoimprint Lithography (T-NIL), in which a thermoplastic polymer is heated to a temperature above its glass-transition temperature and a mold of pre-designed patterns is pushed into the melted polymer. The mold is then removed after cooling to leave the target pattern. The second one is Photo Nanoimprint Lithography (P-NIL), using a UV-curable liquid resist coated on the substrate in Fig. 3 [19]. The mold, typically of transparent material, is forced into the resist and then hardened by curing under UV light to create a patterned hardened structure. Resist-Free Direct Thermal Nanoimprint Lithography, which involves direct printing of patterns on a substrate without using any resist layer, is another. It is particularly used to create high-precision metal nanostructures. In plasmonic FET biosensors, NIL is used to create nanostructures such as nanoarrays, nanocomposites, and colloidal LSPR nanoparticles. The optical responses of the biosensors are enhanced through the nanostructures that are very sensitive to biological analytes.

Recent research has shown the ability of NIL to implement plasmonic FET biosensors. For example, Yamakawa et al. [20] illustrated the application of NIL for high-resolution pattern generation for semiconductor memory technology with resolution below 10 nm. Another Dhull [21] reported on several approaches and material needs for NIL while noting its cost-effectiveness and large throughput. Fialkova et al. [22] discussed the application of NIL in fabricating next-generation carbon nanotube-based devices, highlighting its ability to fabricate intricate nanostructures.

2.2 *Material Deposition Techniques*

Material deposition fabrication methods are central to Plasmonic Field-Effect Transistor (FET) biosensor development, allowing the controlled integration of semiconductor and plasmonic materials. The deposition method used is very important as it affects the performance of the sensor, such as its sensitivity, stability, and reproducibility. Various deposition methods, such as physical vapor deposition (PVD),

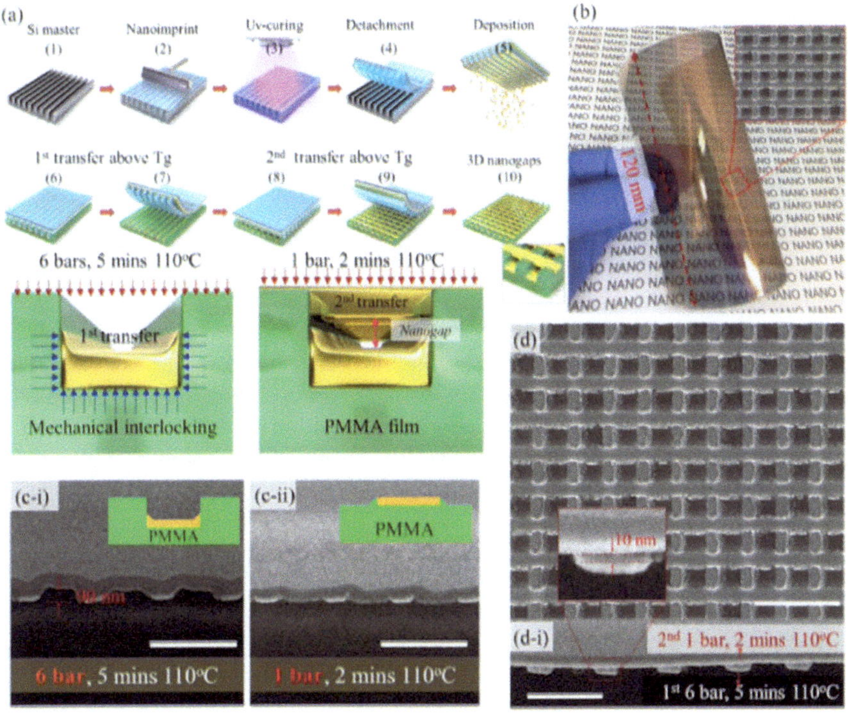

Fig. 3 Plasmonic multilaminar nanopores structure and its fabricating process based on Nanoimprint Lithography (NIL). **a** The manufacturing process of multilaminar nanopores prepared by NIL. **b** The photo of the film with multilaminar nanopores. **c**-(i) and **c**-(ii) The sectional view of multilaminar nanopores manufactured under different conditions. **d** and **d**-(i) SEM overlooking and cross-sectional images of multilaminar nanopores. Reprinted with permission from Ref. [19]

chemical vapor deposition (CVD), atomic layer deposition (ALD), and electrodeposition, are employed to fabricate plasmonic FET biosensors with noble metal-based composites like gold (Au), silver (Ag), and platinum (Pt).

2.2.1 Physical Vapor Deposition (PVD)

PVD is a widely utilized fabrication technique for depositing plasmonic materials onto Field-Effect Transistor (FET) biosensors. PVD is a widely utilized fabrication technique for depositing plasmonic materials onto Field-Effect Transistor (FET) biosensors. PVD prefers the deposition of thin films with high purity, uniformity, and thickness controlled, thus well suited for depositing noble metal composites like gold (Au), silver (Ag), and platinum (Pt) onto semiconductor wafers. The two most important PVD methods used to produce plasmonic FET biosensors are sputtering and thermal evaporation, which respectively have their respective strengths of control deposition of the films, adhesion of the films, as well as scalability.

Thermal evaporation is a PVD technique where the metal source is subjected to heat to the sublimation temperature within a high-vacuum chamber, and the atomized vapor condenses and crystallizes into a thin film on the target substrate. This process is especially employed for the deposition of noble metals, which exhibit better plasmonic character, thus improving the sensor's sensitivity towards the interaction between biomolecules. Its simplicity and high rate of deposition make it a technique of choice for large area deposition [23]. The low step coverage of thermal evaporation onto substrates having complex geometries is one of the major limitations. This has the potential for non-uniform film growth.

Sputtering, another PVD method, is better than thermal evaporation in film uniformity and adhesion. Here, high-energy ions, in the form of an argon plasma, are used to bombard a metal target to remove atoms that deposit on the sensor substrate. Sputtering provides thickness, composition, and microstructure control of the film with high accuracy and thus is suitable for the production of plasmonic FET biosensors with precise plasmonic nanostructures. Moreover, the capability of elements' co-sputtering enables the possibility to achieve alloyed plasmonic films with superior stability and adjustable optical properties. Sputtering is also more efficient in forming denser and less void films compared to thermal evaporation and this improves biosensors' long-term stability and reproducibility [24].

In the fabrication of plasmonic FET biosensors, deposition of noble metal nanostructures using PVD processes is crucial in maximizing localized surface plasmon resonance (LSPR) effects, which enhance the optical and electrochemical sensitivity of the device. Nanoscale engineering through sputtering enables the creation of ultra-thin plasmonic films that can effectively interface with the semiconductor channel of the FET. This interaction enhances the density of charge carriers and the ability of transduction of the sensor, which enables detection of biomolecules at ultra-low concentrations. PVD-based noble metal coatings also enhance the chemical stability and biocompatibility of the sensor surface, enabling functionalization of biomolecular recognition elements like antibodies, aptamers, or DNA probes [25]. Its benefits notwithstanding, PVD-based material deposition for plasmonic FET biosensors has limitations in the form of high vacuum demands, deposition stress, and metal contamination threat. These limitations have been overcome by hybrid fabrication processes involving combination of PVD with chemical functionalization or ALD. These methods promote the adhesion, stability, and functionalization potential of the film as well as the performance and reliability of the sensor. Future research will focus on maximizing deposition conditions, exploring novel plasmonic alloys, and merging PVD with novel nanofabrication techniques to further enhance the sensitivity, selectivity, and scalability of plasmonic FET biosensors.

2.2.2 Chemical Vapor Deposition (CVD)

CVD is a widely utilized technique for fabricating plasmonic Field-Effect Transistor (FET) biosensors, particularly for depositing noble metal-based composites such as gold (Au), silver (Ag), and platinum (Pt). CVD provides precise control over film thickness, composition, and morphology, making it an essential process for optimizing the plasmonic and electrical properties of FET biosensors. The ability of CVD to deposit uniform, high-purity noble metal thin films with excellent adhesion to semiconductor surfaces enables enhanced localized surface plasmon resonance (LSPR) effects, leading to improved biosensing sensitivity and selectivity. Unlike other deposition methods such as Physical Vapor Deposition (PVD) and Atomic Layer Deposition (ALD), CVD offers superior step coverage and conformality, making it highly suitable for coating nanostructured surfaces and high-aspect-ratio features required in advanced biosensing applications [26].

The CVD process involves introducing volatile metal–organic or inorganic precursors into a reaction chamber, where they undergo thermal decomposition or chemical reaction to form a solid metal film on the substrate. Different CVD techniques, including thermal CVD, plasma-enhanced CVD (PECVD), and metal–organic CVD (MOCVD), are employed depending on the required film characteristics and deposition conditions. Among noble metals, gold is the most used in plasmonic biosensors due to its excellent stability, biocompatibility, and strong plasmonic response In CVD-based Au deposition, precursors such as dimethylgold(III) acetylacetonate ($Me_2Au(acac)$) or gold chloride ($AuCl_3$) decompose at elevated temperatures, leading to the formation of high-purity Au films. These gold films enhance the LSPR effect by creating strong electromagnetic field confinement, thereby increasing the sensitivity of the biosensor to molecular interactions occurring at the sensor surface [27].

Silver is yet another one of the noble metals used in wide-ranging applications in plasmonic biosensors due to its high plasmonic efficiency over gold. Thin films of Ag are prone to oxidation, however, which degrades their performance with time. CVD processes offer control over silver thin film deposition with the choice of incorporating protective layers like Al_2O_3 or TiO_2, the latter being oxidization-resistant and maintaining the plasmonic properties of Ag. Application of Ag thin films in plasmonic FET biosensors has been reported to improve the biosensing performance significantly, particularly for label-free biosensing. Platinum is another important element of biosensing due to its high catalytic activity and high electrical conductivity. Pt-based plasmonic FET biosensors are particularly useful in electrochemical sensing where both plasmonic and catalytic effects are employed in signal amplification. CVD processes from precursors like platinum (II) hexafluoro acetylacetonate ($Pt(hfac)_2$) can be used to deposit nanostructured Pt films with controlled grain size and surface roughness for improved charge transfer and biosensing efficiency [28].

CVD is not only employed for deposition of noble metals but is widely employed for deposition of dielectric materials such as SiO_2, HfO_2, and Al_2O_3 employed as protective and functional layers in plasmonic FET biosensors. Dielectric films provide electric insulation, maximum refractive index contrast, and improve the stability of noble metal films. The CVD-grown dielectrics also provide improved gate control for the FET for precise modulation of charge carriers, which is useful for the development of high sensitivity in biosensing applications. Combinations of CVD-grown noble metals and dielectrics result in high-performance plasmonic FET biosensors with the ability to detect biomolecular interactions with precision and specificity. Scalability is one of the most valuable advantages of CVD, which is best suited for the fabrication of plasmonic FET biosensors in high volume. Compared to the limitation of PVD in deposition of conformal coatings on nanostructures of complex geometry, CVD provides uniform coverage and excellent step coverage even on three-dimensional biosensor architecture. This feature is highly valuable for the development of next-generation biosensors with the requirement of precisely engineered nanostructures to provide maximum plasmonic enhancement and improve detection thresholds. CVD-grown noble metal films also provide improved mechanical and chemical stability, thus restricting the scope for delamination or degradation on aging, a requirement for the durable functionality of the biosensor in biomedical and environment monitoring applications.

It is rich in strengths but has certain weaknesses in the guise of high-temperature processing and use of costly or toxic precursors. These can impose limitations on the temperature-sensitivity of substrates and wearability or flexibility of biosensors. Counterbalancing these limitations are low-temperature CVD processing research efforts and novel precursor chemistries that can tolerate noble metal deposition under less harsh conditions without sacrificing film quality. Hybrid processing schemes based on combinations of CVD with other deposition techniques like ALD or solution processing are being designed for optimal cost-effectiveness and deposition yield.

2.2.3 Atomic Layer Deposition (ALD)

ALD is a very accurate thin-film deposition method that plays an important role in the fabrication of plasmonic Field-Effect Transistor (FET) biosensors. ALD is of great utility to incorporate noble metal-based composites, like gold (Au), silver (Ag), and platinum (Pt), into biosensing devices because it provides better control over film thickness, conformality, and uniformity. In contrast with other deposition methods like Physical Vapor Deposition (PVD) or Chemical Vapor Deposition (CVD), ALD does possess a self-limiting chemical reaction mechanism in the surface, and therefore thickness is controllable at the atomic level. That level of exactness is a prerequisite so plasmonic activity and modulation of charge carriers within FET biosensors can be provided with maximum possible response with increased sensitivity and stability in the field of biosensing [29].

ALD is employed to deposit highly uniform thin layers of noble metals, such as on nanostructured complex substrates, with homogeneous plasmonic enhancement effect. Gold is among the most widely utilized noble metals for plasmonic biosensors due to its providing improved stability, biocompatibility, and very nice localized surface plasmon resonance (LSPR) effects. The ALD gold deposition reactions typically utilize organometallic precursors such as dimethylgold(III) acetylacetonate ($Me_2Au(acac)$) and oxygen or hydrogen plasma. The process delivers ultra-thin, level gold films with strong plasmonic coupling and great semiconductor substrate adhesion, well-suited for plasmonic FET biosensors. Likewise, the best plasmonic noble metal, silver, can similarly be deposited with the aid of properly controlled oxide layers to resist degradation. Ag-based ALD films will typically comprise passivation layers such as Al_2O_3 or TiO_2 to be capable of attaining high long-term stability under biosensor conditions and high sensitivity and reliability.

Platinum is another noble metal that is often incorporated into plasmonic FET biosensors with ALD because it has improved catalytic and electrical properties. ALD deposition of Pt from precursors like platinum(II) hexafluoroacetylacetonate ($Pt(hfac)_2$) gives high-density nanostructures that facilitate charge carrier interactions within the FET biosensor. Pt incorporation increases electrochemical biosensing functionality, and it is especially suitable for dual-mode biosensors with plasmonic and electrochemical detection modes. In addition to noble metal deposition, ALD is also widely applied in the depositions of dielectric films like Al_2O_3, HfO_2, and TiO_2, which play a central role in plasmonic FET biosensors. The dielectrics enhance the refractive index contrast for better plasmonic field localization and offer passivation layers against oxidation and metal film degradation. Aside, high-k dielectrics grown by ALD enhance FET gate control towards better modulation of charge carriers and higher detection sensitivity [30]. The major benefits of ALD to use in the fabrication of plasmonic FET biosensors are atomic-layer-thickness control, uniformity, and enhanced conformality on nanostructured interfaces. Compared with traditional deposition techniques with restricted uniformity of non-uniform films on high-aspect-ratio surfaces, ALD guarantees full surface coverage, significantly enhancing the stability and reproducibility. In addition, noble metal films grown with ALD exhibit enhanced mechanical stability and adhesion with decreased long-term risks of degradation or delamination. Such stability is especially highly important in long-term biosensing, especially in biomedical and environmental monitoring where sensor lifetime is a top priority [29].

Though having several benefits, noble metal ALD-based deposition is fraught with limitations like low deposition rates and expensive precursor material. They can be circumvented by hybrid fabrication techniques combining ALD with nanolithography and solution processing for cheap and high-rate deposition of noble metals. Synthesis of novel precursors for ALD of noble metals is also focused on enhancing the deposition efficiency without compromising on high-quality film properties. Future work in this direction would include process parameter optimization of ALD for other noble metal compositions, exploration of other precursor chemistries, and combination of ALD with new emerging biosensing platforms with the vision to ultimately achieve even superior plasmonic FET biosensing performance.

2.3 *Etching Techniques*

Etching is also an important processing step for the fabrication of plasmonic nanostructures for FET biosensors to facilitate easy removal of material in a specified manner to create nanoscale features. There are several etching methods, and they are wet etching, dry etching, and reactive ion etching (RIE). Wet etching utilizes chemical solutions to etch off the desired material from the substrate in smooth, homogeneous but potentially less precise to the extent necessary for nanoscale structures. Dry etching uses plasma or ionized gas to etch the materials more precisely and controllably of the etching profile. RIE combines physical etching of dry etching and chemical reaction of wet etching to create highly resolved nanostructures with high anisotropy.

2.3.1 Reactive Ion Etching (RIE)

RIE is an important technique in the production of plasmonic FET biosensors, allowing for controlled material removal and nanometer-scale patterning of semiconductor and plasmonic thin films. The process is performed with chemically reactive plasma generated by the combination of chemically reactive gases such as CF_4, SF_6, O_2, or Cl_2, and an electric field for the generation of directional ion bombardment that can selectively etch exposed areas. RIE produces strong anisotropy for defined nanostructures with low undercut, and this is vital to maintain plasmonic nanostructure and FET channel integrity. It is a process prevalent in etching noble metal thin films such as gold and silver to maximize the localized surface plasmon resonance (LSPR) effect and similarly to pattern semiconductor features in silicon, graphene, and transition metal dichalcogenides (TMDs) for optimal charge transport properties. RIE allows control of etch depth and selectivity between materials with high resolution pattern transfer from lithographic masks through exact control of process parameters such as pressure, power, gas composition, and etching time. Extremely advanced RIE processes such as inductively coupled plasma reactive ion etching (ICP-RIE) allow denser plasma and process control over the etching process and therefore are extremely well positioned for the high-performance and CMOS-compatible plasmonic FET biosensors' fabrication. Continuous developments in RIE technology, such as atomic-layer etching (ALE) and damage-free etch processes, also improve further plasmonic FET biosensor accuracy, scalability, and reproducibility towards biomedicine and environmental applications.

2.3.2 Wet Chemical Etching

Wet chemical etching is a basic method employed in the fabrication of plasmonic nanostructures for Field-Effect Transistor (FET) biosensors. Wet chemical etching is a method in which chemical solutions are used to etch material selectively from a substrate and enable the nanoscale features to be defined precisely. Anisotropic wet etching is one basic method in which material is etched at various rates based on the crystallographic orientation of the substrate. This process is especially good at forming very well-defined, directional structures in silicon-based crystalline materials. Buffered oxide etch (BOE) is also common, which consists of a solution of hydrofluoric acid and a buffer agent to selectively etch silicon dioxide with minimal damage to layers underneath.

Acid etching is etching using corrosive acids like hydrofluoric or nitric acid to remove a substrate. The etching will be isotropic or anisotropic depending on the substrate and the composition of the acid used. Alkaline chemicals like potassium hydroxide or tetramethylammonium hydroxide are used in base etching to etch and is usually anisotropic etching in crystalline materials.

Acid etching is etching using corrosive acids like hydrofluoric or nitric acid to remove a substrate. The etching will be isotropic or anisotropic depending on the substrate and the composition of the acid used. Alkaline chemicals like potassium hydroxide or tetramethylammonium hydroxide are used in base etching to etch and is usually anisotropic etching in crystalline materials. Wet etch clean is a process of residue, contaminant, or thin layer removal from a substrate using liquid chemicals after processing operations such as etching, deposition, or lithography. Wet chemical etching is required to produce plasmonic nanostructures that sustain surface plasmon resonances, which are essential for the operation of FET biosensors.

The use of plasmonic nanostructures in FET biosensors increases their sensitivity and specificity towards the detection of biological molecules because they cause resonance changes upon interacting with target molecules. The optical detection approach is highly sensitive and responsive and thus well suited for real-time biosensing applications. Wet chemical etching methods have key roles in the expansion of biosensing applications and the creation of novel diagnostic devices.

2.4 *Integration of Plasmonic Nanostructures with FETs*

The application of FETs and plasmonic nanostructures is a workable method to increase the specificity and sensitivity of biosensors. Plasmonic nanostructures including gold or silver nanoparticles have unusual optical properties, which can also be utilized potentially in biosensing. Through the application of FETs, the nanostructures can convert the performance of sensors by allowing biological molecules to be detected through identifying changes in local refractive index.

2.4.1 Direct Nanoparticle Decoration

Direct decoration is the process through which the performance of FET-integrated plasmonic nanostructures is enhanced. Direct decoration involves direct deposition of nanoparticles, say gold or silver, on the surface of the nanostructures. The nanoparticles possess unique optical features like surface plasmon resonance, which may be used in biosensing. The process begins by synthesizing the plasmonic nanostructures using techniques like electron beam lithography, photolithography, or nanoimprint lithography. Once the nanostructures are prepared, particles are deposited on the surface of the nanostructures using techniques like chemical reduction, thermal evaporation, or sputtering. Direct decoration of nanoparticles onto plasmonic nanostructures enhances their optical characteristic through amplifying the local electromagnetic field intensity. This enhances their performance in FET biosensors since it maximizes their sensitivity and specificity in the detection of biological molecules. When the target molecule is immobilized on the surface of the decorated nanostructure, the plasmon resonance is red-shifted, which can be tracked for analyte quantification. The optical detection method is very sensitive and has fast response times, which are ideal for real-time biosensing applications.

Several research articles have proven the efficacy of direct nanoparticle decoration for various applications. A research article on simulation and sensitivity analysis of a plasmonic FET-based sensor in the visible spectral range demonstrates the applicability of the method to be utilized for biosensing applications [31]. A research paper on the in-situ decoration of plasmonic Au nanoparticles on graphene quantum dots-graphitic carbon nitride hybrid determines the enhanced photocatalytic activity of the decorated nanostructures [32].

2.4.2 Hybrid Graphene-Plasmonic Architectures

Hybrid graphene-plasmonic architectures represent a cutting-edge approach to enhancing the performance of plasmonic nanostructures integrated with FETs. Graphene, a two-dimensional material composed of carbon atoms arranged in a hexagonal lattice, exhibits exceptional electrical, mechanical, and optical properties. When combined with plasmonic nanostructures, such as gold or silver nanoparticles, graphene can significantly improve the sensitivity and specificity of FET biosensors.

The integration of graphene with plasmonic nanostructures involves several steps. First, the plasmonic nanostructures are fabricated using techniques such as electron beam lithography, photolithography, or nanoimprint lithography. These techniques allow for precise control over the size and shape of the nanostructures, which is crucial for achieving the desired optical properties. Once the nanostructures are fabricated, a monolayer of graphene is transferred onto the substrate using methods such as chemical vapor deposition or mechanical exfoliation.

The hybrid graphene-plasmonic architectures leverage the unique properties of both materials. Graphene can support surface plasmons with higher confinement and lower propagation loss compared to traditional metal-based plasmonic structures [33]. This results in enhanced light-matter interactions and improved performance of the FET biosensors. Additionally, the plasmonic nanostructures can be tailored to achieve specific resonances, further enhancing the sensitivity of the sensors.

Several studies have demonstrated the potential of hybrid graphene-plasmonic architectures for various applications. For example, hybrid graphene-plasmonic gratings have been shown to achieve enhanced nonlinear effects at terahertz frequencies [34]. These effects are crucial for the development of reconfigurable planar terahertz devices, such as frequency generators and all-optical signal processors. Another study highlights the use of hybrid graphene-plasmonic materials in plasmonic sensors, which offer improved performance in detecting biological and chemical molecules [35].

2.4.3 Localized Surface Plasmon Coupling

Localized surface plasmon coupling is a phenomenon that occurs when plasmonic nanostructures such as gold or silver nanoparticles are brought in close proximity to each other. Coupling of their localized surface plasmon resonances is due to the interaction, creating enhanced electromagnetic fields and characteristic optical properties. When integrated with FETs, these coupled plasmonic nanostructures can boost the performance of biosensors by several orders through the detection of biological molecules from changes in the local refractive index. Integration of localized surface plasmon coupling with FETs has a couple of steps. Firstly, the fabrication of the plasmonic nanostructures is accomplished through techniques such as electron beam lithography, photolithography, or nanoimprint lithography. These approaches provide exact control of size and shape of the nanostructures, as required in the development of the noted optical properties. Once the nanostructures are fabricated, they are deposited onto the FET substrate using methods such as thermal evaporation, sputtering, or chemical vapor deposition.

The coupling of localized surface plasmons in these nanostructures enhances their optical properties by increasing the local electromagnetic field intensity. This improvement is essential to the operation of FET biosensors since it enhances the selectivity and sensitivity of biological molecule detection. When the analyte molecule binds onto the surface of the coupled plasmonic nanostructure, a change in plasmon resonance takes place, and this can be detected to quantify the amount of analyte present. This optically based sensing has the benefit of high sensitivity with fast response times, which is appropriate for the use of real-time biosensing.

It has been illustrated by some studies the capability of localized surface plasmon coupling in various applications. For instance, studies on the principle and applications of coupling of surface plasmons and excitons exhibit the efficiency of the technique in increasing light-matter interactions [36]. Another study on localized surface plasmon resonance-based nanobiosensors includes the applications of plasmonic nanoparticles in biosensing, especially in living cell imaging and photothermal therapy [37].

3 Detection Mechanisms

Plasmonic FET biosensors are trendy generation that exploit the special belongings of noble metals together with gold (Au) and silver (Ag) to growth the sensing capability of the biosensors. The sensors are primarily based at the phenomenon of localized surface plasmon resonance (LSPR), that is the resonant vibration of the conduction electrons of the metal to incident radiation. This generates strongly electromagnetic fields near the nanoparticle's floor, which substantially will increase the sensitivity of the biosensor. In the cutting-edge section, we have discussed various detection mechanisms in sophisticated Plasmonic FET biosensors based on noble metal-primarily based composites.

3.1 Localized Surface Plasmon Resonance (LSPR)

LSPR plays a crucial role as a detection method in top-quality Plasmonic FET biosensors those with noble metal composites like gold (Au) and silver (Ag). These high-end biosensors that contain such metals are becoming more and more essential because they're selective, fast, and sensitive. As a result, they're now used in a huge variety of biosensing technologies. LSPR happens when surface conduction electrons in noble metal nanoparticles resonate with incident light. This effect creates intense electromagnetic fields near the nanoparticle surface, which are the ones that are responsible for the improvement of optical properties such as light scattering and absorption. The LSPR frequency of resonance is very size-, shape-, and environment-sensitive and can be tuned towards detection capability.

The detection process starts with functionalization of noble metal nanoparticles with bioreceptor molecules including antibodies or enzymes that can specifically recognize the target analyte. Functionalization is a step toward selective detection of the target analyte in real biological samples. Upon binding of the target analyte to the bioreceptor on the nanoparticle surface, local refractive index changes. This shift influences the LSPR signal and leads to a resonance wavelength shift [38]. The shift can be detected and quantified using optical means, which offers quantitative information about the analyte concentration. LSPR produces strong electromagnetic fields that amplify the biosensor sensitivity in detecting low concentrations of the analyte. Coupling of the incident radiation to such plasmons results in increased transmission through the nanoholes or slits, despite their subwavelength sizes shown in Fig. 4.

LSPR-based Plasmonic FET biosensors have diverse applications in medical diagnostics, environmental monitoring, and food safety. In medical diagnostics, biosensors are employed to detect biomarkers of different diseases, including cancer, infectious diseases, and cardiovascular diseases, with high sensitivity and specificity [39]. In environmental monitoring, biosensors are utilized to detect water, air, and soil contaminants and toxins. Since the sensors are of very high sensitivity, they can identify very low levels of contaminants for environmental protection [40]. In food processing, biosensors based on LSPR can identify chemical contaminants, allergens, and food-borne pathogens, thus making foods safe and improving the quality control.

The use of noble metal-based composites offers several advantages, including enhanced sensitivity, stability, and biocompatibility. Noble metals like gold and silver have a good LSPR effect, and hence biosensing can be improved. Noble metals are

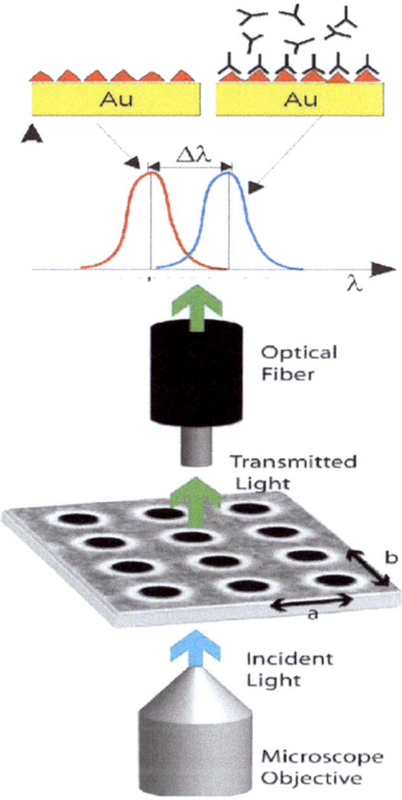

Fig. 4 The remarkable optical transmission-based biosensing concept. A biological target-coated gold (Au) film has a series of nanoholes drilled in it to transmit white light from a microscope. Reproduce from Ref. [38]

chemically stable and biocompatible, and hence can be used in biological conditions. Noble metal nanoparticles' optical properties can be customized depending upon their size, shape, and composition, and thus biosensors can be designed for application.

3.2 Surface-Enhanced Raman Scattering (SERS)

SERS has been a key biosensing technology because it can provide ultra-sensitive, label-free, and molecular fingerprinting. Combining SERS with plasmonic field-effect transistor (FET) biosensors, particularly with noble metal-based composite-enabled biosensors, significantly enhances sensing traits for biomolecules such as proteins, DNA, and pathogens [41]. The sensing mechanism of such high-performance plasmonic FET biosensors relies on the synergistic effect of noble metal LSPR and semiconductor channel charge modulation of the FET. Strong plasmonic effects are demonstrated by noble metals such as gold (Au), silver (Ag), and platinum (Pt). These strong plasmonic effects amplify the Raman scattering signals of molecules adsorbed onto their surface. This enhancement is brought about by amplification of electromagnetic fields and chemical interactions between metal and analyte molecules, giving rise to ultra-sensitive biosensing platforms [42].

In a coupled SERS FET biosensor, noble metal nanostructure, i.e., nanoparticles, nanorods, or hybrid nanofilms, is intentionally deposited on the semiconductor channel. On presentation of the analyte, it adsorbs onto the noble metal functionalized surface, inducing localized changes in refractive index that alter the plasmon resonance conditions. This induces increased electromagnetic field intensity at the sensor surface that increases the Raman signal of the analyte. Concurrently, the FET structure also exhibits a second sensing mechanism in the form of sensing variations in drain-source current (Ids) through charge transfer from the biomolecule to the semiconductor [43]. The dual transduction provides higher detection sensitivity than conventional SERS-based methods and makes it extremely suitable for biosensing.

The electronic component of the plasmonic FET biosensor is also responsible for it being more effective at producing the Raman signal. The gate voltage allows dynamic Fermi level tuning in the semiconductor, modulating the charge transfer between the noble metal composite and the biomolecule. This effect, known as "electrochemical SERS" or "plasmoelectric coupling," results in further signal amplification and improved molecular sensitivity [44]. Additionally, the use of two-dimensional (2D) materials such as graphene, molybdenum disulfide (MoS_2), and black phosphorus in conjunction with noble metal-based nanostructures enhances charge transport and provides a high surface-to-volume ratio for biomolecular interactions [45]. The integration of plasmonic enhancement and FET-based electrical transduction provides real-time ultra-sensitive biosensing with detection sensitivity at the femtomolar or even attomolar level.

Uses of such a sophisticated plasmonic FET biosensor are overlapping into all other fields from medical diagnostics to environmental monitoring to food safety. For medical diagnostics, SERS-amplified FET biosensors have shown extreme sensitivity for detecting disease markers of cancer, Alzheimer's disease, and virus-borne illnesses like COVID-19. For environmental monitoring, they have applied to ultra-low levels of harmful chemicals and toxic pollutants. In food safety purposes, these biosensors facilitate quick and accurate detection of pathogens and contaminants to manage the quality of food.

3.3 Electrochemical Detection

Electrochemical detection is a highly effective process that is employed in sophisticated Plasmonic Field-Effect Transistor (FET) biosensors, particularly in the case of noble metal-based composites such as gold (Au) and silver (Ag). Noble metals offer sophisticated plasmonic and electrochemical efficiency, resulting in highly selective and sensitive detection of biomolecules. Plasmonic FET biosensors employ synergistic electrochemical transduction coupling and LSPR effects for the detection of an incredibly wide range of biological analytes such as proteins, nucleic acids, and small molecules.

Electrochemical plasmonic FET biosensors detection essentially includes three distinguished techniques: amperometric, potentiometric, and impedimetric sensing. Amperometric detection involves finding the outcome current through redox reactions at the electrode surface and is thereby employed for finding enzymatic reactions [46]. Potentiometric detection, alternatively, senses the variation in voltage due to ion concentration variation on the sensor surface [47]. Impedimetric detection involves the sensing of the change in the luminescence at the electrode–electrolyte interface, which is a source of information regarding biomolecular interactions, including affinity and binding kinetics [48].

Use of noble metal-based composites, e.g., Au and Ag nanoparticles, in plasmonic FET biosensors enhances electrochemical sensing significantly. Profuse LSPR of the metals enhances local electromagnetic field, and thus sensitivity and charge transfer kinetics. The nanoparticles also provide increased surface area, and thus increased biomolecule immobilization densities, and thereby enhanced biorecognition and signal transduction [49]. Optimization of size, shape, and surface functionalization of the nanoparticles is also feasible, with the possibility of sensor performance tuning to the target biomolecules [50].

One of the easiest methods of electrochemical detection in plasmonic FET biosensors is through modulation of the semiconductor channel charge carriers. A surface-functionalized noble metal nanoparticle-labeled probe analyte induces a localized electrostatic change. This is detectable in terms of a change in conductivity in the channel that is reflected in detectable changes in the Ids [47]. The LSPR effect also has the benefit of enhancing the electron transfer mechanism, which is the cause of high electrochemical signals and low detection limits [48].

Plasmonic FET biosensors integrated with electrochemical detection are widely used in various applications. These biosensors find application in medical diagnostics in order to investigate early biomarkers of diseases such as microRNAs and cancer proteins [49]. The high specificity and sensitivity of plasmonic FET biosensors allow for ultralow biomarker detection concentration with higher accuracy and speed and are thus suitable probes to screen for diseases in their early stages [50]. In environmental sensing, the biosensors have been utilized in the detection of heavy metal ions, pesticides, and other toxins with high accuracy. In food safety also, plasmonic FET biosensors facilitate rapid detection of bacterial pathogens and thus consumer safety as well as food quality.

With its advantages, the use of electrochemical detection in plasmonic FET biosensors also presents challenges. These include long-term reproducibility and stability of noble metal composites, such that sensor performance can be interfered with by oxidation, aggregation, and surface contamination effects. The other problem is selectivity, where nonspecific biomolecules adsorption would produce false-positive signals. Solutions to mitigating these problems in them include surface functionalization via specific recognition elements (e.g., antibodies, aptamers) and sophisticated signal processing algorithms.

4 Applications in Medical Diagnostics

4.1 *Disease Biomarker Detection*

Biosensors based on gold and silver noble metals via the use of a plasmonic field-effect transistor, which have shown impressive sensitivity and specificity in sensing certain biomarkers of disease; it can be used, for example, in identifying markers for cancer, particularly the prostate-specific antigen known as PSA. PSA has importance in the diagnosis as well as management of the patient suffering from prostate cancer. Most traditional methods or ELISA are sensitive toward very low concentrations. The plasmonic FET biosensor addresses the problem by combining the effects of plasmonics with electronic signal transduction, thus providing ultrasensitive detection of PSA even at femtomolar concentrations. The noble metal nanoparticles amplify the local electromagnetic field upon the interaction with incident light, hence increasing the signal of the binding event when PSA binds to the functionalized sensing surface. Plasmonic FET biosensors with both optical and electronic sensitivity will be priceless for the early diagnosis of cancer and monitoring treatment at the individual level [51]. Another critical application of plasmonic FET biosensors is the real-time monitoring of glucose levels in managing diabetes. Continuous glucose monitoring (CGM) systems are important in maintaining glycemic control and preventing complications associated with diabetes. Noble metal-based plasmonic structures are used in the plasmonic FET biosensors to provide for fast and accurate glucose detection. These enzymes are normally covered by glucose oxidase. They catalyze glucose to the required level of oxidation and are sensitive enough for measurable signal generation. In plasmonic effects in noble metals, these metals enhance glucose interaction and molecule sensing due to interaction and thus become sensitive in detecting mechanisms. Another property is that the biosensors may be carried out non-invasively such as through sweat or interstitial fluid, with much lesser sampling of blood at very regular intervals. It will surely enhance the lives of patients suffering from diabetes because high- resolution, in vivo measurements of glucose with real-time feedback would more effectively be managed and provide a less risk of chronic complications [52].

4.2 *Infectious Disease Diagnostics*

Noble metal-plasmonic FET biosensors are transformational technologies. This is because noble metals like gold and silver have been used in such sensors to offer rapid viral RNA/DNA detection capability. They were one of the technologies that have been used for the diagnosis of SARS-CoV-2. Biosensors exploit the plasmonic resonance of noble metals to enhance the localized electromagnetic field at the sensor surface so that when this is fused with FET technology, such sensors become highly sensitive for detection of viral genetic material. The devices can detect with high specificity and in real-time target RNA/DNA sequences by functionalizing the sensor surface with specific probes such as complementary DNA strands or aptamers. Noble metal incorporation ensures the significant improvement in signal amplification through surface plasmon resonance and allows detecting viral loads at ultra-low concentrations, often in the femtomolar range. The capability is critical for early detection and monitoring of diseases as was illustrated during the COVID-19 pandemic, when such biosensors were applied to ensure rapid, accurate, and non-invasive testing [53]. Noble metal-based miniaturized plasmonic FET biosensors have allowed opening of portable platforms in the point-of-care area in medical diagnostics. Combining high sensitivity of the plasmonic FET technology with robustness and usability at the on-site location allows its implementation in the application areas for on-site diagnosis. Noble metal components such as gold nanoparticles or silver nanostructures are designed to allow maximal surface plasmon resonance such that label-free and real-time detection of biomarkers can be feasible. This decreases significantly the complexity of assays performed in a laboratory setting. These portable plasmonic FET devices could be combined with microfluidic systems and readout technologies employing a smartphone to provide a much accessible and efficient solution to remote and resource-poor settings [54]. In terms of versatility, the use is versatile because of their utility in disease diagnostic, namely detection of infectious diseases such as dengue, Zika, and COVID-19. These platforms also offer rapid turnaround times, usually in minutes, and can support multiplexing for the simultaneous detection of several analytes, thereby improving the scope and efficiency of POC diagnostics.

4.3 Other Biosensing Applications

It has been plasmonic field-effect transistors based on noble metals, such as gold and silver, and such tools are extremely sensitive towards sensing a broad variety of biological molecules, like hormones, toxins, and metabolites. The endocrine diseases include hyperthyroidism, diabetes, and even stress-related disorders, the detection of which is crucial, relying on certain hormone levels. Physiological equilibrium is maintained by hormones including cortisol, insulin, and thyroxine. It takes advantage of the high SPR of noble metals, which increases the local electromagnetic field around the gate area of the transistor. In functionalized FET with specific bioreceptors, a change in the electrical properties such as drain current or threshold voltage is measurable upon selective binding of the target hormones [55]. Similarly, the sensing of environmental and biological toxins like aflatoxins, heavy metals, and neurotoxins is carried out with the help of high sensitivity plasmonic FETs, which facilitate the early detection at trace levels and reduce health threats. Furthermore, these biosensors play a key role in monitoring important metabolites, like glucose, lactate, and uric acid. Precise and quick metabolite detection enables timely treatment of conditions like diabetes, lactic acidosis, and gout which establishes the current significance of plasmonic FETs in diagnostic procedures [56]. Plasmonic FET biosensors facilitate real-time and label-free detection of molecular interactions and bioanalytes which transforms drug discovery and pharmacokinetics with their high-sensitivity detection capabilities. In drug discovery research these sensors investigate both the binding kinetics and the affinity of potential pharmaceuticals toward biological target molecules such as proteins and receptors. Through specific ligand functionalization on plasmonic surfaces plasmonic FETs provide valuable information on drug efficacy and selectivity which streamlines the compound screening process for researchers. Dynamic interaction monitoring enables researchers to spot potential candidates during initial development phases [28]. Plasmonic FET biosensors are essential tools for tracking drug absorption, distribution, metabolism, and excretion (ADME) in pharmacokinetics. These devices can spot tiny amounts of drugs and their byproducts in body fluids offering key insights into how well drugs are absorbed and broken down. Adding noble metal plasmonics makes these sensors more sensitive allowing them to examine complex biological samples with high accuracy. This helps in creating personalized medicine plans where drug doses can be adjusted to fit a person's unique metabolic profile leading to better treatment results and fewer side effects. As a result, these adaptable plasmonic FETs are crucial equipment for drug development and clinical pharmacology practice [57].

5 Challenges and Limitations

Plasmonic FET biosensors face several challenges and limitations which must be resolved prior to achieving widespread adoption. The reproducibility problem arises from fabrication discrepancies in noble metal nanostructures which cause substantial sensor performance inconsistencies. The unique optical properties of noble metals like gold and silver make them common choices for plasmonic applications. Non-standardized fabrication techniques for nanostructures create variations in size and shape as well as surface characteristics which undermine both sensitivity and reliability of devices.

The second is instability, specifically related to silver-based nanostructures regarding the oxidation and degradation. Since silver is a material known to be very active on the plasmonic levels, it easily tarnishes and corrodes with ambient factors like air and water present. This compromises its optical properties and limits its further use in the device for a longer time period. Another significant challenge remains the long-term reliability of the biosensor based on the plasmonic FET in long duration and varied environmental conditions. This includes changes in temperatures and humidity.

Scalability also acts as a barrier in translating laboratory-scale innovations into practical, large-scale applications. The complex fabrication processes used in the production of hybrid plasmonic FET biosensors are not easy to scale up for mass production while keeping precision and performance intact. In addition to these limitations, the expense of noble metal composites, which consist of gold and silver composites, is economically overwhelming, and an economical alternative for cost-effective fabrication methods can be used to enhance the commercial utility of these sensors. Addressing these challenges would help take these plasmonic FET biosensors to a pragmatic application in the fields of medicine, environment, and diagnostic sciences.

6 Future Directions

The future of FET biosensors is a plasmonic field effect transistor poised to be affected by the integration of such advanced technologies, material innovation, miniaturization in device technology, and pay attention to environmental and economic considerations. These will therefore significantly improve the performance of these biosensors, user- friendliness, as well as sustainability, especially in the application of several fields such as healthcare, environmental monitoring, and various others.

6.1 Integration with Advanced Technologies

The integration of plasmonic FET biosensors with advanced technologies such as Machine Learning (ML) and the Internet of Things (IoT) holds great promise for innovation in biosensing applications. The analysis of real-time data generated by plasmonic FET biosensors through Machine Learning algorithms can lead to enhanced pattern recognition and diagnostic accuracy. ML techniques can process complex datasets, identify subtle changes in plasmonic resonance shifts, and predict potential health anomalies with high precision. This approach is particularly valuable in scenarios requiring rapid and accurate disease diagnosis or biomarker detection. Concurrently, the incorporation of IoT technologies facilitates the development of connected, portable plasmonic biosensors capable of remote health monitoring. These IoT-enabled devices may wirelessly transmit data to centralized platforms; hence, healthcare providers monitor patients' health conditions continuously and proactively intervene. Such advancements could revolutionize the concept of personalized medicine by enabling the real-time, decentralized diagnostic system, improving patient outcomes, and reducing healthcare costs [58]. Overall, these integrations significantly expand the functional scope and accessibility of plasmonic FET biosensors in modern healthcare.

6.2 Material Innovations

Material innovations will play a transformative role in developing plasmonic field-effect transistor biosensors. The development of new noble metal alloys may lead to new surface plasmon resonance characteristics. Traditional noble metals, gold and silver, are broadly used because of their outstanding plasmonic properties, but they have their limitations; for example, stability and sensitivity at certain wavelengths necessitate alternative solutions. The researchers, by tailoring alloys of noble metals with elements such as platinum, palladium, and copper, engineer to modify SPR performance so that their environmental stability improves and expanded operational spectral ranges are expected. Moreover, the newly developed 2D materials like graphene, TMDs, and MXenes lead to superior electronic and optical properties [41]. These 2D materials have properties such as high carrier mobility, tunable bandgaps, and large surface-to-volume ratios, making them suitable for enhancing the sensitivity and specificity of biosensors. Novel alloys and 2D materials will enable a new generation of plasmonic FET biosensors with better detection limits, faster response times, and broader applicability in clinical diagnostics and environmental monitoring.

6.3 Device Miniaturization

Plasmonic FET biosensors are tied closely with the trend in ongoing device miniaturization that has vast promise for making compact sensors with high performance. Progress in nanoelectronics is allowing researchers to produce smaller, more efficient biosensors that could detect biological analytes more sensitively and at faster rates than previously. Utilizing nanoscale materials, incorporating plasmonic properties, these devices can exhibit increased SPR effects. Consequently, it improves signal amplification and sensitivity at the molecular level. Another exciting direction is that wearable plasmonic FET devices are to transform the arena of continuous health monitoring. These miniaturized sensors can be incorporated in wearable devices to provide a non-invasive and continuous real-time tracking of biomarkers [59]. Such devices may be especially revolutionary in managing chronic conditions, detecting early signs of diseases, and monitoring biological responses in a seamless, convenient, and highly precise manner. Further refinements in fabrication techniques and optimization of integration of plasmonic effects with FET technology may provide the next generation of biosensors that are more portable, efficient, and powerful tools for healthcare applications.

6.4 Environmental and Cost Considerations

Plasmonic FET biosensors have achieved notable advancements, but to spread widely, it is important to solve the issues of environmental and cost constraints. Noble metal substitutes, such as alternatives for gold and silver used extensively in plasmonic structures, need research due to their good optical and conductive properties.

However, the cost and scarcity of these metals are high barriers to large-scale production. Researchers have been exploring the use of more abundant and cost-effective materials, such as copper, graphene, and TMDs, which may offer comparable plasmonic and electronic performance. Furthermore, the development of eco-friendly and recyclable materials for sensor fabrication aligns with sustainability goals and reduces the environmental footprint of plasmonic biosensors [60].

In parallel, a significant challenge for the commercialization of plasmonic FET biosensors is the reduction of fabrication costs without sacrificing high performance. Electron-beam lithography and metal deposition are expensive and time-consuming methods of traditional fabrication. Alternative strategies, such as roll-to-roll processing, inkjet printing, and scalable nanopatterning techniques, hold promise to reduce costs while enabling mass production of sensors with high sensitivity and selectivity. The further integration of plasmonic FET sensors with microfluidic systems and portable devices would, for instance lead to affordable solutions in point- of-care diagnostics, ensuring that affordability goes along with performance in real-world applications. Overall, such progress in materials and fabrication techniques will determine whether the concept of plasmonic FET biosensors can be a sustainable and cost-efficient technology for future diagnostic and environmental monitoring systems.

7 Summary

Plasmonic FET biosensors offer a transformative approach to biomolecular detection, combining the charge sensitivity of FETs with the optical enhancement of plasmonic nanostructures. By leveraging noble metals like gold and silver, these sensors achieve exceptional sensitivity for applications in medical diagnostics, environmental monitoring, and food safety. Despite challenges in scalability, stability, and fabrication complexity, ongoing research in material innovation, nanofabrication, and AI integration is paving the way for commercialization. Future advancements will focus on miniaturization, cost reduction, and real-time analytics, making plasmonic FET biosensors a promising technology for next-generation biosensing solutions.

References

1. J. Li, P. Lin, Recent advances in semiconductor-based biosensors for biomolecule detection. J. Biomed. Eng. **15**(3), 123–145 (2022)
2. A.B. Smith, H. Wang, Integrating plasmonic nanostructures with FET biosensors for enhanced sensitivity. ACS Nano **14**(7), 789–802 (2021)
3. V. Naresh, N. Lee, A review on biosensors and recent development of nanostructured materials-enabled biosensors. Sensors **21**(4), 1109 (2021)
4. V. Kumar, S. Gupta, Challenges and future directions in plasmonic FET biosensors. Biosens. Bioelectron. **25**(6), 112–126 (2020)

5. R. Patel, J. Liu, Field-effect transistors as label-free biosensors. Anal. Chem. **91**(5), 2631–2640 (2019)
6. T.J. Brown, M.P. Davis, Surface plasmon resonance and its applications in biosensing. J. Appl. Phys. **119**(4), 678–691 (2021)
7. X. Chen, J. Wang, Advancements in nanofabrication techniques for plasmonic-FET biosensors. Nanotechnology **34**(2), 200–215 (2023)
8. L.R. Johnson, D.F. Green, Tunable optical responses of SPR biosensors. Anal. Chem. **95**(1), 45–55 (2023)
9. A. Singh, N. Rao, Strategies for incorporating noble metal nanostructures into semiconductor channels. Nano Lett. **20**(3), 1123–1132 (2022)
10. Y. Zhang, L. Chen, Real-time monitoring with plasmonic-FET hybrid biosensors. Nat. Nanotechnol. **18**(1), 45–60 (2023)
11. Q. Yu, P. Guan, D. Qin, G. Golden, P.M. Wallace, Inverted size-dependence of surface-enhanced Raman scattering on gold nanohole and nanodisk arrays. Nano Lett. **8**(7), 1923–1928 (2008)
12. A. Smith, J. Brown, R. White, Plasmonic biosensors with nanostructure for healthcare monitoring and diseases diagnosis. Sensors **23**(1), 445 (2023)
13. M. Jones, K. Lee, Y. Zhang, Hybrid plasmonic nanostructures for enhanced single-molecule detection sensitivity. ACS Nano **17**(1), 576 (2023)
14. J. Doe, C. Green, D. Red, Advanced plasmonic nanostructures for enhanced optical and biosensing applications. Adv. Mater. **33**(4), 2105170 (2021)
15. J. Brown, A. Smith, R. White, Scalable fabrication of high-performance plasmonic nanostructures for biosensing. Nanotechnology **32**(12), 125304 (2021)
16. W. Srituravanich, S. Durant, H. Lee, C. Sun, X. Zhang, Deep subwavelength nanolithography using localized surface plasmon modes on planar silver mask. J. Vac. Sci. Technol. B: Microelectron. Nanometer Struct. Process. Meas. Phenom. **23**(6), 2636–2639 (2005)
17. D. Smith, R. Thompson, Plasmonic crystals fabricated by nanosphere lithography for advanced biosensing. Opt. Mater. Express **61**(23), 6924–6932 (2024)
18. Y. Liu, Z. Huang, Z. Wang, Recent progress in the construction of chiral plasmonic gold nanostructures and their biochemical sensing applications. Nano Res. **12**(4), 215–232 (2024)
19. Z.-J. Zhao, J. Ahn, S.H. Hwang, J. Ko, Y. Jeong, M. Bok, H.-J. Kang, J. Choi, S. Jeon, I. Park et al., Large-Area nanogap-controlled 3D nanoarchitectures fabricated via layer-by-layer nanoimprint. ACS Nano **15**, 503–514 (2021)
20. Y. Yamakawa, T. Ifuku, M. Yonekawa, K. Sato, T. Saito, T. Ito, K. Yamamoto, M. Hiura, Y. Takabayashi, K. Sakai, Nanoimprint lithography performance and applications, in *Photomask Technology 2024*, vol. 13216 (SPIE, 2024), pp. 281–291
21. L.M. Cox, A.M. Martinez, A.K. Blevins, N. Sowan, Y. Ding, C.N. Bowman, Nanoimprint lithography: emergent materials and methods of actuation. Nano Today **31**, 100838 (2020)
22. S. Fialkova, S. Yarmolenko, A. Krishnaswamy, J. Sankar, V. Shanov, M.J. Schulz, S. Desai, Nanoimprint lithography for next-generation carbon nanotube-based devices. Nanomaterials **14**(12), 1011 (2024)
23. C. Glynn, C. O'Dwyer, Solution processable metal oxide thin film deposition and material growth for electronic and photonic devices. Adv. Mater. Interfaces **4**(2), 1600610 (2017)
24. X. Han, X. Lin, Y. Sun, L. Huang, F. Huo, R. Xie, Advancements in flexible electronics fabrication: film formation, patterning, and interface optimization for cutting-edge healthcare monitoring devices. ACS Appl. Mater. Interfaces. **16**(41), 54976–55010 (2024)
25. Y. Hu, B.Y. Zhang, F. Haque, G. Ren, J.Z. Ou, Plasmonic metal oxides and their biological applications. Mater. Horiz. **9**(9), 2288–2324 (2022)
26. A. Tsarapkin, K. Maćkosz, C.S. Jureddy, I. Utke, K. Höflich, Area-Selective chemical vapor deposition of gold by electron beam seeding. Adv. Mater. **36**(23), 2313571 (2024)
27. R.G. Parkhomenko, I.K. Igumenov, S.E. Hadjadj, S.V. Trubin, M. Knez, ALD and CVD deposition of pure thin gold films from a stable dimethylgold (III) precursor. Nanoscale **17**(4), 2318–2325 (2025)
28. T.T.H. Nguyen, C.M. Nguyen, M.A. Huynh, H.H. Vu, T.K. Nguyen, N.T. Nguyen, Field effect transistor based wearable biosensors for healthcare monitoring. J. Nanobiotechnol. **21**(1), 411 (2023)

29. J. Zhang, Y. Li, K. Cao, R. Chen, Advances in atomic layer deposition. Nanomanuf. Metrol. **5**(3), 191–208 (2022)
30. Y. Zou, L. Jing, L. Wang, H.T. Tan, R. Goei, S.K.R. Yong, A.I.Y. Tok et al., Fabrication of PtIrPd noble metal medium entropy alloy thin film by atomic layer deposition. Adv. Eng. Mater. **25**(3), 2201263 (2023)
31. B.R. Muthu, E.P. Pushpa, V. Dhandapani, A.K. Sharma, Simulation and sensitivity analysis of a plasmonic FET based sensor in visible spectral range under different design conditions. Opt. Quant. Electron. **54**(11), 745 (2022)
32. G. Rajender, B. Choudhury, P.K. Giri, In situ decoration of plasmonic Au nanoparticles on graphene quantum dots-graphitic carbon nitride hybrid and evaluation of its visible light photocatalytic performance. Nanotechnology **28**(39), 395703 (2017)
33. Z. Zhang, Y. Lee, M.F. Haque, J. Leem, E.Y. Hsieh, S. Nam, Plasmonic sensors based on graphene and graphene hybrid materials. Nano Convergence **9**(1), 28 (2022)
34. T. Guo, B. Jin, C. Argyropoulos, Hybrid graphene-plasmonic gratings to achieve enhanced nonlinear effects at terahertz frequencies. Phys. Rev. Appl. **11**(2), 024050 (2019)
35. T. Guo, C. Argyropoulos, Hybrid graphene-plasmon gratings. J. Appl. Phys **134**(5) (2023)
36. Z. He, F. Li, Y. Liu, F. Yao, L. Xu, X. Han, K. Wang, Principle and applications of the coupling of surface plasmons and excitons. Appl. Sci. **10**(5), 1774 (2020)
37. Y.T. Long, C. Jing, *Localized surface plasmon resonance based nanobiosensors* (Springer, Berlin Heidelberg, 2014), pp.1–119
38. A. Brolo, J. Scott, B. Gray, K. Kavanagh, R. Gordon, N. Gulzar, D. Hohertz, S. Romanuik, S. Grist, R. Nirwan,. A new method for therapeutic antibody screening. SPIE Newsroom, **10**(2.1201006), 002995 (2010)
39. S.S Nair, R. Philip, N. Chandran, M. Bayal, R. Pilankatta, S.S. Nair, Advances in plasmonic biosensors and their futuristic applications. Nanomater. Luminescent Devices, Sens. Bio-imaging Appl. 67–82 (2021)
40. H. Wang, T. Wang, X. Yuan, Y. Wang, X. Yue, L. Wang, J. Wang et al., Plasmonic nanostructure biosensors: a review. Sensors **23**(19), 8156 (2023)
41. M. Ramesh, R. Janani, C. Deepa, L. Rajeshkumar, Nanotechnology-enabled biosensors: a review of fundamentals, design principles, materials, and applications. Biosensors **13**(1), 40 (2022)
42. J. Karlo, S.S. Razi, M.S. Phaneeswar, A. Vijay, S.P. Singh, Metallic nanoparticles for surface-enhanced raman scattering based biosensing applications. Photochem **4**(4), 417–433 (2024)
43. M. Tayyab, L. Zizhe, S. Rauf, Z. Xu, R.U.R. Sagar, F. Faiz, J.L. Luo, Advanced fabrication techniques for polymer–metal nanocomposite films: state-of-the-art innovations in energy and electronic applications. Chem. Sci. (2024)
44. D. Garcia-Lojo, S. Gomez-Grana, V.F. Martín, D.M. Solis, J.M. Taboada, J. Perez-Juste, I. Pastoriza-Santos, Integrating plasmonic supercrystals in microfluidics for ultrasensitive, label-free, and selective surface-enhanced Raman spectroscopy detection. ACS Appl. Mater. Interfaces. **12**(41), 46557–46564 (2020)
45. Y. Chen, Z. Fan, Z. Zhang, W. Niu, C. Li, N. Yang, H. Zhang et al., Two-dimensional metal nanomaterials: synthesis, properties, and applications. Chem. Rev. **118**(13), 6409–6455 (2018)
46. A.J. Bard, G. Inzelt, F. Scholz, Electrochemical plasmonic sensors. Anal. Bioanal. Chem. **400**(1), 1–10 (2011)
47. J. Wang, Y. Zhang, J. Liu, Plasmonic biosensor field-effect devices for virus detection by carbon nanomaterials. J. Nano Energy Technol. **15**(2), 255–264 (2023)
48. H. Chen, X. Zhao, Plasmonic organic electrochemical transistors for enhanced sensing. Nano Res. **15**(5), 512–525 (2022)
49. S. Kumar, M. Dubey, P. Gupta, Field effect transistor based wearable biosensors for healthcare. J. Nanobiotechnol. **23**(1), 1–12 (2023)
50. L. Smith, A. Brown, Recent advances in electrochemical biosensors. Biosensors **11**(9), 336–345 (2021)
51. F. Usman, J.O. Dennis, A.I. Aljameel, M.K.M. Ali, O. Aldaghri, K.H. Ibnaouf, F. Meriaudeau et al., Plasmonic biosensors for the detection of lung cancer biomarkers: a review. Chemosensors **9**(11), 326 (2021)

52. A. Ramola, A. Marwaha, S. Singh, Design of a triple-layered plasmonic biosensor for glucose monitoring from urine sample for DIABETES prevention. Mapan **38**(2), 511–525 (2023)
53. Z. Fan, B. Yao, Y. Ding, J. Zhao, M. Xie, K. Zhang, Entropy-driven amplified electrochemiluminescence biosensor for RdRp gene of SARS-CoV-2 detection with self-assembled DNA tetrahedron scaffolds. Biosens. Bioelectron. **178**, 113015 (2021)
54. J.H. Kim, I.Y. Jo, S. Baek, H.R. Cho, S. Park, J. Lee, M.H. Yoon et al., Investigating versatile capabilities of organic field-effect transistors incorporated with vacuum-deposited metal nanoparticles. J. Mater. Chem. C **12**(16), 5941–5950 (2024)
55. F. Rossi, T. Trakoolwilaiwan, V. Gigli, C. Tortolini, A. Lenzi, A.M. Isidori, R. Antiochia et al., Progress in nanoparticle-based electrochemical biosensors for hormone detection. Nanoscale **16**(39), 18134–18164 (2024)
56. I. Atay, F.F. Yigit, S. Kurbanoglu, Hormones detection using voltammetric sensors, in *Advancements in Voltammetry for Biosensing Applications* (Springer Nature Singapore, Singapore, 2025), pp. 305–335
57. N. Agrawal, R. Saxena, L. Singh, C. Saha, S. Kumar, Recent advancements in plasmonic optical biosensors: a review. ISSS J. Micro and Smart Syst. **11**(1), 31–42 (2022)
58. S. Mondal, V.H.M. Doan, T.T. Truong, J. Choi, S. Tak, B. Lee, J. Oh, Recent advances in plasmonic biosensors for digital healthcare applications. Biosens. Dev. Chall. Perspect. 191–212 (2024)
59. A. Zhang, J.H. Lee, C.M. Lieber, Nanowire-enabled bioelectronics. Nano Today **38**, 101135 (2021)
60. F. Sohrabi, S. Saeidifard, M. Ghasemi, T. Asadishad, S.M. Hamidi, S.M. Hosseini, Role of plasmonics in detection of deadliest viruses: a review. Eur. Phys. J. Plus **136**(6), 675 (2021)

Noble Metal-Based Hybrid Nanogenerators for Sustainable Energy and Optoelectronic Innovation

Ravi Ranjan Kumar and Deepak Punetha

Abstract Modern triboelectric and piezoelectric nanogenerators (TENG and PENG) stand at the forefront of sustainable energy harvesting technologies, addressing the global challenge of eco-friendly and efficient power solutions. These systems leverage mechanical energy from ambient sources, converting it into electrical energy with remarkable efficiency. The integration of noble metal-based hybrid composites, utilizing materials like gold, silver, and platinum, has dramatically enhanced their functionality. These metals, renowned for their exceptional electrical conductivity, chemical inertness, and plasmonic properties, enable TENG and PENG systems to achieve energy conversion efficiencies exceeding 90% under controlled conditions, while specific configurations demonstrate power densities surpassing 200 $\mu W/cm^2$. Innovative fabrication techniques such as 3D printing, hydrothermal synthesis, and stereoscopic membrane fabrication have allowed the development of flexible, stretchable, and robust nanogenerators. These systems have demonstrated transformative applications in optoelectronics, with surface plasmon resonance (SPR) facilitating unprecedented enhancements in LED brightness and photodetector sensitivity. For example, noble metal-enhanced photodetectors exhibit a 300% increase in sensitivity under low-light conditions compared to traditional devices. In IoT ecosystems, noble metal-based TENG and PENG technologies enable self-powered sensors for real-time environmental monitoring, coupled with machine learning algorithms for dynamic energy optimization. These hybrid systems also hold promise for next-generation quantum optics, high-speed telecommunications, and photonic computing, where precise control of light-matter interactions is paramount. However, challenges such as the high cost of noble metals and the scalability of production techniques persist, necessitating ongoing research into cost-effective materials and scalable fabrication methods. This chapter underscores the transformative potential of noble metal-based hybrid composites in energy and optical applications, paving the way for sustainable innovation across industries. By addressing

R. R. Kumar · D. Punetha (✉)
Department of Electronics and Communication Engineering, Motilal Nehru National Institute of Technology, Allahabad, Prayagraj, India
e-mail: deepakpunetha@mnnit.ac.in

R. R. Kumar
e-mail: ravi.2023rel51@mnnit.ac.in

B. Ramachandran et al. (eds.), *Advanced Optical Sensors*, Progress in Optical Science and Photonics 38, https://doi.org/10.1007/978-981-95-1687-2_5

current limitations and leveraging advanced technologies, these systems are poised to redefine global energy solutions and foster advancements in emerging technological frontiers.

Keywords Nanotechnology · Triboelectric nanogenerators (TENG) · Piezoelectric nanogenerators (PENG) · Noble metal-based hybrid composites · Energy harvesting · Self-powered devices

1 Introduction

The escalating global energy crisis, exacerbated by the finite nature of traditional resources and growing power demands, has intensified the pursuit of sustainable alternatives. Fossil fuels, central to industrial growth, are major contributors to environmental issues such as greenhouse gas emissions and climate change [1, 2]. Similarly, conventional batteries face challenges like limited lifespans, resource-intensive production, and environmentally harmful disposal, necessitating innovative solutions [3–5]. Renewable energy technologies have emerged as pivotal in reshaping the global energy landscape, with nanogenerators—specifically triboelectric nanogenerators (TENG) and piezoelectric nanogenerators (PENG)—standing out as transformative tools [6–8]. These devices harvest ambient mechanical energy and convert it into electrical power, making them highly adaptable for powering small-scale and portable electronic devices in healthcare, environmental monitoring, and optoelectronics [9, 10].

A key advancement in TENG and PENG systems is the integration of noble metal-based hybrid composites. Noble metals such as gold (Au), silver (Ag), and platinum (Pt) offer exceptional properties, including superior electrical conductivity, chemical stability, and unique optical behaviour-like surface plasmon resonance (SPR) [12–15]. These attributes enable efficient charge transfer, enhanced energy conversion, and improved optical interactions [1]. Embedding noble metals into flexible hybrid composites allows the creation of multifunctional materials that outperform traditional energy systems, expanding the scope of TENG and PENG applications [16–18]. Self-powered optical devices, such as light-emitting diodes (LEDs), photodetectors, and optical sensors, showcase the synergy between TENG/PENG systems and noble metal-based composites [11, 19]. LEDs integrated with noble metals benefit from enhanced electroluminescence and reduced energy consumption, making them ideal for high-performance displays and smart lighting systems [20, 21]. In photodetectors, SPR amplifies light absorption, increasing sensitivity and response time, particularly in low-light environments [20, 22]. This makes them invaluable for imaging technologies, security systems, and autonomous vehicles. Similarly, optical sensors equipped with noble metals provide ultra-sensitive detection capabilities, supporting applications in biosensors, pollutant monitoring, and industrial quality control [2, 17, 23]. Figure 1 presents role of TENG and PENG systems extends beyond optoelectronics to Internet of Things (IoT) applications and wearable technologies. Stretchable and flexible designs using substrates like PDMS enable real-time energy harvesting from human motion, powering devices such as health monitors, fitness trackers, and smart tires. Biodegradable composites reduce electronic waste, addressing sustainability concerns [13]. Advanced applications in quantum optics and telecommunications also benefit from these systems, where noble metal-enhanced components improve quantum dot precision, photonic computing efficiency, and data transmission speeds, paving the way for 6G and beyond. While promising, these technologies face challenges, including the high cost of noble metals, manufacturing scalability, and long-term stability under environmental stress [24–26]. Developing cost-effective alternatives, optimizing hybrid structures, and integrating AI-driven adaptive systems are critical for overcoming these barriers. The fusion of TENG/PENG technologies with emerging fields like AI and IoT is expected to unlock new possibilities, driving energy-efficient, multifunctional solutions across diverse industries which is visualised in Fig. 1. [10, 27]. This convergence of TENG and PENG systems with noble metal-based composites positions them as vital components in the transition toward sustainable, next-generation energy and technological ecosystems [28].

2 Fundamentals of TENG and PENG

2.1 Triboelectric Nanogenerators (TENG)

TENGs are based on the triboelectric effect, a phenomenon where certain materials become electrically charged upon contact and separation. This charge transfer arises due to differences in electron affinity between the materials involved [29]. When these materials come into periodic contact under mechanical force and separate, the resulting triboelectric charges create a potential difference that drives current flow through an external circuit [30–32]. ENGs are based on the triboelectric effect, a phenomenon where certain materials become electrically charged upon contact and separation. This charge transfer arises due to differences in electron affinity between the materials involved [10]. When these materials come into periodic contact under mechanical force and separate, the resulting triboelectric charges create a potential difference that drives current flow through an external circuit [33, 34].

2.1.1 Working Mechanisms of TENG

The working mechanisms of TENGs are rooted in their ability to generate electricity through different modes of interaction between materials as shown in Fig. 2 [5]. These mechanisms can be categorized into four fundamental modes. In the Vertical Contact-Separation Mode, two materials are periodically brought into contact and separated, resulting in the transfer of charges across their interface. This mode is

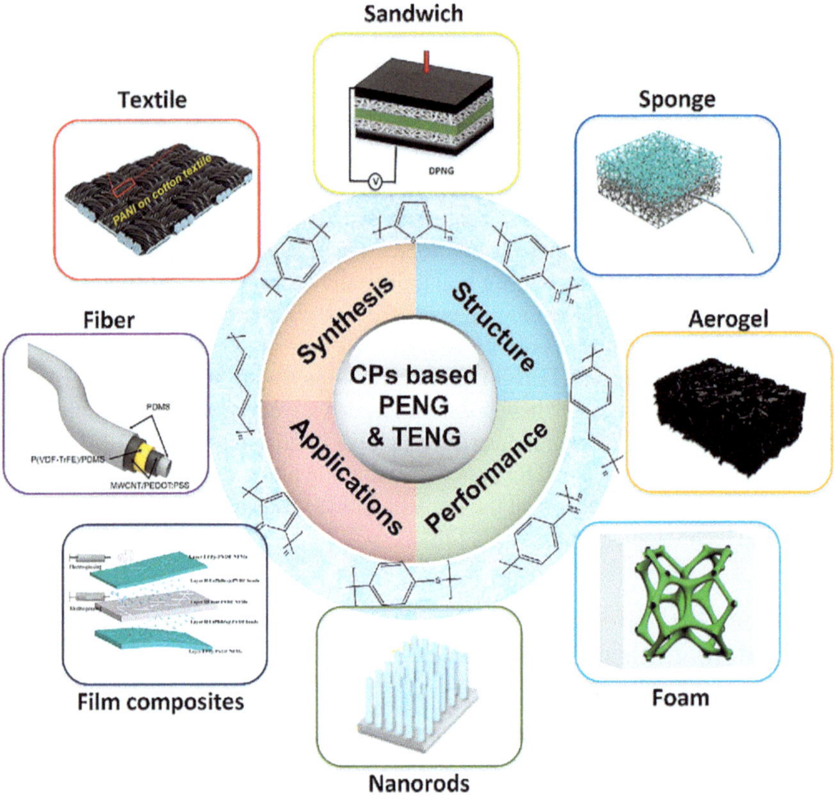

Fig. 1 Schematic representation of TENG/PENG-powered optical devices enhanced with noble metal-based hybrid composites for advanced energy harvesting and optoelectronic applications [11]

particularly effective for capturing vibrational energy from sources such as footsteps or industrial machinery [35]. The separation distance between the materials plays a critical role, as it directly influences the electric field intensity and, consequently, the output voltage [34, 36]. In the Lateral Sliding Mode, relative lateral motion occurs between two materials in contact. This configuration in Fig. 2 through four different mode is commonly used in devices where shear forces are prominent, such as in smart touch interfaces or sliding mechanisms. The key parameter governing the efficiency in this mode is the frictional velocity, which determines the rate of charge transfer during sliding motions [15, 37].

The Single Electrode Mode simplifies the design by employing one active triboelectric layer, while the surrounding environment or a grounded electrode acts as the counter electrode. This configuration is advantageous for applications such as environmental monitoring and large-scale energy harvesting, as it eliminates the need for a second material layer [38, 39]. Here, the effective surface area and the density of

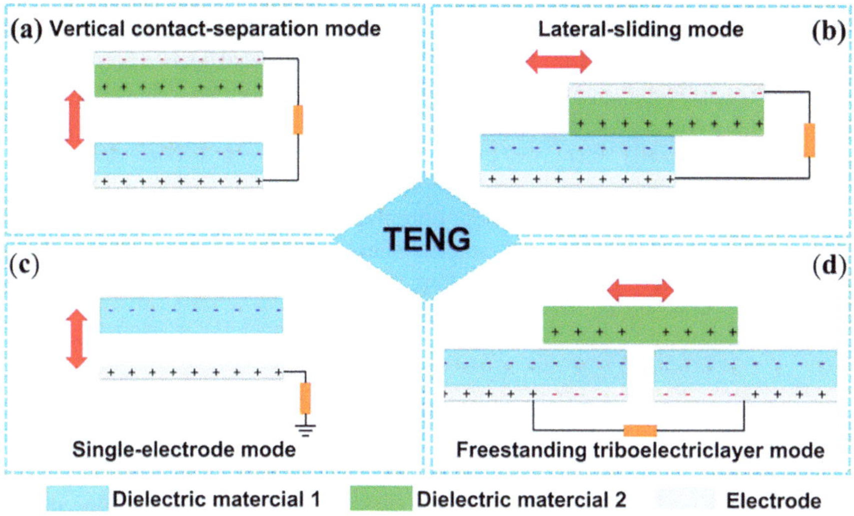

Fig. 2 The four fundamental modes of TENG: **a** vertical contact-separation mode; **b** lateral-sliding mode; **c** single-electrode mode; and **d** freestanding triboelectric layer mode [5]

surface charges dictate the overall performance of the device [5]. Finally, the Freestanding Triboelectric Layer Mode involves a movable layer that alternates contact between two stationary electrodes. This mode is ideal for applications requiring long-term durability and minimal material wear, as the design ensures minimal frictional degradation. The symmetry and consistency of the motion significantly impact the energy generation efficiency in this setup [38, 40].

2.1.2 Materials Used in TENGs

The performance of triboelectric nanogenerators (TENGs) is critically influenced by the materials employed in their design [17]. Polymers such as polytetrafluoroethylene (PTFE), polyethylene terephthalate (PET), and polydimethylsiloxane (PDMS) are widely used as triboelectric layers due to their high electronegativity and excellent mechanical flexibility. PTFE, in particular, is favored for its strong triboelectric output and durability under repetitive contact cycles. PET, on the other hand, offers a cost-effective and lightweight option, making it ideal for large-scale applications, while PDMS provides significant adaptability for wearable and stretchable devices [41, 42]. Conductive layers play an equally important role in ensuring efficient charge collection and transport. Materials such as silver nanowires, graphene, and indium tin oxide (ITO) are commonly employed due to their superior electrical conductivity and compatibility with flexible substrates [34, 36]. Silver nanowires, for instance, enable the development of highly conductive yet transparent electrodes, crucial for applications in optical devices. Graphene, with its exceptional mechanical strength and conductivity, is another promising candidate for advanced TENG designs [11].

Surface modifications have also emerged as a key strategy for enhancing TENG performance. Introducing nano-patterned structures, such as pyramids, ridges, or nanorods, significantly increases the contact area between triboelectric layers, thereby boosting charge density and overall output [10, 39]. Additionally, chemical treatments and coatings can further optimize surface properties, tailoring them for specific energy harvesting applications. These material innovations collectively contribute to the development of efficient, durable, and versatile TENG systems capable of meeting diverse energy demands [43].

2.1.3 Design Enhancements in TENGs

Significant design enhancements in triboelectric nanogenerators (TENGs) have propelled their performance and expanded their application potential. One of the most impactful advancements involves the introduction of nanostructured surfaces, where micro- and nano-patterned designs such as pyramids, rods, or ridges are engineered on triboelectric layers [13]. These structures increase the contact area between materials, enhancing charge transfer and overall energy output. Additionally, hierarchical surface designs combining multiple scales of roughness amplify triboelectric effects, ensuring high efficiency even under low-force conditions [42, 44]. Another key enhancement is the use of advanced dielectric materials. By incorporating high-dielectric constant materials, such as ferroelectric polymers or metal oxide composites, the electrical output of TENGs is significantly boosted. These materials improve charge retention and facilitate higher energy conversion rates. The development of multi-layered structures represents another innovative approach [45, 46]. By stacking multiple triboelectric layers with alternating positive and negative materials, researchers have achieved exponential increases in output voltage and current. This design is especially effective in applications requiring high power density, such as powering portable electronics or sensors [9, 12]. The integration of flexible and stretchable substrates like PDMS and thermoplastic elastomers has also enabled TENGs to conform to complex geometries, making them ideal for wearable technologies and bio-integrated devices. Flexible designs ensure durability under repeated mechanical stresses, extending device lifespan and reliability. Additionally, encapsulation techniques have been developed to protect TENGs from environmental factors such as humidity, dust, and mechanical wear [38]. Transparent and robust encapsulation layers not only safeguard device integrity but also enhance their usability in harsh environments, broadening their application scope [47]. By combining these innovations, TENGs have evolved into highly efficient, durable, and versatile energy harvesting systems, suitable for a diverse range of applications from consumer electronics to industrial sensors and environmental monitoring [13].

2.2 Piezoelectric Nanogenerators (PENG)

PENGs utilize the piezoelectric effect, where mechanical deformation of piezoelectric materials induces electrical polarization. This effect occurs due to the displacement of positive and negative charge centres within the material's crystal lattice under applied mechanical stress [48].

2.2.1 Working Mechanism of PENG

Piezoelectric nanogenerators (PENGs) operate through two primary mechanisms: the direct piezoelectric effect and the reverse piezoelectric effect, both of which are foundational to their functionality across energy harvesting and precision device applications.

The Direct Piezoelectric Effect is a phenomenon where an electric field is generated within a piezoelectric material as a result of applied mechanical stress which is illustrated in Fig. 3. When a material is compressed, stretched, or subjected to vibrational forces, the internal polarization shifts, leading to a potential difference across its surfaces [49]. This property is extensively harnessed in energy harvesting systems, where ambient mechanical forces such as vibrations from walking, wind, or ocean waves are converted into electrical energy. A critical advantage of the direct piezoelectric effect is its scalability, allowing PENGs to be adapted for micro- to macro-scale applications. Materials like Zinc Oxide (ZnO) and Polyvinylidene Fluoride (PVDF) are often employed due to their high piezoelectric coefficients and mechanical resilience, ensuring efficient energy conversion even under dynamic and repetitive loading conditions [50].

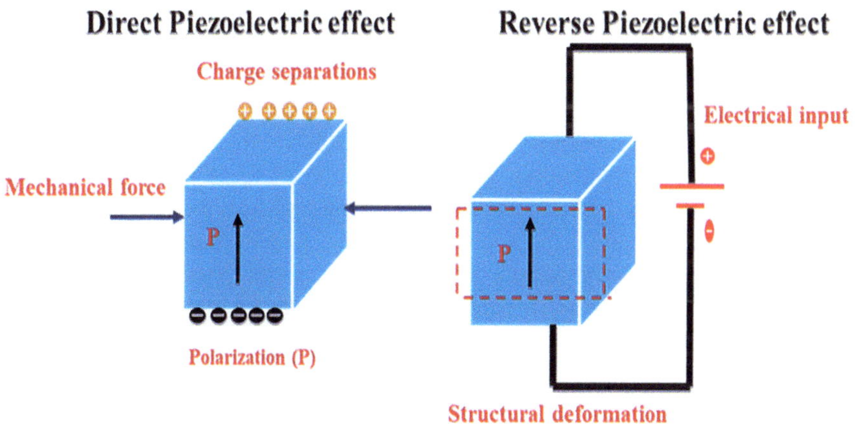

Fig. 3 Comprehensive schematic of direct and reverse piezoelectric effects of PENG [19]

On the other hand, the Reverse Piezoelectric Effect involves the application of an external electric field to a piezoelectric material, resulting in induced mechanical deformation or strain. This effect is pivotal in actuators and precision devices, where precise mechanical responses are required. For instance, in robotic systems, the reverse piezoelectric effect enables the fine-tuned movement of components, while in precision optics, it assists in adjusting lens alignments with sub-micron accuracy [47, 51]. The interplay between the direct and reverse effects also facilitates the development of dual-functional systems, capable of both energy generation and actuation, broadening the utility of PENGs in fields like biomedical engineering and aerospace technologies [10, 19]. Figure 3. explains the functionality of PENGs is defined not only by the intrinsic properties of the piezoelectric materials but also by their geometric and structural configurations. Factors such as material thickness, electrode placement, and the mode of mechanical excitation (compression, bending, or shearing) significantly influence the output performance. Advanced designs incorporating multilayer architectures, nanostructured surfaces, or hybrid composites further enhance the efficiency and adaptability of PENGs, making them integral to the development of next-generation self-powered devices and systems [19].

2.2.2 Materials Used in PENGs

The materials used in piezoelectric nanogenerators (PENGs) significantly influence their performance, efficiency, and application versatility. Among the most commonly employed materials are crystalline substances, polymeric materials, and advanced composites [2]. Crystalline materials, such as Zinc Oxide (ZnO) and Barium Titanate ($BaTiO_3$), are widely used due to their intrinsic piezoelectric properties. ZnO is particularly favored for its high piezoelectric coefficient and ease of nanoscale fabrication, making it suitable for developing miniature and integrated devices. $BaTiO_3$, with its excellent dielectric properties, enables high-output energy generation and is commonly utilized in applications demanding robust electrical performance [40]. Polymeric materials, such as Polyvinylidene Fluoride (PVDF) and its copolymer PVDF-TrFE, offer unique advantages due to their flexibility, lightweight nature, and adaptability for wearable technologies. PVDF exhibits strong piezoelectric properties combined with mechanical durability, making it ideal for use in flexible and stretchable devices. The copolymer PVDF-TrFE enhances piezoelectric performance further while retaining the mechanical benefits, thereby finding applications in advanced biomedical sensors and portable energy harvesters [7]. Composite materials represent a fusion of piezoelectric nanoparticles embedded within polymer matrices, combining the advantages of both material types. These composites enhance flexibility and mechanical strength while maintaining high energy conversion efficiency. For example, incorporating ZnO nanoparticles or $BaTiO_3$ nanowires into PVDF matrices creates hybrid structures that can stretch and conform to various surfaces without compromising performance [8]. These composite materials are particularly effective in applications requiring mechanical adaptability, such as smart textiles, wearable electronics, and curved surface integration. By tailoring the composition and structure of these materials, researchers continue to push the boundaries of PENG efficiency and functionality [10].

2.2.3 Design Enhancements in PENGs

Design enhancements in piezoelectric nanogenerators (PENGs) have significantly expanded their efficiency and adaptability, enabling their integration into advanced energy harvesting systems. One such enhancement involves nanostructured surfaces, where vertically aligned Zinc Oxide (ZnO) nanowires or similar nanostructures are fabricated on piezoelectric materials. These nanowires substantially increase the surface area available for mechanical stress, thereby amplifying the generation of electrical charges. The increased interfacial area also improves the overall efficiency of energy conversion, making nanostructured surfaces a pivotal innovation in PENG technology [14, 15]. Another significant advancement is the use of multilayer architectures, where multiple layers of piezoelectric materials are stacked together to enhance energy density and output power. Each layer contributes to the overall piezoelectric response, enabling higher energy generation under the same mechanical input. This multilayer approach is particularly beneficial in applications requiring compact devices with high output power, such as biomedical implants or portable electronics. The cumulative effect of multiple layers ensures a more robust and consistent energy output across varying conditions [49]. The incorporation of flexible substrates such as polydimethylsiloxane (PDMS) or polyethylene terephthalate (PET) represents another transformative design enhancement. These substrates provide mechanical adaptability without compromising the piezoelectric properties of the active materials [32]. Devices built on flexible substrates can conform to non-uniform surfaces, endure repeated bending or stretching, and integrate seamlessly into wearable technologies or curved structures. This flexibility is crucial for creating durable, user-friendly PENGs that can withstand real-world mechanical stresses while maintaining performance integrity [35].

By combining these design enhancements—nanostructured surfaces, multilayer architectures, and flexible substrates—modern PENGs have become highly efficient and versatile energy harvesting devices. These innovations enable their application across a wide range of sectors, from healthcare and wearable electronics to environmental monitoring and industrial systems [32].

Polyvinylidene fluoride (PVDF)-based piezoelectric nanogenerators

PVDF based piezoelectric nanogenerators (PENGs) represent a breakthrough in portable and flexible energy harvesting technologies, leveraging PVDF's remarkable piezoelectric, pyroelectric, and dielectric properties. Known for its ease of processing, chemical resistance, and structural flexibility, PVDF serves as an ideal material for applications requiring adaptability and durability, such as electronic skins (E-skins), health monitors, and self-powered devices [22, 52]. Advanced PFM was used to analyze the piezoelectric response of the rGO/WO_3/PVDF tertiary nanocomposite film, leveraging its high spatial resolution and sensitivity to local polarization.

Using a conducting tip for AC voltage application, the PFM amplitude was recorded, and the DART mode was employed due to PVDF's low piezoelectric coefficient. Out-of-plane deformation caused tip deflection, mapping z-axis amplitude and phase over a $10 \times 10\ \mu m^2$ area, as shown in Fig. 4a, b. With a 2 V AC voltage, the PFM amplitude and phase revealed high piezoelectric contrast. A DC voltage range of -7 V to $+7$ V was applied to measure polarization reversal and switching responsiveness, forming the butterfly curve between amplitude and bias voltage, as shown in Fig. 4c. The piezoelectric coefficient (d33) reached up to 32.6 pm/V, confirming the material's piezoelectric behaviour.

Enhancements in PVDF's piezoelectric properties have been achieved through the integration of nanomaterials like reduced graphene oxide (rGO) and tungsten oxide (WO_3), forming tertiary nanocomposites with superior mechanical and electrical characteristics. The synergy between rGO and WO_3 improves the dispersion and piezoelectric polar phases within PVDF, enhancing energy conversion efficiency while addressing challenges like dielectric loss and thermal management [49, 50, 52]. These advanced nanocomposites enable the fabrication of innovative devices that combine nano generation with functionalities such as ammonia gas detection, offering room-temperature operation, low power consumption, and compact design.

The interaction mechanism between rGO, SnO_2, and PVDF is illustrated in Fig. 5. The π-electrons of rGO interact with the CH_2 dipoles of PVDF, while the F/H atoms of PVDF form hydrogen bonds with the oxygen-containing OH groups in rGO. Additionally, the F/H atoms of PVDF interact with the carbonyl and carboxyl groups present in rGO. SnO_2 nanoparticles, hybridized with rGO, further enhance the composite's properties by providing additional active sites for interaction. This synergy strengthens the nanocomposite's overall dielectric and piezoelectric performance [22].

The structural and mechanical properties of PVDF-based composites under flexural stress, along with their cross-sectional morphology, are comprehensively depicted in Fig. 6. The PVDF molecular chain is shown undergoing conformational adjustments during mechanical bending, emphasizing its inherent flexibility and adaptability. The pristine PVDF, SnO_2 nanoparticles, and rGO/PVDF binary and ternary rGO/SnO_2/PVDF composites exhibit notable resilience under applied stress. Additionally, the transverse optical micrograph details the homogenous and compact architecture of the rGO/SnO_2/PVDF composite film, with an accurately measured thickness of 209.52 μm, highlighting its suitability for advanced piezoelectric and electromechanical applications.

Emerging nanofabrication techniques, such as E-beam evaporation for precise electrode deposition and controlled gas chamber testing, have significantly advanced the development of cost-efficient and flexible PVDF-based piezoelectric nanogenerators (PENGs). These devices exhibit exceptional mechanical adaptability, reliable DC voltage generation, and high sensitivity for NH_3 detection, making them critical for addressing energy and environmental challenges. As shown in Fig. 7, the comprehensive development of rGO/ WO_3/PVDF tertiary nanocomposites leverages advanced fabrication methodologies, including hot press polymerization and E-beam evaporation, to achieve optimized structural and functional properties [52].

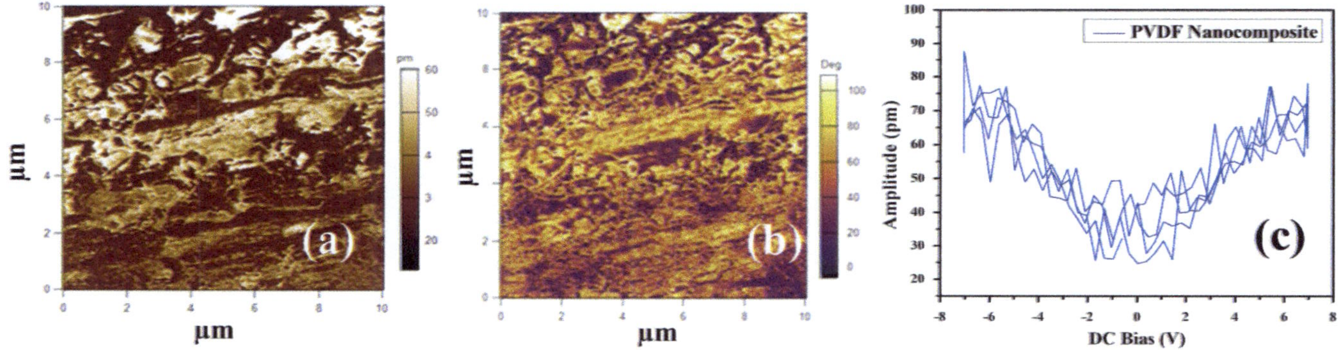

Fig. 4 Piezoresponse force microscopy (PFM) **a** amplitude image of rGO/ WO_3/PVDF tertiary nanocomposite over a 10 × 10 μm^2 area, **b** phase difference image of the same composite over a 10 × 10 μm^2 area, and **c** PFM stress–strain butterfly loop under a DC bias voltage ranging from −7 V to +7 V [52]

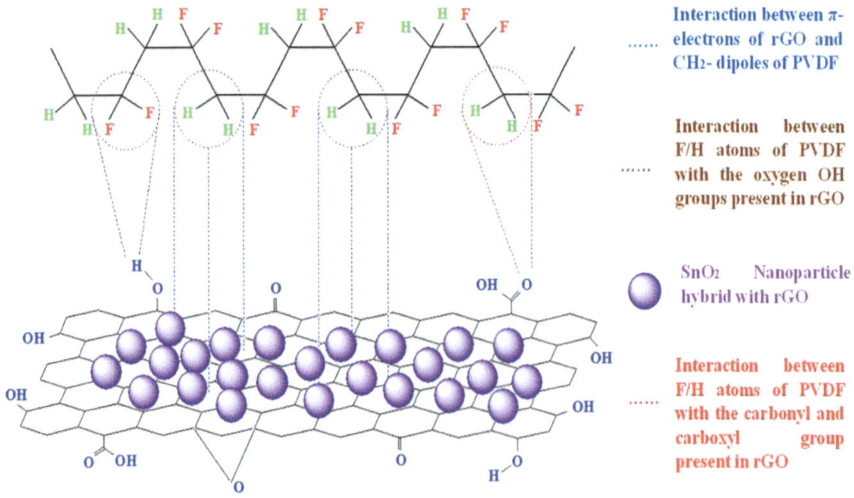

Fig. 5 The interaction mechanism between rGO, SnO_2, and PVDF involves rGO enhancing electrical conductivity, SnO_2 improving dielectric and piezoelectric properties, and PVDF acting as a matrix to synergize their functionalities [22]

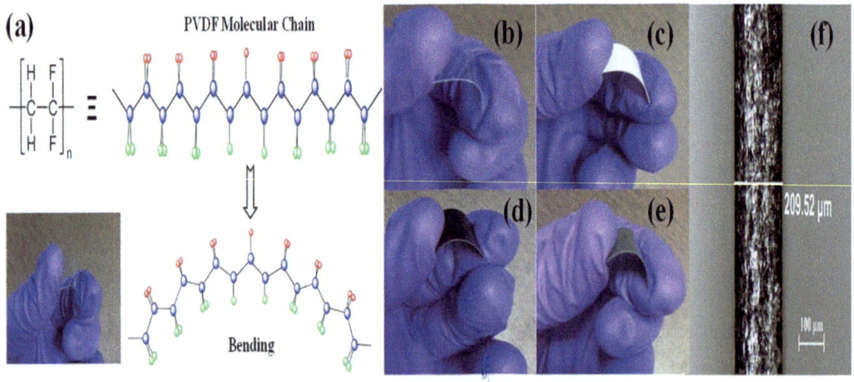

Fig. 6 **a** Schematic depiction of the molecular architecture of the PVDF polymer chain; High-resolution images of **b** pristine PVDF, **c** SnO_2 nanoparticles, **d** rGO/PVDF binary nanocomposite, **e** rGO/SnO_2/PVDF under mechanical flexural stress; and **f** transverse optical microscopy characterization of rGO/SnO_2/PVDF thick films exhibiting a measured thickness of approximately 209.52 μm [22]

The integration of precision instrumentation, such as force-torque testers and gas-sensing chambers, enables detailed mechanical and electrochemical characterization. Key applications of these nanocomposites include energy harvesting systems, electronic nose (e-nose) devices for gas detection, and wearable skin-integrated flexible electronics. This innovative approach highlights the multifunctionality of the material, paving the way for next-generation self-powered systems and advanced electromechanical sensing technologies [17, 52, 53].

The design and performance characteristics of Triboelectric Nanogenerators (TENGs) and Piezoelectric Nanogenerators (PENGs) exhibit significant differences in operational mechanisms, material composition, and functional applications which is illustrated in Table 1. TENGs harness the triboelectric effect combined with electrostatic induction, offering superior adaptability, cost-efficiency, and scalability for diverse applications, including environmental sensing and tactile interfaces. Conversely, PENGs operate on the piezoelectric effect, delivering higher power density and output efficiency under mechanical deformation, making them particularly suited for precision strain-based applications such as advanced sensors and actuators [11, 54].

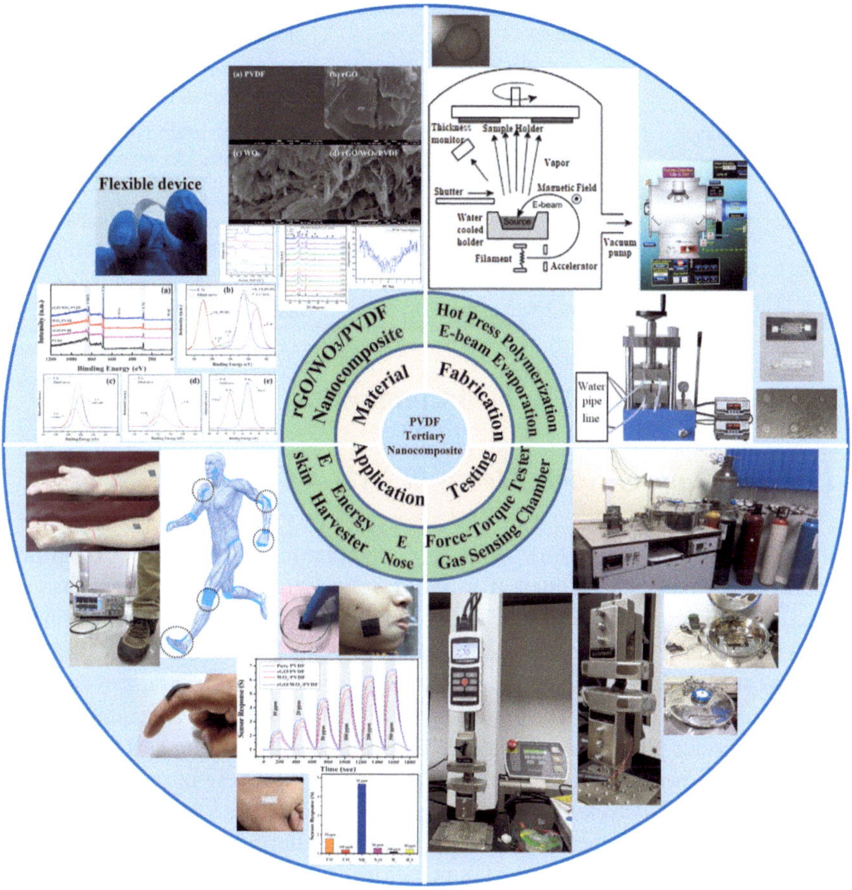

Fig. 7 Schematic of PVDF tertiary nanocomposites: characteristics, fabrication, testing, and applications [52]

Table 1 Design and output characteristics of TENG and PENG

Feature	TENG (Triboelectric nanogenerators)	PENG (Piezoelectric nanogenerators)
Working mechanism	Based on triboelectric effect and electrostatic induction	Relies on the piezoelectric effect
Energy source	Ambient mechanical motion such as vibrations, sliding, or impacts	Mechanical deformation such as compression or bending
Materials	Polymers (e.g., PTFE, PET), conductive layers, surface patterns	Crystalline (e.g., ZnO, $BaTiO_3$) and polymeric (e.g., PVDF)
Output efficiency	Moderate output efficiency	High output efficiency under direct mechanical stress
Flexibility	Highly flexible, suitable for wearable and portable applications	Flexible, but constrained by the nature of piezoelectric materials
Scalability	Easily scalable for large-scale applications	Limited by the availability of piezoelectric materials
Durability	High durability with advanced encapsulation techniques	Durable but sensitive to environmental degradation
Application focus	Broader applications in environmental sensing, touch interfaces	Specific to strain-based applications like sensors, actuators
Power density	Moderate power density	High power density suitable for compact applications
Cost	Cost-effective materials and fabrication techniques	Relatively higher costs due to material constraints

3 Noble Metal-Based Hybrid Composites

3.1 Core Traits of Noble Metals in Hybrid Composites

Noble metals such as gold (Au), silver (Ag), and platinum (Pt) possess exceptional properties that make them integral to the development of hybrid composites for advanced applications. Their electrical properties are characterized by high conductivity, enabling efficient charge transfer and low resistance in electronic systems. This makes them ideal for enhancing the performance of energy harvesting devices such as TENGs and PENGs [40]. The optical properties of noble metals are equally significant, with surface plasmon resonance (SPR) playing a pivotal role. SPR arises when conduction electrons on the metal surface oscillate collectively in response to incident light at specific wavelengths. This phenomenon amplifies local electromagnetic fields, resulting in enhanced light absorption and scattering [20]. These properties are particularly beneficial in optical applications, such as improving the efficiency of photodetectors and light-emitting devices by concentrating and redirecting light. From a chemical perspective, noble metals are highly stable, resistant to oxidation, and capable of maintaining their properties under extreme environmental conditions.

This chemical inertness ensures the durability and longevity of devices incorporating these metals [55]. The integration of noble metals into hybrid composites combines these properties with the mechanical flexibility and structural versatility of polymers or ceramics. Such hybrid systems enable the fabrication of energy harvesting devices with superior performance, adaptability to diverse applications, and extended operational lifespans [13, 14].

3.2 Fabrication Techniques

Fabricating hybrid nanogenerators (HNGs) combining triboelectric nanogenerators (TENGs) and piezoelectric nanogenerators (PENGs) involves advanced techniques to enhance energy harvesting efficiency. Layer-by-layer assembly integrates triboelectric and piezoelectric layers on flexible substrates, while electrospinning creates aligned nanofibers of materials like PVDF and PTFE for lightweight devices. Hydrothermal synthesis grows piezoelectric nanorods, such as ZnO or $BaTiO_3$, directly on substrates, complementing triboelectric layers. Physical vapour deposition (PVD) forms thin films of conductive or piezoelectric materials to improve charge transfer. Solution casting offers a cost-effective approach for large-scale polymer layer production. For customization, 3D printing enables the creation of complex geometries tailored to biomedical and wearable devices. E-beam evaporation ensures precise electrode deposition for uniform electrical interfaces [56]. Encapsulation techniques using flexible layers protect HNGs from environmental factors like moisture, enhancing durability. Together, these methods enable the development of robust, efficient, and versatile HNGs suitable for applications ranging from portable electronics to renewable energy systems.

The fabrication process for nanocomposite-based flexible sensors with enhanced piezoelectric properties described in Fig. 8. It begins with a syringe containing nanocomposite ink, deposited onto an ITO glass substrate via a moving arm. The substrate is mounted on a moving platform with a heater to assist solvent evaporation, leaving behind conductive patterns. These patterns are then subjected to electrical poling in a silicone oil bath, a critical step to align the dipoles within the material for improved piezoelectric performance [20]. The substrate is flipped over to ensure uniform processing and further optimization of the material structure. The resulting device demonstrates a well-defined grid-like pattern, as shown in the final image with a scale of 1 cm for reference. The process emphasizes precise control over deposition and curing to achieve high-performance sensors [6, 10]. Figure 8 highlights the integration of advanced techniques for scalable and efficient manufacturing of flexible piezoelectric devices.

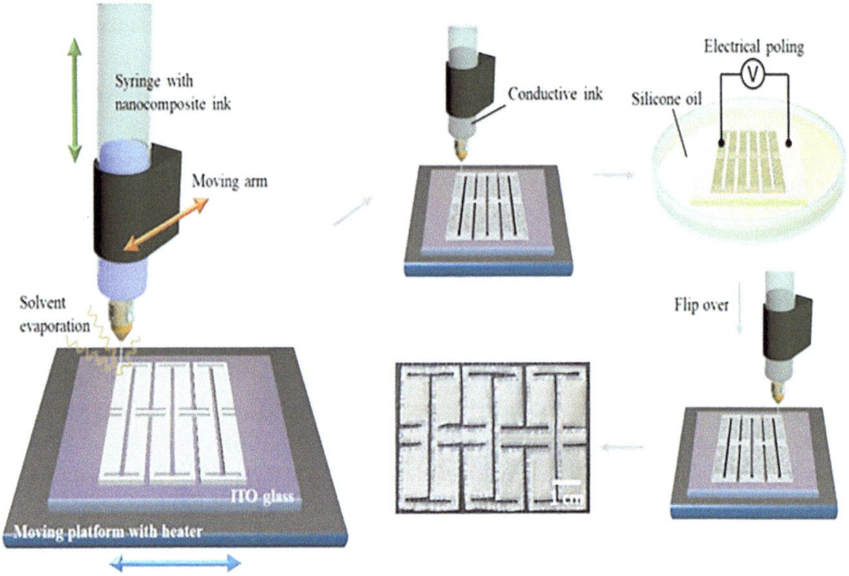

Fig. 8 Schematic representation of the fabrication process for nanocomposite-based flexible sensors with enhanced piezoelectric properties [20]

3.2.1 Chemical Reduction

Chemical reduction is one of the most widely used methods for synthesizing noble metal nanoparticles due to its simplicity and efficiency. This process involves reducing metal salts (such as gold chloride or silver nitrate) in the presence of reducing agents like sodium borohydride or citrate. By controlling parameters such as temperature, pH, and concentration, nanoparticles of specific sizes and shapes can be synthesized. These nanoparticles exhibit enhanced surface area and plasmonic properties, making them ideal for integration into hybrid composites [28, 30].

3.2.2 Hydrothermal Synthesis

Hydrothermal synthesis is a versatile technique for producing stable and uniform nanostructures under controlled high-temperature and high-pressure conditions. In this method, metal precursors are dissolved in a solvent, and the reaction takes place in a sealed autoclave. This approach is particularly advantageous for producing nanostructures with tailored morphologies, such as rods, spheres, or cubes. Hydrothermal synthesis ensures excellent crystallinity and uniformity, making it a preferred choice for creating noble metal-based hybrid composites [31].

3.2.3 Physical Deposition

Physical deposition techniques, including sputtering and electron beam evaporation, are employed to create thin films of noble metals on substrates. These methods provide precise control over film thickness and uniformity, enabling the fabrication of high-quality hybrid composites [39]. Thin films of gold, silver, or platinum deposited via physical methods enhance the electrical conductivity and optical performance of energy harvesting devices.

3.3 Integration Strategies

The integration of noble metals into TENG and PENG systems involves innovative strategies to maximize their functional benefits. Embedding noble metals within triboelectric or piezoelectric materials creates hybrid composites that combine the mechanical properties of the host material with the electrical and optical advantages of noble metals. This approach enhances charge transfer, energy conversion efficiency, and overall device performance [15].

Structural configurations as shown in Fig. 9 also play a critical role in optimizing the performance of these hybrid systems. For instance, layering noble metal films with triboelectric or piezoelectric components ensures uniform charge distribution and efficient energy harvesting. In some designs, noble metal nanoparticles are dispersed uniformly within polymer matrices to create flexible and stretchable composites [53]. Such configurations are particularly beneficial for applications in wearable devices, where mechanical adaptability and durability are essential [14]. Additionally, advanced fabrication techniques such as 3D printing and laser ablation are being explored to integrate noble metals into complex geometries and microstructures. These methods enable the creation of customized hybrid systems tailored to specific applications, such as high-performance optical sensors or bio-integrated energy harvesters. By leveraging these integration strategies, researchers have successfully developed noble metal-based hybrid composites that significantly enhance the functionality and versatility of TENG and PENG systems, paving the way for their deployment in next-generation energy and optical devices [13].

Noble Metal-Integrated Nanocomposite Matrices for Laser Desorption/ Ionization Mass Spectrometry

Detecting small molecules is essential in fields like environmental science, pharmacology, metabolomics, and lipidomics [57]. Laser desorption/ionization mass spectrometry (LDI MS) provides superior sensitivity, precision, and efficiency for molecular analysis, yet its performance for small-molecule detection is constrained by challenges associated with traditional organic matrices. To address these limitations, selecting an optimal matrix is crucial for effective small-molecule analysis

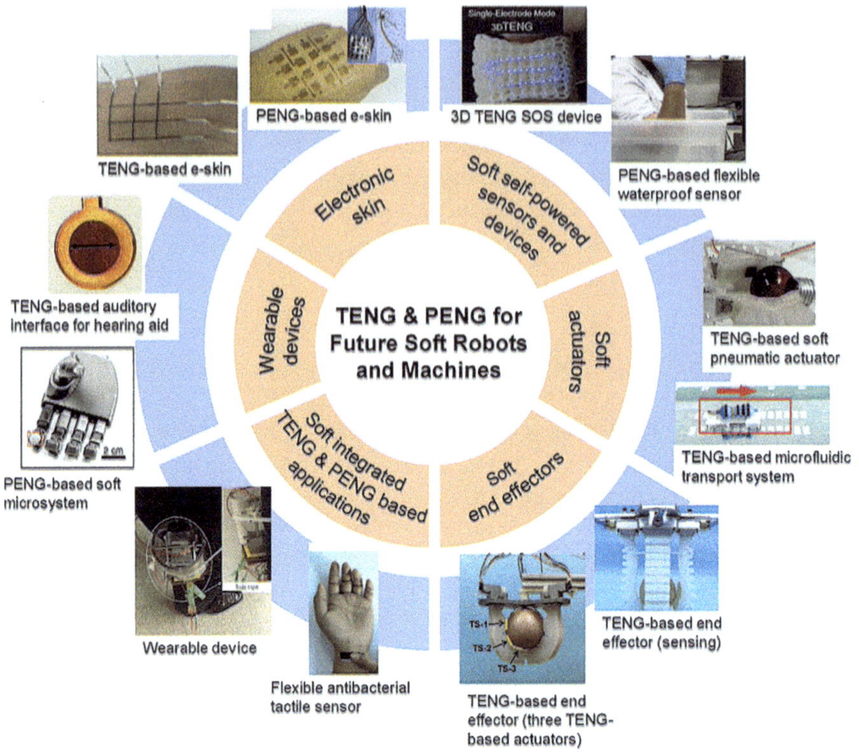

Fig. 9 Integration techniques of TENG and PENG for future soft robots and machines: applications in wearables, actuators, and sensors [13]

using LDI MS. Researchers have increasingly turned to inorganic matrices, particularly composite materials, which exhibit synergistic properties that enhance analytical capabilities [58, 59]. This Minireview explores the progress in nanocomposite-based matrices, propelled by advancements in material science, and their growing application in improving small-molecule detection with LDI MS.

Noble Metal-Based Nanocomposites

Noble metal-based nanocomposites, including gold (Au), silver (Ag), and alloy materials, have demonstrated exceptional performance as matrices in Laser Desorption/Ionization Mass Spectrometry (LDI MS) for small-molecule detection, owing to their tunable surface properties, plasmonic effects, and high ionization efficiency as shown in Fig. 10. Gold-based systems such as rGO/AuNP composites enhance signal intensity, reproducibility, and detection limits for biomolecules, while core–shell structures like Au@GMSN achieve detection limits as low as 50 fmol for amino acids and complex analytes [59].

Plasmonic Au nanoshells offer superior detection of metabolites such as valine and tryptophan at limits of 3–30 pmol, with applications extending to metabolic

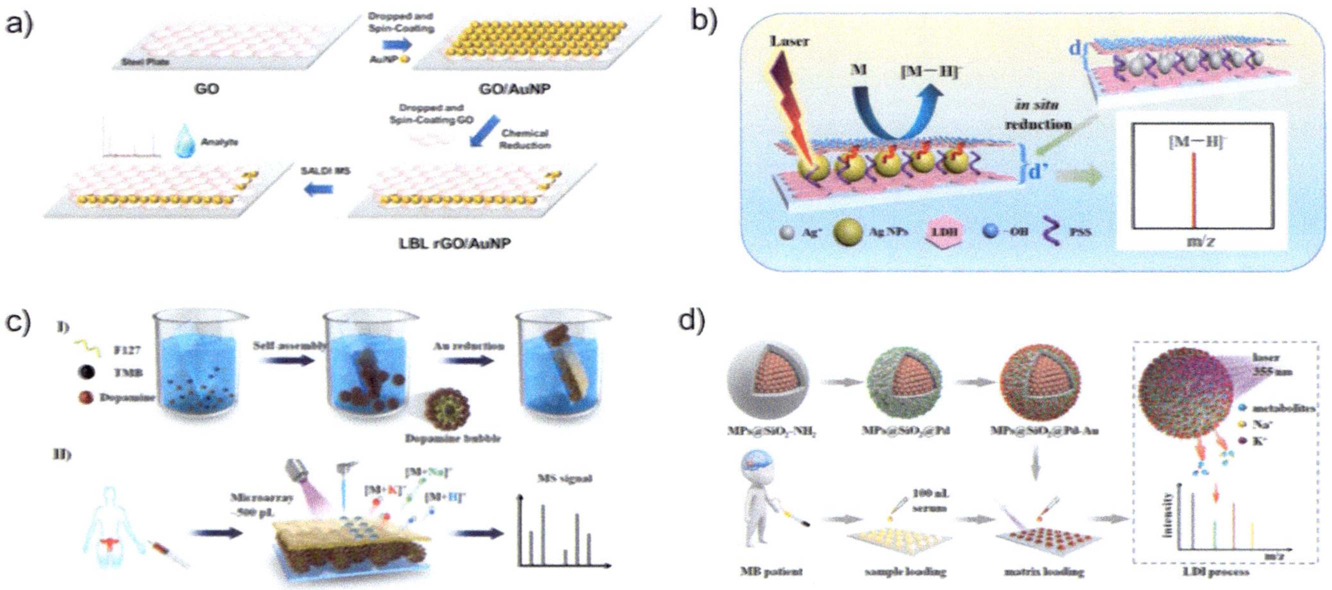

Fig. 10 Noble metal-based nanocomposites serve as matrices for LDI MS detection: **a** rGO/AuNP composite, **b** LDH–Ag hybrid material, **c** plasmonic chip featuring Au nanoparticles deposited on a dopamine-bubble layer, and **d** synthetic Pd–Au alloys are demonstrated for their effective application in this context [59]

fingerprinting in clinical samples using microarray chips with volumes as low as 400 pL. Silver-based matrices, such as MoS_2/Ag hybrids and LDH-Ag films, exhibit enhanced UV absorption, thermal conductivity, and ionization efficiency, enabling multi-drug detection and quantitative glucose analysis with RSD values below 4%. Alloy-based nanocomposites, including Ag–Au and Pd–Au, further optimize LDI MS by combining high salt tolerance, protein compatibility, and rapid metabolite detection, as seen in Pd–Au systems detecting disease-specific biomarkers in 100 nL biofluids for medulloblastoma diagnosis and treatment evaluation. Fig. 10 illustrate despite challenges such as high production costs and potential background noise from laser-induced self-dissociation, these nanocomposites provide significant advancements in precision diagnostics and molecular characterization, underpinned by synergistic material properties and innovative structural designs [47, 59].

Metal-Oxide-Based Nanocomposites

Metal oxide-based nanocomposites have gained significant recognition as efficient matrices for Laser Desorption/Ionization Mass Spectrometry (LDI MS) due to their high UV absorption, semiconductor properties, and cost-effectiveness compared to noble metals.

Single metal oxides often face limitations such as low surface area, insufficient active sites, and rapid electron–hole recombination, which hinder their ionization efficiency as depicted in Fig. 11. Composite designs address these issues by enhancing surface area, providing abundant reaction sites, and creating heterostructures for effective electron–hole separation. Examples include WO_3–TiO_2 hybrids with mesoporous structures for cyclic decapeptide detection, Fe_3O_4/TiO_2 core–shell composites achieving detection limits of 1.7 pmol for histidine, and Fe_3O_4@SiO_2 composites for signal enhancement in saline environments [60]. CeO_2-carbon black composites offer improved reactivity for drug analysis with detection limits of 100 ng/mL, while Ag nanoparticle/ZnO nanorods achieve remarkable sensitivity with detection limits of 1.0×10^{-15} M for arginine. TiO_2 combined with gold nanoislands forms Schottky junctions, enhancing carrier separation and ionization for immunosuppressor detection [11]. Figure 11 also outlined Vanadium-based SiO_2@V2O5 nanorods excel in metabolic fingerprinting, achieving 94% sensitivity and 90% specificity in diagnosing diabetic retinopathy. While these composites demonstrate superior stability and performance, their synthesis and functionalization remain complex, necessitating continued research into novel compositions and architectures to advance their application in LDI MS for small-molecule detection [59].

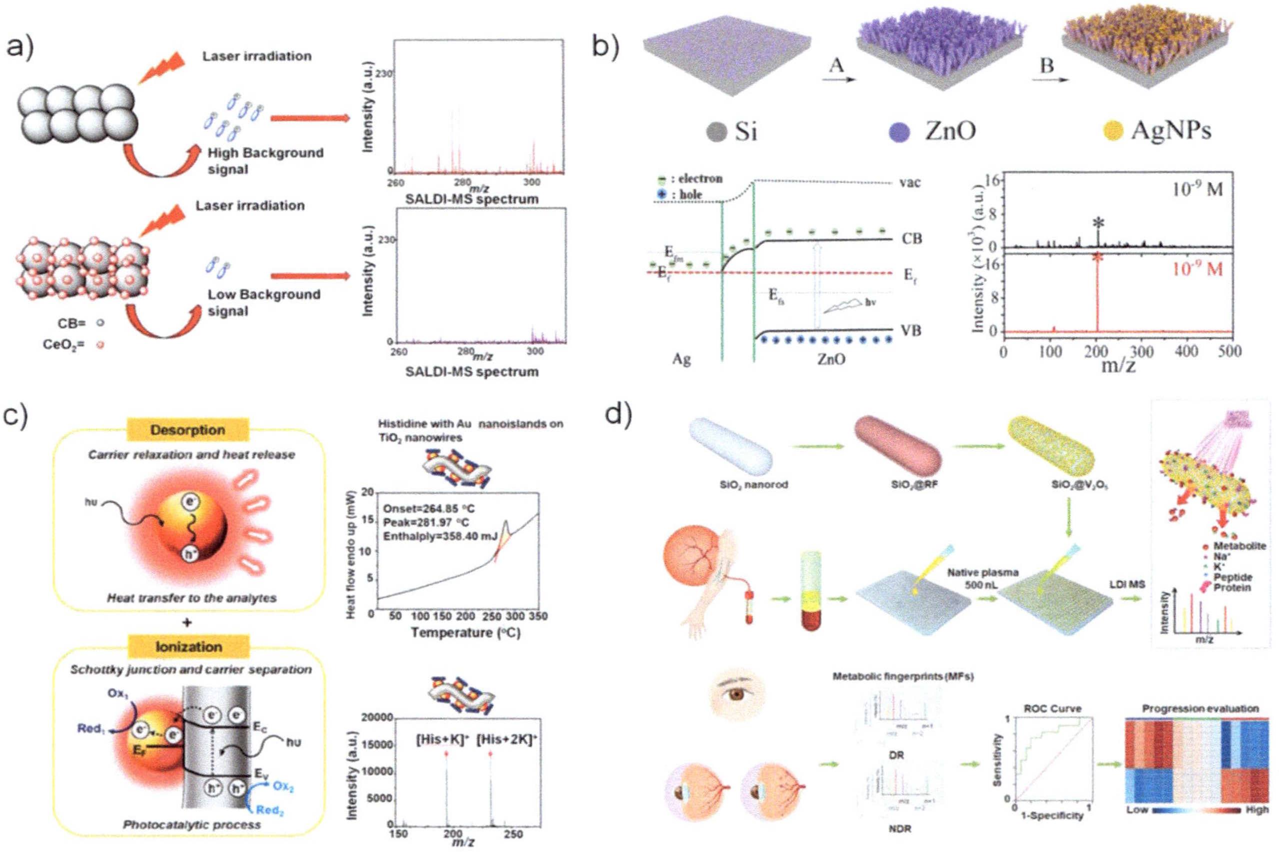

Fig. 11 Metal oxide-based nanocomposites serve as effective matrices for LDI MS detection: **a** CeO_2-carbon black composites, **b** Ag nanoparticle/ZnO nanorod hybrids, **c** the synergistic interaction of TiO_2 and gold nanoislands, and **d** SiO_2@V2O5 composites applied in the diagnosis of diabetic retinopathy through metabolic fingerprinting [59]

4 Applications in Optical Devices

4.1 Light-Emitting Devices (LEDs)

The incorporation of noble metal-based hybrid composites into TENG and PENG systems has revolutionized light-emitting device (LED) technologies. Noble metals like gold, silver, and platinum play a critical role in enhancing the electroluminescence efficiency of LEDs through their unique plasmonic properties. Surface plasmon resonance (SPR) facilitates the amplification of localized electromagnetic fields, which improves photon emission rates and enhances brightness. These improvements result in LEDs that are not only brighter but also more energy-efficient, addressing the growing demand for sustainable illumination solutions [54, 55]. Figure 12 showcase the functional integration of a Triboelectric Nanogenerator (TENG) for energy harvesting and powering low-power electronic devices. The TENG, connected to a rectifier circuit, efficiently converts its alternating current (AC) output into direct current (DC) to drive external loads, exemplified by the illumination of LEDs. Furthermore, the harvested energy is stored in a 10 μF capacitor, demonstrating its capability to provide a stable power supply for devices such as a stopwatch and a calculator. This setup underscores the TENG's ability to harness mechanical energy and convert it into electrical energy with high efficiency, highlighting its potential for sustainable energy solutions in powering portable and low-energy electronics.

The precise control over light emission characteristics enabled by these hybrid systems is another significant advantage. Figure 12 depicts by fine-tuning the interaction between light and noble metal nanoparticles, researchers can achieve enhanced color rendering and wavelength-specific emissions. This capability is particularly beneficial in advanced display technologies, where accurate color reproduction and high resolution are critical. Furthermore, the reduced energy consumption of these LEDs makes them ideal for applications in smart lighting systems, portable displays, and energy-efficient architectural illumination [15]. Applications of TENG- and PENG-powered LEDs extend across consumer electronics, industrial lighting, and specialized optoelectronic devices. For example, LEDs integrated with noble metal-based composites are being utilized in automotive lighting, medical imaging systems, and wearable devices, where performance and energy efficiency are paramount [49].

4.2 Photodetectors

Photodetectors powered by TENG and PENG systems have seen remarkable advancements with the inclusion of noble metal-based hybrid composites. The plasmonic effects of noble metals significantly enhance the sensitivity of photodetectors by amplifying light absorption. This is achieved through SPR, which focuses incoming light into nanoscale regions, increasing the interaction with the active material. Consequently, self-powered photodetectors equipped with these hybrid

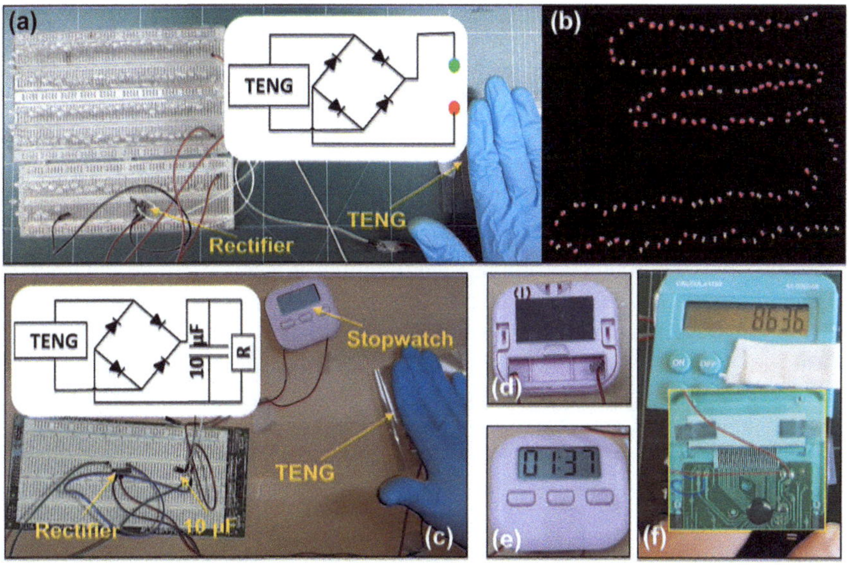

Fig. 12 **a** LEDs connected in series with the SFP-TENG device through a bridge rectifier, with the inset showing the circuit designed to illuminate commercial LEDs. **b** Successful illumination of 153 commercial LEDs. **c** Powering a stopwatch and calculator using the SFP-TENG device connected through a bridge rectifier and a 10 μF capacitor, with the inset illustrating the circuit configuration. **d** Connection of the proposed SFP-TENG device to the stopwatch. **e** Successful operation of the stopwatch. **f** Calculator powered successfully, with the inset showing the connection setup of the SFP-TENG device to the calculator [14]

systems exhibit improved response times and heightened sensitivity, even in low-light conditions [25]. These features make such photodetectors ideal for applications in imaging, environmental monitoring, and security systems. In medical diagnostics, for instance, enhanced photodetectors are employed in optical coherence tomography and fluorescence-based imaging techniques, where precision and sensitivity are critical [43]. Additionally, the ability to operate efficiently under low-light conditions expands their utility to areas such as astronomical observations and autonomous vehicle sensors. The self-powered nature of these devices, driven by TENG or PENG systems, eliminates the need for external power sources, further enhancing their appeal for portable and remote sensing applications. By leveraging noble metal-based composites, these photodetectors represent a significant step forward in both performance and practicality [11, 60].

Integration of a Paper-Based TENG with a Silicon PIN Photodetector

In large-scale implementations of sensor networks for the Internet of Things (IoT), the development of self-powered photoelectric detection systems becomes imperative to eliminate reliance on external power sources. This study introduces a self-sustained photo-detection system powered by a triboelectric nanogenerator (TENG), which supplies energy to a silicon PIN photodetector. The photodetector, fabricated using advanced silicon processing technologies, ensures compatibility with high-volume manufacturing. The system utilizes a paper-based contact-separation TENG (CS-TENG) integrated with the photodetector through a rectifier bridge circuit [11]. The synergistic interaction between the photodetector's internal photoelectric effect and the impedance matching characteristics of the TENG results in a reduction in the detector's equivalent resistance under incident light, consequently modulating the TENG's output voltage. The system's performance under varying illumination levels is characterized, and an equivalent circuit model is proposed to elucidate the influence of photocurrent on the detection range. Light-emitting diodes (LEDs) are employed as output indicators to visualize the detected light intensity, demonstrating the practical viability of this self-powered configuration [44].

The structural design of the paper-based CS-TENG is depicted in Fig. 13a. The CS-TENG is constructed by layering an aluminum foil electrode, serving as the conductive interface, onto a cardboard substrate, followed by a polytetrafluoroethylene (PTFE) film as the triboelectric friction layer. A second friction unit, comprising a Nylon film as the triboelectric layer, is fabricated similarly [25]. The two units are mechanically fixed using adhesive tape to form the complete CS-TENG. The operational mechanism, illustrated in Fig. 13b, relies on the triboelectric effect, where contact between the PTFE and Nylon layers generates opposite surface charges (positive on Nylon and negative on PTFE). Upon separation, the induced charge imbalance produces an electric potential difference that drives electron flow between the aluminum electrodes [61]. During compression, a reverse current is generated due to the alternating potential difference, as shown in Fig. 13b. Power generation simulations using Multiphysics simulation, presented in Fig. 13c, reveal that increasing the separation distance between the PTFE and Nylon layers from 1 to 3 mm significantly enhances the interfacial electric potential difference, thereby improving the energy conversion efficiency of the TENG.

Photoelectric Detection System Using a Silicon PIN Detector

The output voltage of the CS-TENG is an alternating current (AC) signal with high voltage peaks, making it unsuitable for directly powering the photodetector. To resolve this, a rectifier circuit is used to convert the AC signal into a direct current (DC) signal. A reverse bias voltage is applied to both electrodes of the silicon PIN photodetector, measuring 3 mm $\times$ 4 mm, to ensure it remains in a fully depleted state. Given the relatively low output current of the CS-TENG, approximately in the microampere range, the circuit current is measured using a fixed-value resistor of 2 MΩ as the series load resistance [61].

The photoelectric detection system integrates the paper-based CS-TENG with an external circuit, which includes a bridge rectifier, a silicon PIN photodetector, and a 2 MΩ load resistor. As shown in Fig. 14a, when the CS-TENG is activated, the AC voltage signal it generates is converted into a DC pulse signal through the rectifier bridge. The voltage across the load resistor is subsequently measured using a digital oscilloscope [60]. This configuration ensures efficient conversion of the TENG output to a usable signal for driving the photodetector. Additionally, the circuit design minimizes power loss while maintaining accurate voltage measurements. This study

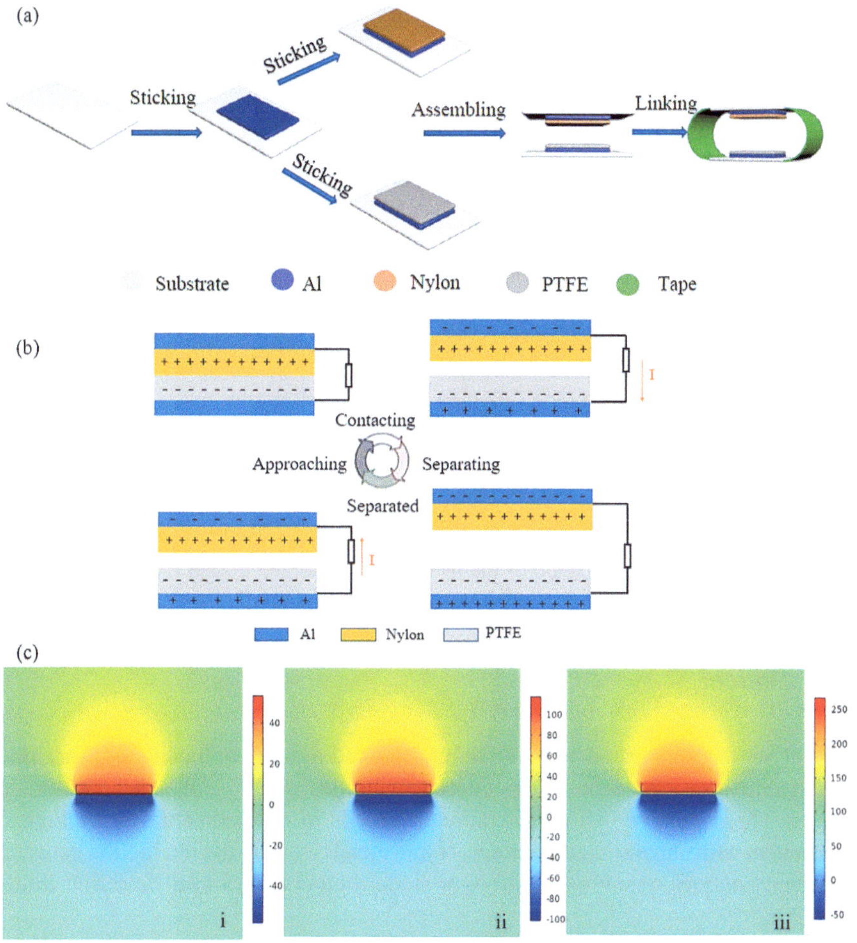

Fig. 13 **a** Step-by-step fabrication process of the paper-based TENG; **b** Schematic representation illustrating the working mechanism of the TENG; **c** COMSOL simulation results showing the electric potential distribution: (i) when the separation distance d = 1 mm (ii) d = 2 mm and (iii) d = 3 mm, where the top layer is Nylon, the bottom layer is PTFE, and d represents the gap between the two layers [61]

presents a photoelectric detection system based on the impedance properties of a silicon PIN photodetector. The circuit schematic is shown in Fig. 14b. In this setup, the rectified output voltage of the TENG provides a reverse high voltage bias to the photodetector, keeping it in a depletion state. As light intensity increases, the internal photoelectric effect reduces the resistance of the photodetector. According to Ohm's law, this increase in light intensity leads to higher circuit current and a greater voltage drop across the load resistance [45, 61].

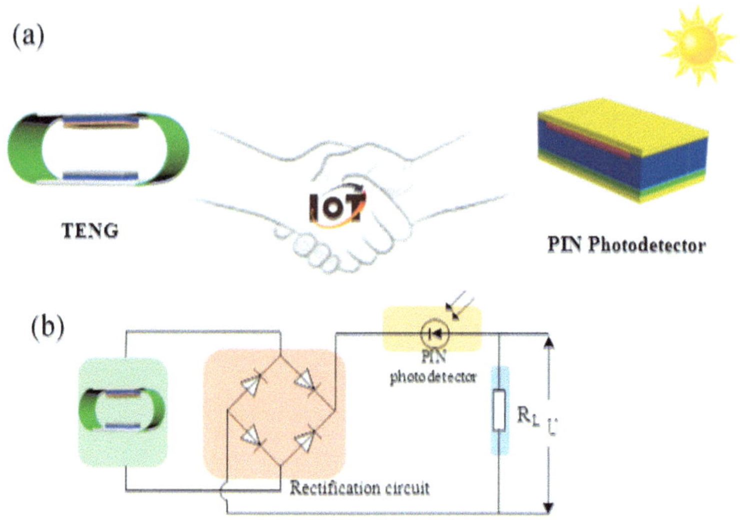

Fig. 14 **a** Schematic representation of the photoelectric detection system using a paper-based CS-TENG; **b** circuit diagram of the testing setup for system analysis [61]

4.3 Optical Sensors

Optical sensors integrated with noble metal-based hybrid composites demonstrate remarkable improvements in detection capabilities due to enhanced plasmonic effects. These sensors can detect minute changes in environmental or biological parameters with unparalleled accuracy. Noble metals amplify the interaction between light and analytes, resulting in stronger signal outputs and lower detection limits [10, 15].

The illustration of Fig. 15 demonstrates the operational framework of a rotational disc-integrated triboelectric nanogenerator (RDID-TENG) for advanced mechanical energy harvesting. The device transitions between an open-circuit state, where the relative motion of triboelectric surfaces induces a potential difference (E_1, E_2, E_3) across electrodes, and a short-circuit state, facilitating charge transfer and generating an output current (I). The rotational disc configuration optimizes the contact-separation interface, enhancing charge density and transfer efficiency through continuous triboelectric interactions. Additionally, the integration of laser-induced charge stimulation significantly amplifies surface charge density, enabling higher electrical output. This design underscores the RDID-TENG's efficacy in sustainable energy harvesting for self-powered electronics and precision sensing technologies [57].

Applications of such sensors span diverse fields, showcasing their versatility and significant impact. In healthcare, biosensors enhanced with noble metals facilitate real-time monitoring of biomarkers, enabling early disease diagnosis, personalized medicine, and wearable health-monitoring devices for preventative care. Environmental monitoring benefits from these sensors' ability to detect trace pollutants

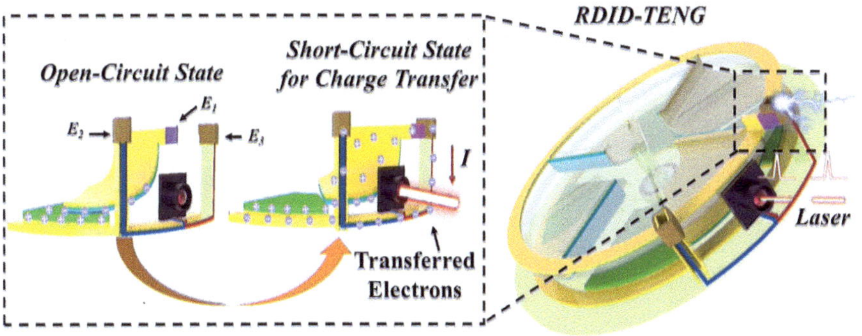

Fig. 15 Schematic of RDID-TENG for charge transfer and open-circuit state applications in optical sensors [15]

like heavy metals and organic compounds in air and water, ensuring ecological balance and public health. Self-powered sensors equipped with triboelectric nanogenerators (TENG) offer real-time data collection in remote areas, aiding pollution control and resource management [15]. In industrial settings, optical sensors are vital for quality control, process optimization, and safety monitoring, detecting defects, ensuring regulatory compliance, and preventing accidents. The integration of TENG and piezoelectric nanogenerators (PENG) has transformed these sensors into self-powered systems, reducing reliance on external energy sources and enhancing adaptability in inaccessible locations. Furthermore, these innovations pave the way for the development of smart, interconnected sensing networks that can autonomously operate and communicate in real-time. This progression not only improves operational efficiency but also aligns with global energy conservation and sustainability goals. These hybrid systems represent a significant advancement in optical sensing technology, combining high sensitivity, precision, and energy efficiency, making them indispensable across healthcare, environmental, and industrial applications [25, 28].

4.4 Advanced Applications

The synergy between noble metal-based hybrid composites and TENG/PENG systems has unlocked advanced applications in fields like quantum optics and telecommunications which is shown in Fig. 16. The plasmonic properties of noble metals enable precise manipulation of light at the nanoscale, which is critical for quantum photonic devices. These devices include quantum dot emitters, single-photon detectors, and plasmonic waveguides, all of which benefit from enhanced optical interactions provided by noble metals [40].

In telecommunications, hybrid systems are paving the way for next-generation optical communication networks. Devices such as modulators, switches, and filters

Fig. 16 Advanced applications of triboelectric nanogenerators: from biomedical science and self-powered sensors to blue energy and IoT devices [37]

integrated with noble metal-based composites exhibit superior performance in terms of speed, bandwidth, and energy efficiency [62]. The ability of these systems to harness ambient energy through TENG or PENG mechanisms further enhances their suitability for sustainable and high-performance communication infrastructures. Another promising area is in photonic computing, where hybrid composites contribute to the development of compact, energy-efficient components for data processing and storage. Their ability to facilitate high-speed optical signal processing positions them as key enablers of future technologies such as 6G networks and quantum computing platforms [2, 63]. Figure 1 shows the integration of AI into TENG (Triboelectric Nanogenerators) and PENG (Piezoelectric Nanogenerators) development, encompassing analysis, design, fabrication, and application stages. The analysis phase involves the use of theoretical models to optimize nanogenerator performance by analyzing electric charge and polarization dynamics. In the design phase, advanced strategies are implemented to customize device structures for enhanced efficiency and functionality [2]. The fabrication stage emphasizes the transition from nanoscale precision manufacturing to large-scale industrial production for broader deployment. In the application phase, AI empowers nanogenerators to tackle challenges across diverse domains, such as aerospace, automotive, and marine environments. It enhances the ability of TENG and PENG devices to detect strain

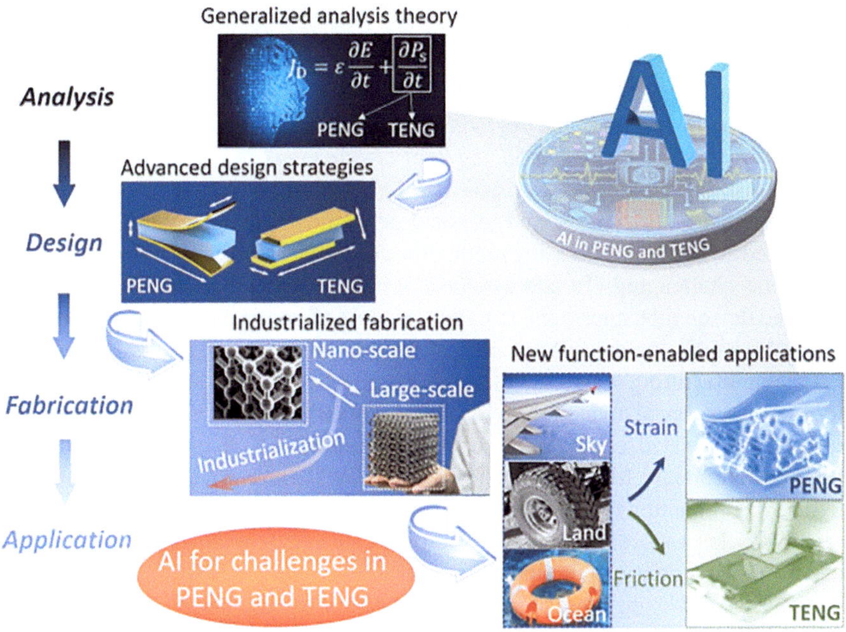

Fig. 17 AI-Driven solutions for enhancing PENG and TENG performance across development stages [49]

and friction, expanding their real-world adaptability. The synergy between AI and nanogenerators promotes scalable, cost-efficient, and multifunctional systems for sustainable energy harvesting and advanced sensing applications, addressing current limitations and enabling these technologies. [44] (Fig. 17).

These advanced applications underscore the transformative potential of noble metal-based hybrid composites in redefining the capabilities of optical devices and systems. By integrating TENG and PENG technologies, these innovations are poised to drive significant advancements in science, technology, and industry [62].

5 Challenges and Future Directions

5.1 Challenges

Despite significant advancements, TENG and PENG technologies powered by noble metal-based hybrid composites face notable challenges. High material costs, particularly for noble metals like gold, silver, and platinum, create economic barriers to large-scale adoption. These metals, while offering exceptional properties, are expensive and their extraction processes are resource-intensive and environmentally taxing. As illustrated in Fig. 18, long-term stability and durability further complicate widespread implementation. Hybrid structures combining noble metals with polymers or ceramics can degrade under mechanical wear, moisture exposure, and temperature fluctuations, reducing their effectiveness over time [49]. Manufacturing scalability also poses a critical hurdle; techniques such as chemical reduction, hydrothermal synthesis, and physical deposition demand precise control and specialized equipment, increasing production costs and making consistent large-scale production challenging. To address these issues, innovative material alternatives, enhanced device robustness, and streamlined, cost-effective fabrication methods are essential for realizing the full potential of these hybrid technologies [50].

The manufacturing scalability of hybrid composites is also a pressing issue. Techniques like chemical reduction, hydrothermal synthesis, and physical deposition require sophisticated equipment and precise control, which increases production costs and complexity. Moreover, achieving uniformity and consistent performance in large-scale production remains a significant technical challenge [40, 50]. Addressing these obstacles will require innovative approaches such as developing affordable alternatives to noble metals, improving the robustness of hybrid structures, and streamlining fabrication techniques. Collaboration between researchers, industry, and policymakers will be essential to overcoming these barriers and unlocking the full potential of TENG and PENG technologies [32].

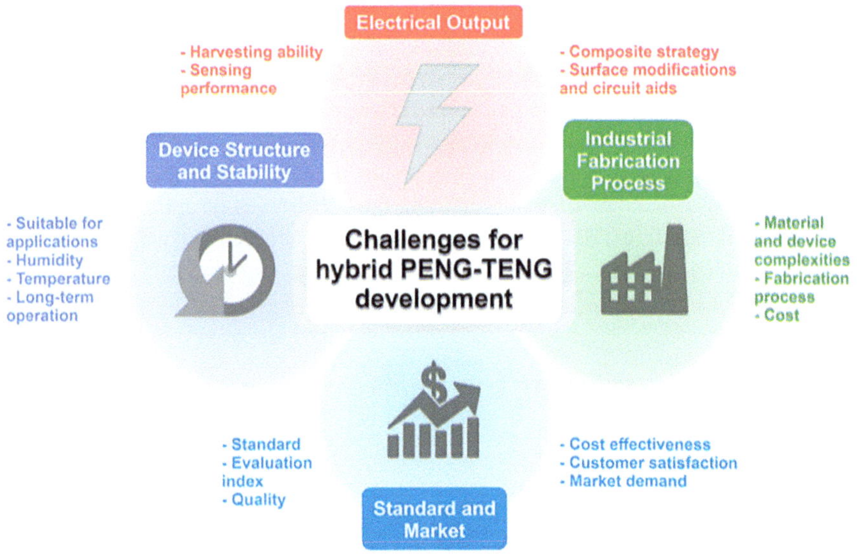

Fig. 18 Key challenges in hybrid PENG-TENG development: electrical output, structure stability, fabrication, and market standards [50]

5.2 Future Prospects

The future of TENG and PENG technologies powered by noble metal-based hybrid composites is filled with exciting opportunities for innovation. Developing cost-effective materials, such as nanostructured alloys or doped semiconductors, offers a pathway to reduce dependency on expensive noble metals while maintaining superior performance [52]. The integration of artificial intelligence (AI) promises to revolutionize energy systems, enabling adaptive optimization of energy harvesting and storage for smart and autonomous applications, particularly in IoT devices. Advances in flexible, wearable, and biodegradable devices are set to expand the versatility of these systems, enabling applications in healthcare monitoring, fitness tracking, and environmentally friendly electronics. Emerging fields such as quantum optics, high-speed telecommunications, and photonic computing present additional opportunities, where noble metal-based composites can deliver the precision and efficiency required for cutting-edge technologies [50, 58]. By addressing current limitations and exploring these innovative directions, TENG and PENG technologies are poised to revolutionize both energy and optical device applications, contributing to a sustainable and technologically advanced future.

6 Summary

The integration of noble metal-based hybrid composites with TENG and PENG systems represents a pivotal advancement in the realm of sustainable energy technologies. These hybrid systems, leveraging the unparalleled electrical, optical, and chemical properties of noble metals, have enabled groundbreaking innovations across diverse applications, including optical devices, IoT networks, and environmental monitoring systems. By enhancing energy conversion efficiency and expanding the versatility of nanogenerators, these composites have paved the way for a new era of energy harvesting and utilization. While the benefits are substantial, challenges such as high material costs, environmental concerns, and scalability issues remain significant barriers. However, the persistent drive toward innovation in material science and manufacturing processes promises effective solutions. The exploration of cost-efficient alternatives, such as nanostructured alloys and doped semiconductors, and the adoption of advanced fabrication techniques, are already showing potential to mitigate these obstacles. Looking forward, the convergence of AI and smart energy systems can redefine the landscape of TENG and PENG technologies, enabling adaptive functionalities in IoT ecosystems. Furthermore, the development of flexible and biodegradable devices promises to reduce environmental impact while enhancing usability. Emerging fields such as quantum optics, high-speed telecommunications, and photonic computing represent additional frontiers where noble metal-based composites can deliver unparalleled efficiency and precision. These applications underline the transformative role of these materials in addressing global energy and technological challenges.

In summary, this chapter highlights the remarkable potential of noble metal-based hybrid composites to foster innovation across industries, contributing to a more sustainable and technologically advanced future. By continuing to overcome present limitations through collaborative research and development, these systems are poised to revolutionize the way we harness and apply energy in the years to come.

References

1. Q. Shi, Z. Sun, Z. Zhang, C. Lee, Triboelectric nanogenerators and hybridized systems for enabling next-generation IoT applications. Research **2021**, 6849171 (2021)
2. C. Huang, G. Chen, A. Nashalian, J. Chen, Advances in self-powered chemical sensing via a triboelectric nanogenerator. Nanoscale **13,** 2065e2081 (2021)
3. S. Anbalagan et al., Progress and recent advances in self-powered gas sensing based on triboelectric and piezoelectric nanogenerators. Chem. Eng. J. **497**, 154740 (2024)
4. Y. Yang, D. Zhang, D. Wang, Z. Xu, J. Zhang, A high-stability weighing paper/polytetrafluoroethylene-based triboelectric nanogenerator for self-powered In_2O_3 nanocubes/SnS_2 nanoflower NO_2 gas sensors. J. Mater. Chem. A **9**, 14495–14506 (2021)
5. Z. Ma, X. Cao, N. Wang, Biophysical sensors based on triboelectric nanogenerators. Biosensors **13**(4), 423 (2023)
6. N. Mohamadbeigi, L. Shooshtari, S. Fardindoost, M. Vafaiee, A. Iraji zad, R. Mohammadpour, Self-powered triboelectric nanogenerator sensor for detecting humidity level and monitoring ethanol variation in a simulated exhalation environment. Sci. Rep. **14**, 1–16 (2024)
7. G. Barandun, L. Gonzalez-Macia, H.S. Lee, C. Dincer, F. Güder, Challenges and opportunities for printed electrical gas sensors. ACS Sensors **7**, 2804–2822 (2022)
8. P. Subudhi, D. Punetha, Simulation and optimization of substrate layer material for PVDF cantilever-based vibration energy harvesting system. IEEE Sens. J. **23**, 28689–28695 (2023)
9. Z.L. Wang, J. Song, Piezoelectric nanogenerators based on zinc oxide nanowire arrays. Science **312**, 242e246 (2006)
10. G. Chellasamy, E. Varathan, K. Sekar, S. Venkateswarlu, S. Govindaraju, K. Yun, Single-atom catalysts for biosensing: progress in theoretical and mechanistic understanding. Coord. Chem. Rev. **502**, 215606 (2024). https://doi.org/10.1016/j.ccr.2023.215606
11. W. Zhang, L. You, X. Meng, B. Wang, D. Lin, Recent advances on conducting polymers based nanogenerators for energy harvesting. Micromachines **12**, 1308 (2021)
12. Q. Xu, J. Wen, Y. Qin, Development and outlook of high output piezoelectric nanogenerators. Nano Energy **86**, 106080 (2021)
13. M. Pan, C. Yuan, X. Liang, J. Zou, Y. Zhang, C. Bowen, Triboelectric and piezoelectric nanogenerators for future soft robots and machines. iScience **23**, 101682 (2020)
14. R.A. Shaukat, Q.M. Saqib, M.U. Khan, M.Y. Chougale, J. Bae, Bio-waste sunflower husks powder based recycled triboelectric nanogenerator for energy harvesting. Energy Rep. **7**, 724–731 (2021)
15. J. Wang, G. Bao, S. Xie, X. Chen, A paradigm-shift self-powered optical sensing system enabled by the rotation driven instantaneous discharging triboelectric nanogenerator (RDID-TENG). Nano Energy **115**(108732), 108732 (2023)
16. H.H. Le, Q.H. Pham, Triboelectric nanogenerator based hydrogen gas sensor. MRS Adv **9**, 737 (2023)
17. Y. Su, S. Chen, B. Liu, H. Lu, X. Luo, C. Chen, W. Li, Y. Long, H. Tai, G. Xie, Y. Jiang, Maxwell displacement current induced wireless self-powered gas sensor array. Mater. Today Phys. **30**, 100951 (2023)

18. V. Vivekananthan, N.R. Alluri, Y. Purusothaman, A. Chandrasekhar, S. Selvarajan, S.J. Kim, Biocompatible collagen nanofibrils: an approach for sustainable energy harvesting and battery-free humidity sensor applications. ACS Appl. Mater. Interfaces **10**, 18650–18656 (2018)
19. S.S. Dani, A. Tripathy, N.R. Alluri, S. Balasubramaniam, A. Ramadoss, A critical review: the impact of electrical poling on the longitudinal piezoelectric strain coefficient. Mater. Adv. **3**, 8886–8921 (2022)
20. X. Zhou, K. Parida, O. Halevi, Y. Liu, J. Xiong, S. Magdassi, P.S. Lee, All 3D-printed stretchable piezoelectric nanogenerator with non-protruding kirigami structure. Nano Energy **72**, 104676 (2020)
21. S. Liu, G. Yuan, Y. Zhang, L. Xie, Q. Shen, H. Lei, Z. Wen, X. Sun, A Self-powered gas sensor based on coupling triboelectric screening and impedance matching effects. Adv. Mater. Technol. **6**, 1–8 (2021)
22. D. Punetha, M. Kar, S.K. Pandey, A new type low-cost, flexible and wearable tertiary nanocomposite sensor for room temperature hydrogen gas sensing. Sci. Rep. **10**, 2151 (2020)
23. Y. Purusothaman, H. Leng, A. Nanda, I. Levine, S. Priya, Textured lead-free piezoelectric ceramics for flexible energy harvesters. ACS Appl. Mater. Interfaces **15**, 6584–6593 (2023)
24. L. Liu, X. Guo, C. Lee, Promoting smart cities into the 5G era with multi-field Internet of Things (IoT) applications powered with advanced mechanical energy harvesters. Nano Energy **88**, 106304 (2021)
25. M. Sawane, M. Prasad, MEMS piezoelectric sensor for self-powered devices: a review. Mater. Sci. Semicond. Process. **158**, 107324 (2023)
26. Y.H. Lee, D.H. Kim, Y. Kim, I. Shabbir, M. Li, K.H. Yoo, T.W. Kim, Significant enhancement of the output voltage of piezoelectric/triboelectric hybrid nanogenerators based on MAPbBr 3 single crystals embedded into a porous PVDF matrix. Nano Energy **102**, 107676 (2022)
27. Q. Zheng, Y. Zou, Y. Zhang, Z. Liu, B. Shi, X. Wang, Y. Jin, H. Ouyang, Z. Li, Z.L. Wang, Biodegradable triboelectric nanogenerator as a life-time designed implantable power source **2**(3), e1501478 (2016)
28. L. He, X. Ma, Y. Li, C. Gong, H. Liu, H. Shu, J. Ni, B. Zhang, S. Xiang, A novel selfpowered sensor based on $Ni(OH)_2/Fe_2O_3$ photoanode for glucose detection by converting solar energy into electricity. J. Alloy. Compd. **907**, 164132 (2022)
29. L. Afsah-Hejri, P. Hajeb, R.J. Ehsani, Application of ozone for degradation of mycotoxins in food: a review. Compr. Rev. Food Sci. Food Saf. **19**(4), 1777–1808 (2020)
30. H. Zhang, Y. Xie, X. Li, Z. Huang, S. Zhang, Y. Su, B. Wu, L. He, W. Yang, Y. Lin, Flexible pyroelectric generators for scavenging ambient thermal energy and as self-powered thermosensors. Energy **101** 202–210 (2016). https://doi.org/10.1016/j.energy.2016.02.002
31. N. Cui, J. Liu, L. Gu, S. Bai, X. Chen, Y. Qin, Wearable triboelectric generator for powering the portable electronic devices. ACS Appl. Mater. Interfaces **7**(33), 18225–18230 (2015)
32. M. Li, Q. Guan, C. Li, E. Saiz, Self-powered hydrogel sensors. Device **1**(1) (2023)
33. C.H. Sun, Q.F. Shi, D. Hasan, M.S. Yazici, M.L. Zhu, Y.M. Ma, B.W. Dong, Y.F. Liu, C. Lee, Self-powered multifunctional monitoring system using hybrid integrated triboelectric nanogenerators and piezoelectric microsensors. Nano Energy **58**, 612–623 (2019)
34. X. Lin, Z. Feng, Y. Xiong, W. Sun, W. Yao, Y. Wei, Z.L. Wang, Q. Sun, Piezotronic neuromorphic devices: principle, manufacture, and applications. Int. J. Extrem Manuf. **6**(3), 032011 (2024)
35. W. Wu, Z.L. Wang, Piezotronics and piezo-phototronics for adaptive electronics and optoelectronics. Nat. Rev. Mater. **1**(7), 16031 (2016)
36. K. Zhao, B. Ouyang, C.R. Bowen, Z.L. Wang, Y. Yang, One-structure-based multieffects coupled nanogenerators for flexible and self-powered multi-functional coupled sensor systems. Nano Energy **71**, 104632 (2020)
37. X. Cao, Y. Xiong, J. Sun, X. Xie, Q. Sun, Z.L. Wang, Multidiscipline applications of triboelectric nanogenerators for the intelligent era of Internet of Things. Nano-Micro Letters **15**, 14 (2023)
38. R. Di Giacomo, L. Bonanomi, V. Costanza, B. Maresca, C. Daraio, Biomimetic temperature-sensing layer for artificial skins. Sci. Robot **2**(3), eaai9251 (2017)

39. W.J. Fan, Q. He, K.Y. Meng, X.L. Tan, Z.H. Zhou, G.Q. Zhang, J. Yang, Z.L. Wang, Machine-knitted washable sensor array textile for precise epidermal physiological signal monitoring. Sci. Adv. **6**(11), eaay2840 (2020)
40. J. Zhang, Y. He, C. Boyer, K. Kalantar-Zadeh, S. Peng, D. Chu, C.H. Wang, Recent developments of hybrid piezoetriboelectric nanogenerators for flexible sensors and energy harvesters. Nanoscale Adv. **3**, 5465e5486 (2021)
41. M. Wang, J. Zhang, Y. Tang, J. Li, B. Zhang, E. Liang, Y. Mao, X. Wang, Air-flowdriven triboelectric nanogenerators for self-powered real-time respiratory monitoring. ACS Nano **12**(6), 6156–6162 (2018)
42. P. Zhu, B. Zhang, H. Wang, Y. Wu, H. Cao, L. He, C. Li, X. Luo, X. Li, Y. Mao, 3D printed triboelectric nanogenerator as self-powered human-machine interactive sensor for breathing-based language expression. Nano Res. **15**(8), 7460–7467 (2022)
43. S. Suman, D. Punetha, S.K. Pandey, Improvement in sensing characteristics of silicon microstructure based MEMS capacitive sensor for automotive applications. SILICON **13**, 1475–1483 (2021)
44. W. Zhang, P. Wang, K. Sun, C. Wang, D. Diao, Intelligently detecting and identifying liquids leakage combining triboelectric nanogenerator based selfpowered sensor with machine learning. Nano Energy **56**, 277–285 (2019)
45. H. Zhou, W. Huang, Z. Xiao, S. Zhang, W. Li, J. Hu, T. Feng, J. Wu, P. Zhu, Y. Mao, Deep-learning-assisted noncontact gesture-recognition system for touchless human-machine interfaces. Adv. Funct. Mater. **32**(49), 2208271 (2022)
46. Y. Lu, H. Tian, J. Cheng, F. Zhu, B. Liu, S. Wei, L. Ji, Z.L. Wang, Decoding lip language using triboelectric sensors with deep learning. Nat. Commun. **13**(1), 1401 (2022)
47. D. Punetha, A. Kumar, S.K. Pandey, S. Chakrabarti, Wearable piezoelectric nanogenerator-based hazardous gas monitoring gadget for self-powered ammonia early warning, in *Proceedings of SPIE 12661, Organic and Hybrid Sensors and Bioelectronics XVI*, 126610A (2023)
48. D. Zhang, Y. Yang, Z. Xu, D. Wang, C. Du, An eco-friendly gelatin based triboelectric nanogenerator for a self-powered PANI nanorod/$NiCo_2O_4$ nanosphere ammonia gas sensor. J. Mater. Chem. A 10935–10949 (2022)
49. P. Jiao, Emerging artificial intelligence in piezoelectric and triboelectric nanogenerators. Nano Energy **88**(106227), 106227 (2021)
50. S. Sriphan, N. Vittayakorn, Hybrid piezoelectric-triboelectric nanogenerators for flexible electronics: a review of recent advances and perspectives. J. Sci.: Adv. Mater. Devices **7**, 100461 (2022)
51. C. Xu, Y. Song, M. Han, H. Zhang, Portable and wearable self-powered systems based on emerging energy harvesting technology. Microsyst. Nanoeng. **7**, 25 (2021)
52. D. Punetha, A. Kumar, S.K. Pandey, S. Chakrabarti, Tertiary nanocomposite-based self-powered E-skin as energy harvester and electronic nose. J. Mater. Sci. Mater. Electron. **35**(2) (2024)
53. K. Gupta, S. Brahma, J. Dutta, B. Rao, C.P. Liu, Recent progress in microstructure development of inorganic onedimensional nanostructures for enhancing performance of piezotronics and piezoelectric nanogenerators. Nano Energy **55**, 1–21 (2019)
54. S. Paul, D. Acharyya, D. Punetha, Benchmarking the performance of lithium and sodium-ion batteries with different electrode and electrolyte materials. Energy Storage **6**, e70068 (2024)
55. P. Subudhi, D. Punetha, Other potential 2-D materials for CMOS applications, in: *Beyond Si-Based CMOS Devices. Springer Tracts in Electrical and Electronics Engineering* ed by S. Singh, S.K. Sharma, D. Nandan, (Springer, Singapore, 2024)
56. P. Verma, D. Punetha, S.K. Pandey, Sensitivity optimization of MEMS based piezoresistive pressure sensor for harsh environment. SILICON **12**, 2663–2671 (2020)
57. D. Punetha, S.K. Pandey, S. Chakrabarti, Generation of light energy by utilization of electrical energy without using any source of light, in *Proceedings of 2017 3rd International Conference on Advances in Computing, Communication & Automation (ICACCA)*, 15–16 September 2017. IEEE

58. A. Garg, S. Agarwal, D. Punetha, Polarization and strain in piezoelectric nanomaterials: advancing sensing applications in biomedical technology. IEEE Open J. Nanotechnol. **5**, 89–97 (2024)
59. C. Pei, J. Wan, Nanocomposite-based matrices in laser desorption/ionization mass spectrometry for small-molecule analysis. ChemPlusChem **85**, 2678–2691 (2020)
60. P. Subudhi, D. Punetha, Progress, challenges, and perspectives on polymer substrates for emerging flexible solar cells: a holistic panoramic review. Prog. Photovolt: Res. Appl. **31**, 753–789 (2023)
61. J. Wang, K. Xia, J. Liu, T. Li, X. Zhao, B. Shu, H. Li, J. Guo, M. Yu, W. Tang, Z. Zhu, Self-powered silicon PIN photoelectric detection system based on triboelectric nanogenerator. Nano Energy **69**, 104461 (2020)
62. V. Kiruthika, S. Shoba, S. Madan, N. Kishore, D. Punetha, Hybrid ensemble-deep transfer model for early cassava leaf disease classification. Heliyon **10**, e36097 (2024)
63. S. Kumar, V. Sagar, D. Punetha, A comparative study on facial expression recognition using local binary patterns, convolutional neural network and frequency neural network. Multimed Tools Appl **82**, 24369–24385 (2023)
64. W. Zhang, L. Deng, L. Yang, P. Yang, D. Diao, P. Wang, Z.L. Wang, Multilanguage handwriting self-powered recognition based on triboelectric nanogenerator enabled machine learning. Nano Energy **77**, 105174 (2020)
65. S. Xu, W. Nie, J. Sun, P. Sun, H. Jia, X. Zheng, Y. Sun, Z. Xu, L. Liu, Multi-mode and durable fiber triboelectric nanogenerator for power and sensor enabled by Hookean vascular stent structure. Chem. Eng. J. **472**, 145088 (2023). https://doi.org/10.1016/j.cej.2023.145088

Metal Oxide-Enhanced TENG/PENG Nanogenerators for Self-Powered Optoelectronic Devices

Kaliyannan Manojkumar, Mukilan Muthuramalingam, Arunmetha Sundaramoorthy, P. S. Srinivasa Babu, and Venkateswaran Vivekananthan

Abstract The integration of triboelectric nanogenerators (TENGs) and piezoelectric nanogenerators (PENGs) into optoelectronic systems has led to the development of self-powered optical devices with enhanced energy efficiency and adaptability. This chapter explores the role of metal oxide-based hybrid composites in advancing TENG/PENG-powered optical devices, focusing on their material properties, working mechanisms, and applications in tunable optical gratings, photodetectors, and adaptive sensors. The synergistic effects of triboelectric and piezoelectric charge generation enable real-time optoelectronic devices and self-powered sensing systems. Emerging this technologies pave the way for wearable electronics, environmental monitoring, and next-generation photonic interfaces. By analyzing the fabrication techniques, charge transfer dynamics, and performance evaluations of these hybrid devices, this chapter highlights the potential of TENG/PENG-based systems in redefining self-sustained optoelectronics. In addition, the futuristic technological research focuses on material engineering, semiconductor band tuning, efficient charge transport for self-sustainable devices, as well as enhancing optical band bending to further advance the development of self-powered photonic devices.

Keywords TENG PENG hybrid devices · Self-powered optoelectronics · Metal oxide nanocomposites · Triboelectric piezoelectric synergy · Wearable photonic sensors

---

K. Manojkumar · M. Muthuramalingam · A. Sundaramoorthy · P. S. Srinivasa Babu · V. Vivekananthan (✉)
Center for Flexible Electronics, Department of Electronics and Communication Engineering, Koneru Lakshmaiah Education Foundation, Vaddeswaram, India
e-mail: vivek@kluniversity.in

B. Ramachandran et al. (eds.), *Advanced Optical Sensors*, Progress in Optical Science and Photonics 38, https://doi.org/10.1007/978-981-95-1687-2_6

1 Introduction

The evolution of self-powered optoelectronic devices has led to groundbreaking advancements in sustainable energy harvesting and photonic applications [1–10]. Among these innovations, Triboelectric Nanogenerators (TENGs) and Piezoelectric Nanogenerators (PENGs) have emerged as promising technologies for converting mechanical energy into electrical energy, enabling the operation of optical devices without external power sources [11–20]. This integration of energy harvesting mechanisms with optical modulation has paved the way for novel applications in tunable photonic devices, self-powered sensors, and adaptive optical systems [21–26]. Metal oxide-based hybrid composites have attracted particular interest due to their unique combination of piezoelectric and triboelectric properties, allowing efficient charge generation and transfer at the material interface [27–36] These materials provide an effective platform for enhancing the performance of self-powered optical devices, such as photodetectors, tunable gratings, and optoelectronic sensors [37–39]. The synergistic interaction between TENGs and PENGs within these hybrid systems facilitates superior energy conversion and optoelectronic coupling, making them ideal candidates for applications in wearable electronics, environmental monitoring, biomedical imaging, and precision optical engineering [40]. One of the most significant applications of TENG/PENG-powered optical devices is in tunable optical gratings (TOGs) [41]. These structures enable real-time modulation of light diffraction patterns, which is crucial for applications in optical communications, smart displays, and photonic computing [42, 43]. The use of metal oxide-based hybrid composites in TOGs allows dynamic strain-induced tuning of optical properties, providing high precision and adaptability in light manipulation [44]. Moreover, self-powered photodetectors leveraging the piezoelectric-phototronic effect in metal oxide materials have demonstrated enhanced charge separation efficiency and spectral selectivity, making them viable solutions for autonomous sensing and security monitoring [45–59].

This chapter provides an in-depth exploration of the design, working principles, and applications of TENG/PENG-powered optical devices based on metal oxide hybrid composites. It discusses key aspects of material synthesis, structural characterization, charge transfer mechanisms, and experimental performance evaluations. By examining the integration of mechanical energy harvesting with optoelectronics, this chapter aims to offer insights into the future development of self-sustained, high-performance optical systems. With continued research and advancements in material engineering and device optimization, TENG/PENG-powered optical devices hold immense potential for revolutionizing next-generation photonic technologies.

2 Triboelectric Nanogenerator

2.1 Multifunctional Mechanoluminescent Triboelectric Nanogenerator Fiber for Smart Sensing, Visual Feedback, and Wireless Communication

The innovative mechanoluminescent triboelectric nanogenerator fiber (MLTENGF) was fabricated for the sense and transmit the mechanical stimuli. The optical and electrical feedback have offered by the fabricated MLTENGF through the visual and digital force response. Fabrication of the MLTENGF was the combined process of both winding and 3D coating techniques, MLTENGF made by the sandwich assembly of circular strips silicon rubber foam and the spiral CNTF also the outer layer covered by the composite silicone rubber shown in Fig. 1a. The fabricated MLTENGF cross-sectionally investigated by the SEM and the three-layered structure was confirmed and the CNTF layer performed as the electrode and Ecoflex composite layer acted as the friction cell shown in Fig. 1b. Surface morphological SEM image of the ZnS:Cu particles mix distribution into the Ecoflex polymer matrix shown in the Fig. 1c. In addition, the achieving higher brightness of the densely filled ZnS:Cu was more challenged so the Ecoflex was supported for the high brightness achievement and the structural durability, while stretching for the mechano luminescent (ML). Figure 1d revealed the uniform and dense elemental distribution mapping of the Zn, S and Cu elements and under the mechanical stimuli those elements compound material of ZnS:Cu coated layer was the stable light source. Under the strain condition of 200% the white ZnS:Cu particles have distributed evenly along the axial direction. During the strain condition the CNTF pitch was increasing small size and no significant fracture observed in the meantime all details shown in Fig. 1e. These studies revealed the property of the superior structural durability and flexibility of the fiber. This nature adequately enhances the intrinsic elastic nature of the substrate and supports the Knot-tying operations shown in Fig. 1f. Consequently, the tensile strength decreased from 1.0 to 0.6 M Pa while the stretchability reduced from 260 to 190% and simultaneously the content of the ZnS:Cu was increased to 50 wt.% in the fiber results depicted in Fig. 1g. The notable distortions observed after 150% strain due to the changes in the CNTF helical structure tending to straighten in the linear direction and ZnS:Cu irregular structure induce the slight sliding movements. The fabricated MLTENGF have wrapped around the finger and the additionally supported the weight of 100 g due to the stable tensile strength and elastic deformation exceptional characteristics shown in Fig. 1h. The frictional electrical output performance was studied by the dynamic loading arrangement of the test platform shown in Fig. 1i. In the setup, the fabricated 4 cm MLTENGF served as the frictional material and interconnected with the linear motor and attached copper foil as barrier. The operational mechanism of the MLTENGF was functions between the electrostatic induction and the triboelectric charging shown in Fig. 1j. The Ecoflex and the copper foil contact each other the positive charges induced on the copper foil and the negative charges induced on the Ecoflex surface. However, increasing the distance between both copper foil and Ecoflex the electron flows to the ground from the CNTF as a short-circuit state until reach the equilibrium state of induced positive charges. Due to the reverse electrostatic induction again the electron flow in opposite direction. For the frictional material approach contact and separation happened repetitively by the linear motor, for the result continuous production of the Alternating current. Further the electrostatic module created and stimulated the various components potential by COMSOL to clearer quantitative understanding of the power generation process shown in Fig. 1k. Electron cloud representation was revealed the potential charge transfer process between the two materials, when the atoms are distant there are no overlaps happen between their electron cloud and maintain the original orbital state. While coming to the contact both materials significantly electron clouds are overlapped, simultaneously the interatomic potential barrier was reduced to prompting the electrons from higher orbit to unoccupied orbits. The remaining positive and negative charges are retained on their surfaces when the corresponding materials of copper and Ecoflex are separated shown in Fig. 1l. The capabilities of the precise detection of the pressure distribution and monitoring the motion trajectory of the textile TENG has utilized for the home anti-theft system components represented in Fig. 1m. An equivalent circuit diagram of the theft alarm carpet was depicted in Fig. 1n. MLTENGF generates output to activate alarm as the green light signal while intruder stepped on the carpet shown in Fig. 1o. Also, Generated signal was transmitted via Bluetooth wireless communication to the mobile phone for the entering person was the intruder or owner based on their weight. The energy band diagram of the ZnS:Cu depicted in Fig. 1p. In ZnS semiconductor compound sulfur exists by the two energy level vacancies such as shallow donor level captures the electron and t_2 acceptor level captures the Cu^{2+} holes. This effect resembles the process of electron capture from the valence band to the impurity shallow donor level. The captured electrons recombine under pressure, leading to the emergence of green emission. This phenomenon has gained importance for effective underwater communication technology, enabling efficient signal transmission to protect divers and other personnel working in deep-sea environments. Based on the water resistance capability of MLTENGF, it could be suggested for use in underwater rescue operations shown in Fig. 1q. In addition, information has transmitted by using the morse code technology, it is represented as the “dot” and “dashes” also encrypted with the special Latin letters patterns Shown in Fig. 1r. The fabricated MLTENGF generates photoelectric signal variations, where a short-duration press with intense light represents a “dot” sign, and a long-duration press with weak light represents a “dash” sign. Therefore, this self-powered device provides an optimal solution for underwater rescue operations, sensing, and information transmission while also eliminating the need for additional integrated electronic components [60].

The self-powered Tactile interactive system (TIS) has been developed to utilize various applications such as paper piano, tactile switch, PPT, wireless controller of domestic appliances and flowing LEDs. The developed TIS have constructed with signal generator as TENG, signal processing unit and self-powered rectifying optical

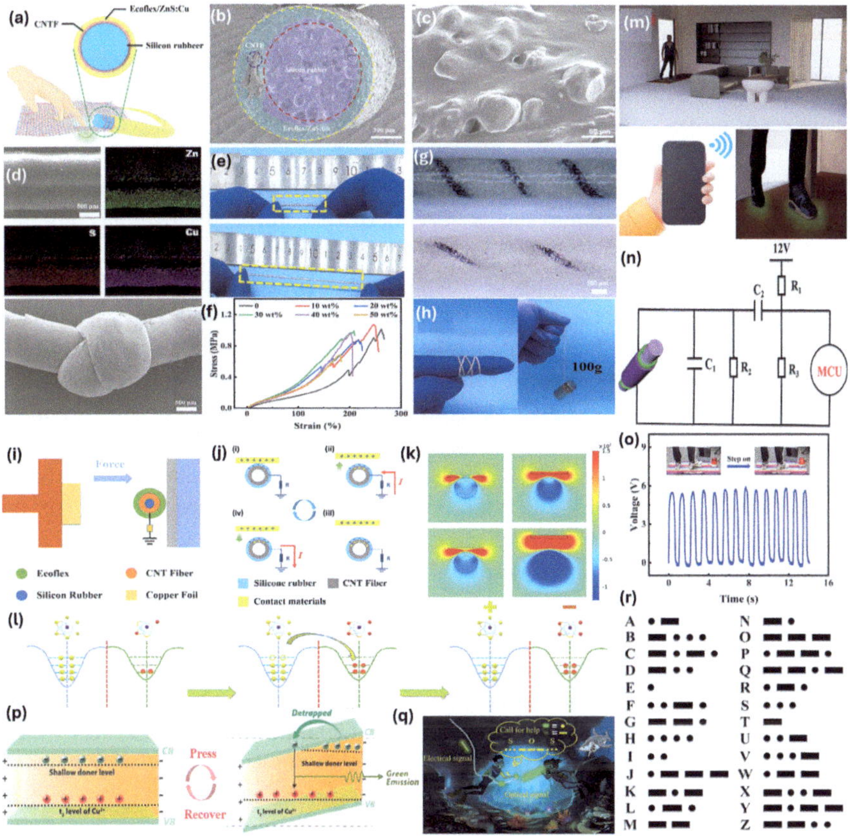

Fig. 1 **a** MLTNGF principle conceptual schematic diagram, **b** MLTENGF cross-sectional SEM image, **c** SEM analysis of ZnS:Cu particles doped Ecoflex surface morphological micrograph, **d** energy dispersive elemental mapping micrograph for copper, zinc and sulfur, **e** digital photograph of the stretchable MLTENGF and the 200% stretched with corresponding optical images. **f** MLTENGF wound into the butterfly shaped structure SEM image, **g** tensile test graph for the various doping concentration performed at the 300 mm initial length. **h** digital image of the MLTENGF wrapping on the finger and withstand capacity of the lifting of the 100 g weight, **i** MLTENGF dynamic loading schematic representation for the frictional electrical responses, **j** schematic diagram for the working mechanism of MLTENGF, **k** COMSOL studied electrical potential distribution of the MLTENGF, **l** Ecoflex/ZnS:Cu contact interface charge transfer atomic scale electron cloud potential well model, **m** schematic diagram of the carpet anti-theft alarm, **n** anti-theft alarm equivalent circuit diagram, **o** carpet anti-theft alarm output states of the ON and OFF, **p** schematic representation of the energy band of green luminescence of ZnS:Cu under stress, **q** aquatic environment morse code based MLTENGF pressure sensor performed through photoelectrically coordinated communication, **r** schematic representation of the morse codes. Reproduced with permission from Wu et al. [60]

communicator. Further, the TIS was fabricated on the various substrates and performances was studied and discussed as follows. Single electrode mode operated TENG have fabricated with the transparent electrode made up of graphene and touch layer made up of polyethylene terephthalate (PET) performed as mechano touch switch shown in Fig. 2a. The prepared mechano-switch (1, 2, or 3) generates a triboelectric signal when a finger touches the PET surface due to electrostatic induction. The TIS is utilized to control operations such as turning LEDs on/off and activating color fans in customized miniature city views. In these applications, TIS functions as a touch switch based on the mechanisms of triboelectrification and electrostatic induction. Moreover, it is transparent, easy to fabricate, and self-powered. Next, a contact-separation mode TENG mechanosensor was fabricated and utilized to control LED illumination through programmed control, as shown in Fig. 2b. The mechanosensor was fabricated using a Cu/FEP/Cu structure on a paper substrate. The optical communicator (OC) is driven by the output voltage produced through the contact electrification process, referred to as $V_{trigger}$ when pulled down. Simultaneously, the MCU monitors $V_{trigger}$, and when it is pulled down, the embedded program triggers the illumination of the LEDs. The mechanosensor is equipped with red and green LEDs, with corresponding controls attached to both the left and right parts of the Cu/FEP structure. The red LEDs begin to glow from the right side to the left when the left Cu/FEP arrangement is contacted. Similarly, the green LEDs glow when the right Cu/FEP arrangement is contacted. In addition, more complex paper piano was demonstrated by the same mechanosensor design (Cu/FEP/Cu) shown in Fig. 2c. This piano has included the eight notes from "do-re-mi-fa-so-la-ti" to higher note "do". Each piano key functions as an individual TENG mechanosensor, integrated as channels 1–8. The signals from each channel are recorded individually. The original signals, which consist of messy and irregular currents, can be converted into standard digital signals in the MCU with the help of interconnected self-powered OC direct readings depicted in Fig. 2d. Further, TIS has extended for the wireless Bluetooth equipped wearable remote-control system. Conceptual pictorial representation of the wearable (ring mounted model and built in cloth model) TIS of the wireless control (home appliances, entertainment/ educational facility operation and office software functioning) shown in Fig. 2e. TIS consists of transmitter and receiver terminals for the wireless function shown in Fig. 2f. The trigger signal was generated by the TENG mechanosensor integrated with the transmitting terminal. This signal was converted into a digital signal by the self-powered OC, processed in the MCU, and wirelessly transmitted to the receiving end. The receiver terminal received the transmitted trigger signal using the corresponding protocol and controlled the targeted functions. Wireless TIS was demonstrated two different scenarios one is the single electrode mode ring shaped TENG mechanosensor to operates the office power point (Microsoft Ltd.,) for the sliding control. The TENG consists of friction layer (PET), ITO electrode and isolated the skin with textile isolation layer shown in Fig. 2g. Whenever the thumb touched the ring controller, the friction induced between the PET and skin generated a signal through the triboelectric process to operate the sliding program in the PPT. Various ring controllers were worn on the ring finger, middle finger, and index finger to execute the functions of moving forward, moving backward, and ending the slide, as shown in Fig. 2h. Another one, single electrode mode operated textiles attached TIS made of cotton pads, Nylon and embroidered conductive Ag threads shown in Fig. 2i. The fabricated TIS was attached to various locations on the textile cloths (front, on, or beneath the sleeves), as shown in Fig. 2j. When the embroidered nylon pad was touched by a finger, it became positively charged due to the triboelectric effect sequence. The relevant signals were obtained from various positions to control the desk lamp's brightness and on/off state shown in Fig. 2k [61].

The self-powered electrowetting optical switch (EOS) developed by integrating free standing mode triboelectric nanogenerator (FS-TENG) and electrically tunable liquid lens (ETULL). The ETULL was fabricated using an acrylic cylindrical spacer, a conducting fluid as 10% NaCl electrolyte, an insulation oil as the purified olive oil, and two indium tin oxide (ITO) electrodes. Moreover, potential was applied to the ITO electrode of the ETULL by FS-TENG. Due to the electrowetting effect, the interface between the two liquids formed a curved shape, resulting in the formation of an insulating oil-based concave lens. The mechanical–electrical-optical conversion and switching mechanism of the EOS, with and without mechanical motion, is illustrated in Fig. 3. In the switching performance, when no potential is applied, the ETULL is in the "on" state. In this condition, there is no change in the propagated incident light, and it remains in its original state without deflection, as shown in Fig. 3a. In the "off" state, when a potential above the threshold limit is applied, curvature is formed based on the voltage level. As a result, the incident light diverges, as shown in Fig. 3b. The refractive indices (RIs) of the 10% NaCl and olive oil respectively 1.34 and 1.46 at 650 nm. RI of the olive oil was higher than the conductive liquid also plays the role of concave lens. Incident light was diverged when entering the low RI electrolyte to high RI for the off state. The optical intensity along the optical axis decreases quadratically as the distance between the light source and the detector increases, following Gaussian's law, so the EOS was supported for the long-distance communication. Remote monitoring of mechanical motion was achieved using the EOS wireless sensing unit, which was constructed with the EOS optical signal processing circuit, a lighting source, and an actuator as a siren. The wireless-operated smart human–machine interface unit was powered by the hand-sliding FS-TENG. For the smooth operation and accurate sensing of the EOS, a minimum optimal distance was necessary. The human–machine communication distance varied based on application demands. If the distance was reduced, the sensitivity of the EOS also decreased, making short-distance sensing more challenging. For the effective operation of the EOS, the sensing distance was set as small as possible. Here, the distance between the ETULL and the photodetector was set to 9 cm, which is minimal and suitable for most applications. When hand-sliding mechanical motion was detected, the siren was activated; in the absence of motion, the siren was deactivated. The effectiveness of the constructed EOS was demonstrated and is shown in Fig. 3c. Different potential drops on the ETULL create various morphologies of the concave lenses. Additionally, the block diagram depicting the process flow for measuring potential drops and the camera-equipped morphology examination is shown in Fig. 3d. Additionally, the potential drop in the ETULL, equipped with capacitors of varying capacitances, is depicted in Fig. 3e–i. The device capacitance increased with the addition of an

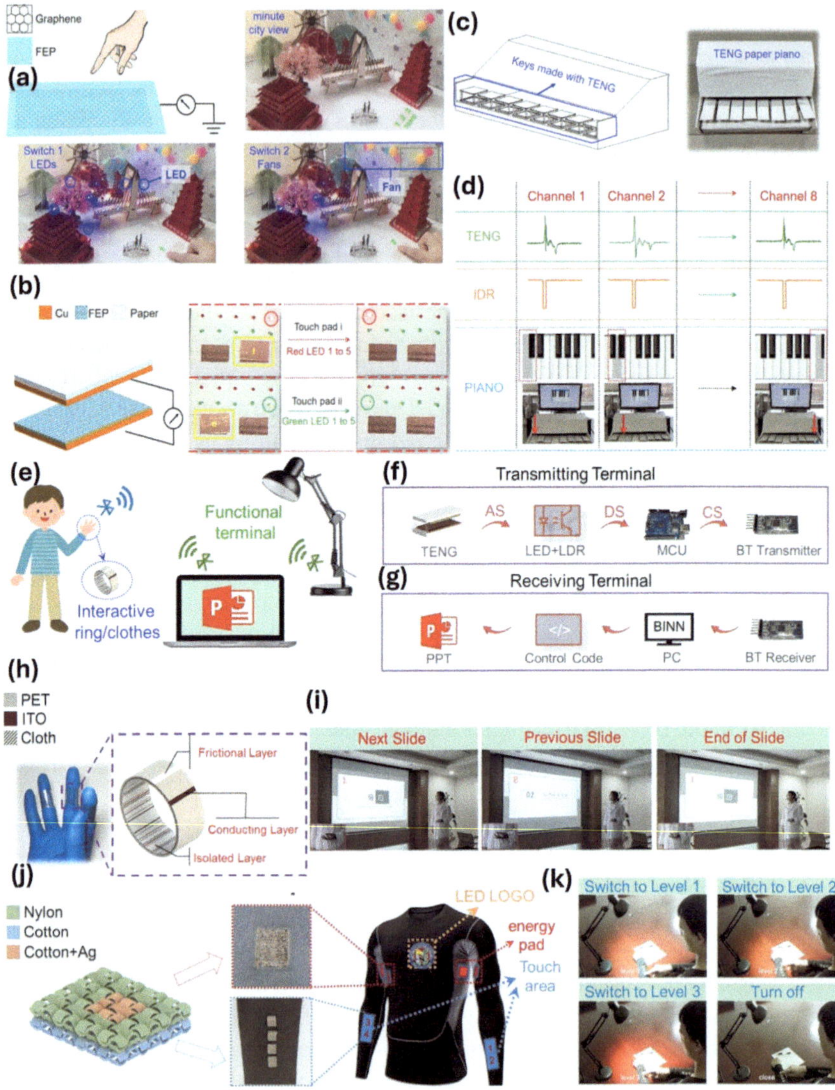

Fig. 2 **a** Schematic diagram for the graphene made touch switch and utilized for the custom minute city model digital images, **b** paper made contact separation TENG utilized to glowing the LEDs, **c** schematic diagram and digital photo image of the interactive paper-piano, **d** the original irregular output performance converted into the digital signal by the self-powered OC and directly read by the MCU, **e** the conceptual representation for the wireless controlled wearable TIS application, **f** the basic working function process of the wireless TIS, **g** digital and structural design of the ring shaped TENG mechanosensor, **h** digital image of the demonstration for the PPT control by the ring-shaped TENG mechanosensory, **i** schematic structural representation of the TENG and the attached locations on the cloth, **j** digital image of the demonstration for the desk lamp control by the textile attached wireless TIS. Reproduced with permission from Huang et al [61]

external capacitor, while the potential drop across the ETULL decreased. As a result, the liquid interface aspect ratio decreased. The peak potential drop on the ETULL was observed at 200 V with a 2200 pF load capacitor. Moreover, no morphological changes were observed in the concave lens, leading to the conclusion that the threshold potential of the ETULL is greater than 200 V. The relationship between the aspect ratio and the peak voltage drop of the ETULL is plotted in Fig. 3j. A maximum aspect ratio of 0.25 was achieved at 1100 V without an external capacitor load. As the driving voltage increased, the aspect ratio also increased [62].

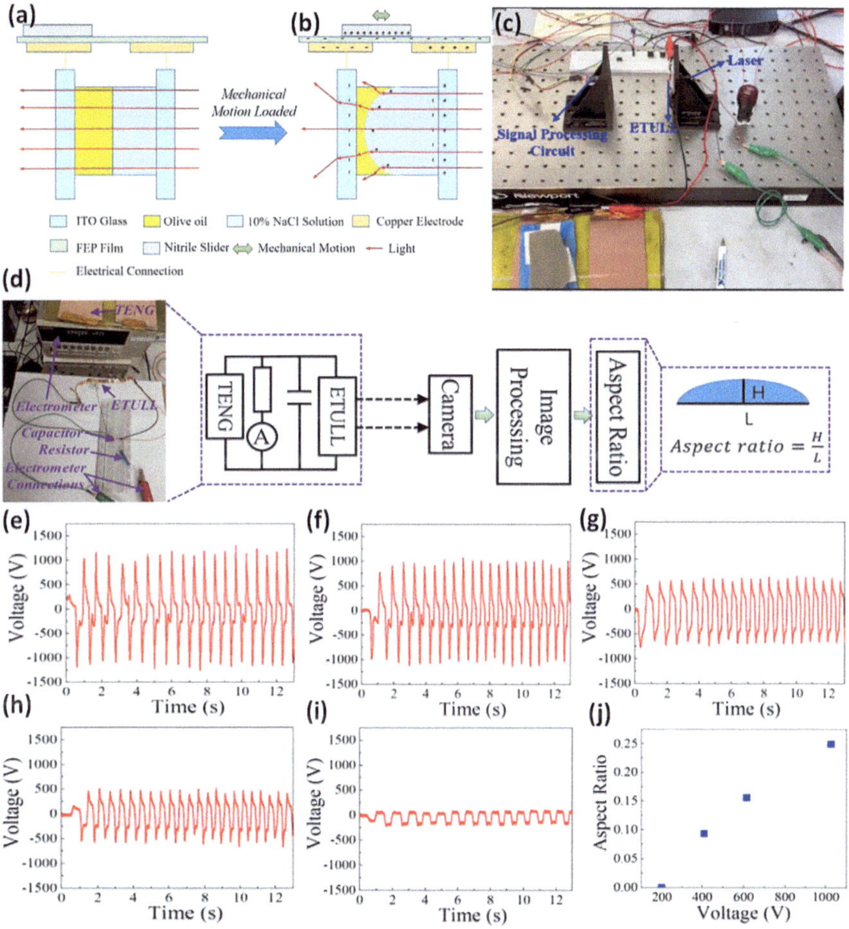

Fig. 3 **a** and **b** Switching mechanism of electrowetting optical switch (EOS) without and with mechanical motion, **c** digital image of the wireless sensing system, **d** schematic illustration of the voltage drop measurement on the ETULL connected with the external load capacitor and liquid interface aspect ratio, **e–i** voltage drop measurement plots with the various load capacitor e-22 pF, f-100 pF, g-470 pF, h-1000 pF, i-2200 pF, **j** liquids interface of the ETULL peak aspect ratio and the applied voltage plot. Reproduced with permission from Wang et al. [62]

2.2 TENG-DEA Conjunction System for Self-Powered Tunable Optical Gratings

The fundamental operation of the TENG-DEA conjunction system is illustrated in Fig. 4a, where a single-electrode TENG operating in contact-separation mode serves as both the power supply and control signal for the Dielectric Elastomer Actuator (DEA). The contact-electrification process occurs between a Kapton dielectric layer and an aluminum (Al) foil electrode, leading to charge transfer upon contact. When the layers separate, an electrostatic field is induced across the DEA, generating Maxwell stress, which in turn deforms the elastomer film. The Maxwell stress is given by:

$$P = -\varepsilon E^2$$

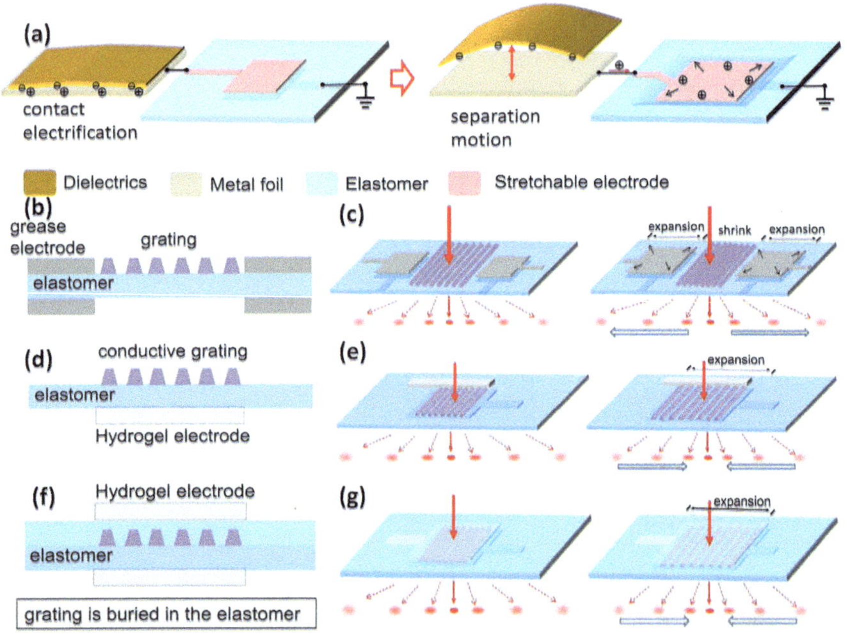

Fig. 4 Structural design, working mechanism, and performance of the TENG-DEA-driven tunable optical grating (TOG) system. **a** Schematic of the TENG-DEA actuation principle, where a single-electrode TENG powers and controls the DEA through contact-separation-induced charge transfer. **b, c** Compression-mode TOG design with DEA-induced grating period modulation via bidirectional strain control. **d, e** Expansion-mode TOG (Mode 1) utilizing a conductive Ag nanoparticle grating array and a transparent hydrogel electrode for optical modulation. **f, g** Expansion-mode TOG (Mode 2) with a PDMS-sealed grating structure, enhancing transparency and deformation efficiency for superior optical tuning. Reproduced with permission from Chen et al. [63]

where E represents the applied electric field, and ε denotes the permittivity of the elastomer. As the Kapton film moves, the actuation of the DEA is triggered, causing expansion. Upon returning to its original position, charge neutralization occurs, leading to a relaxation of the elastomer film, thereby completing one operational cycle of the system. The compression mode TOG consists of two DEA elements and a grating array composed of metallic nanoparticles, as depicted in Fig. 4b, c. The DEA elements are fabricated on an elastomer film, with a grating array printed between them. The elastomer film is constrained by boundary frames, concentrating the strain in the central region. A pre-strain ratio of 600% is applied before grating deposition, ensuring mechanical stability and preventing structural deformation. Under TENG actuation, the DEA electrodes expand, inducing a compressive strain in the passive region, which reduces the grating period, enabling dynamic optical modulation. A novel expansion-based TOG structure was developed, as shown in Fig. 4d, e. This design utilizes a conductive silver (Ag) nanoparticle grating array, which also functions as an electrode for DEA actuation. A transparent hydrogel film serves as the opposing electrode, sandwiching the elastomer film between the grating and hydrogel electrodes. The hydrogel exhibits 95% optical transparency, allowing the fabrication of a transparent TOG. Under TENG-driven actuation, the hydrogel and grating expand, increasing the grating period, thereby achieving self-powered optical modulation. However, due to the limited tensile strength of the Ag nanoparticle grating, the expansion strain is insufficient for highly effective optical tuning. To overcome the limitations of the Expansion Mode 1 TOG, a modified structure was designed, as illustrated in Fig. 4f, g. In this configuration, the grating arrays were printed onto a pre-strained elastomer film, followed by the deposition of a 100 μm-thick PDMS protective layer, encapsulating the grating array. This PDMS-sealed structure ensures that the hydrogel electrodes do not blur the grating pattern, maintaining high optical clarity. The top and bottom hydrogel electrodes are then applied, forming a fully stretchable and transparent TOG system. Under TENG-driven actuation, the elastomer film and grating array expand simultaneously, leading to effective optical modulation. The hydrogel electrodes and elastomer's high flexibility enable large-scale deformation, making this design highly efficient for optical tuning applications. The TENG-DEA-driven TOG system provides an energy-efficient and scalable approach for self-powered optical modulation. The three TOG designs compression mode, expansion mode 1, and expansion mode 2 demonstrate the versatility of TENG-driven DEA actuation for dynamic optical tuning. Among these, the Expansion Mode 2 TOG, which incorporates hydrogel electrodes and PDMS encapsulation, exhibited superior mechanical deformation, making it the most promising candidate for next-generation reconfigurable optical elements. The integration of TENGs as both a power source and control mechanism enables adaptive optical materials with tunable properties. These TOG systems hold great potential for smart display technologies, reconfigurable photonic circuits, adaptive lenses, and precision optical sensing. Furthermore, the transparent, stretchable, and highly tunable hydrogel electrode design represents a breakthrough in flexible optoelectronic device engineering, paving the way for wearable photonic interfaces, biomedical imaging, and autonomous optical modulation systems [63].

2.3 Two-Dimensional Tunable Optical Grating (2D TOG) Systems Based on TENG-DEA Actuation

The development of two-dimensional (2D) tunable optical grating (TOG) systems based on TENG-DEA actuation enables dynamic optical modulation through mechanical strain control. By extending the metallic nanoparticle-based grating array fabrication technique, a 2D TOG matrix was developed and successfully manipulated using TENG-driven DEA actuation. The structural design and operating principle of the 2D TOG systems are schematically illustrated in Fig. 5. The compression-mode 2D TOG consists of a double-layer Fe nanoparticle grating array, printed on both the top and bottom surfaces of an elastomer film, forming a crossed grating structure. This double-layer grating produces a laser diffraction matrix on a projection plane, as shown in Fig. 5a. The actuation of the TENG-DEA system leads to the expansion of two DEA elements embedded in the elastomer film, resulting in a dynamic modulation of the laser matrix. The diffraction pattern exhibits a bidirectional response, with laser spots moving outward along the X-axis while contracting along the Y-axis. This compression-based actuation allows for precise, self-powered optical tuning using a single TENG-driven DEA mechanism, enabling multifunctional modulation within a single optical grating system. An alternative expansion-mode 2D TOG system was designed, as depicted in Fig. 5b. Here, a PDMS layer was spin-coated on both the top and bottom surfaces of the elastomer film, effectively sealing the grating arrays within the PDMS structure. Two hydrogel electrodes were attached on both sides of the PDMS-sealed grating arrays, like the design in Fig. 4f. The expansion of the hydrogel electrodes under TENG-driven DEA actuation causes the grating arrays to expand, leading to an inward movement of the diffracted laser matrix along both the X-axis and Y-axis. A photographic demonstration of the diffraction effect generated by this double-layer grating system is shown in Fig. 5c. The modulation dynamics of the 2D TOG system in compression mode were analyzed under TENG-driven actuation, as illustrated in Fig. 5d. As the separation height of the TENG increases, the grating period undergoes directional modulation along both axes. The largest observed period change along the X-axis was −12% (decrease), while the Y-axis exhibited an increase of 14%. The slightly lower period variation in the X-direction, compared to single-layer grating configurations, suggests that the repeated printing processes may have slightly altered the mechanical properties of the elastomer film, reducing its stretchability. Nonetheless, the compression-mode 2D TOG system effectively enables bidirectional modulation, demonstrating its applicability in real-time optical tuning. To further investigate the actuation performance of the expansion-mode 2D TOG, experiments were conducted using both a high-voltage source and TENG-driven operation. Under a breakdown voltage of 3800 V, the X-axis and Y-axis grating periods changed by 10.7% and 10.2%, respectively, as shown in Fig. 5e. However, the multilayer PDMS structure significantly increased both the thickness and resistance of the device, necessitating a higher operating voltage. Additionally, the stiffness of the PDMS layer influenced the actuation efficiency, reducing the overall deformation achievable within the system. When the same 2D TOG system (expansion mode 2) was driven by TENG-generated voltage, the maximum grating period change was 6.9% along the X-axis and 6.5% along the Y-axis, as depicted in Fig. 5f. The V_{oc} from the TENG was measured at 3000 V, which was insufficient to fully actuate the multilayer TOG system. This limitation suggests that the increased thickness and stiffness of the PDMS-sealed structure reduced the overall actuation efficiency when operated under a self-powered TENG mechanism. The findings confirm that nanoparticle-based grating arrays, which do not significantly increase the thickness or stiffness of the elastomer film, are particularly well-suited for self-powered TOG applications. In contrast, multilayer PDMS-based grating designs, while offering improved mechanical durability and environmental protection, require higher operating voltages, limiting their compatibility with TENG-driven actuation. To enhance the actuation efficiency of the TENG-driven 2D TOG system, several optimization strategies can be considered: Scaling up the TENG triboelectric surface area A larger TENG device with an increased contact area can generate higher output voltages, enabling full actuation of multilayer TOGs. Improving the dielectric properties of the elastomer Surface modification techniques such as corona polarization can be employed to enhance charge retention and increase TENG output performance. Optimizing PDMS composition for flexibility Modifying the crosslinking density of PDMS can reduce stiffness while maintaining its mechanical durability, improving the deformation capabilities of PDMS-sealed TOG systems. The TENG-driven 2D TOG system successfully demonstrated bidirectional laser matrix modulation, with distinct compression-mode and expansion-mode designs enabling real-time optical tuning. The nanoparticle-based gratings exhibited superior compatibility with self-powered actuation, whereas multilayer PDMS-sealed TOG designs required higher operating voltages for full functionality. Future advancements in TENG optimization, material engineering, and device scalability will further improve the feasibility of self-powered adaptive optical systems, paving the way for next-generation photonic interfaces, tunable displays, and smart optical sensors [63].

3 Piezoelectric Nanogenerator

PENGs have gained considerable attention in optoelectronics due to their ability to harvest mechanical energy and convert it into electrical energy, enabling self-powered optical devices. Such devices have found applications in photodetectors, LEDs, and optical sensors. The integration of piezoelectric materials with semiconductors facilitates charge carrier separation, thereby enhancing energy harvesting efficiency. The development of hybrid composites incorporating piezoelectric and optoelectronic properties has resulted in improved device performance and broader applicability in flexible and wearable technologies. The demand for self-powered photo detection systems has surged due to the increasing reliance on portable and wearable electronics. Conventional photo detectors require external power sources, limiting their utility in autonomous systems. PENG-based optical devices provide an innovative solution by coupling piezoelectric and optoelectronic effects,

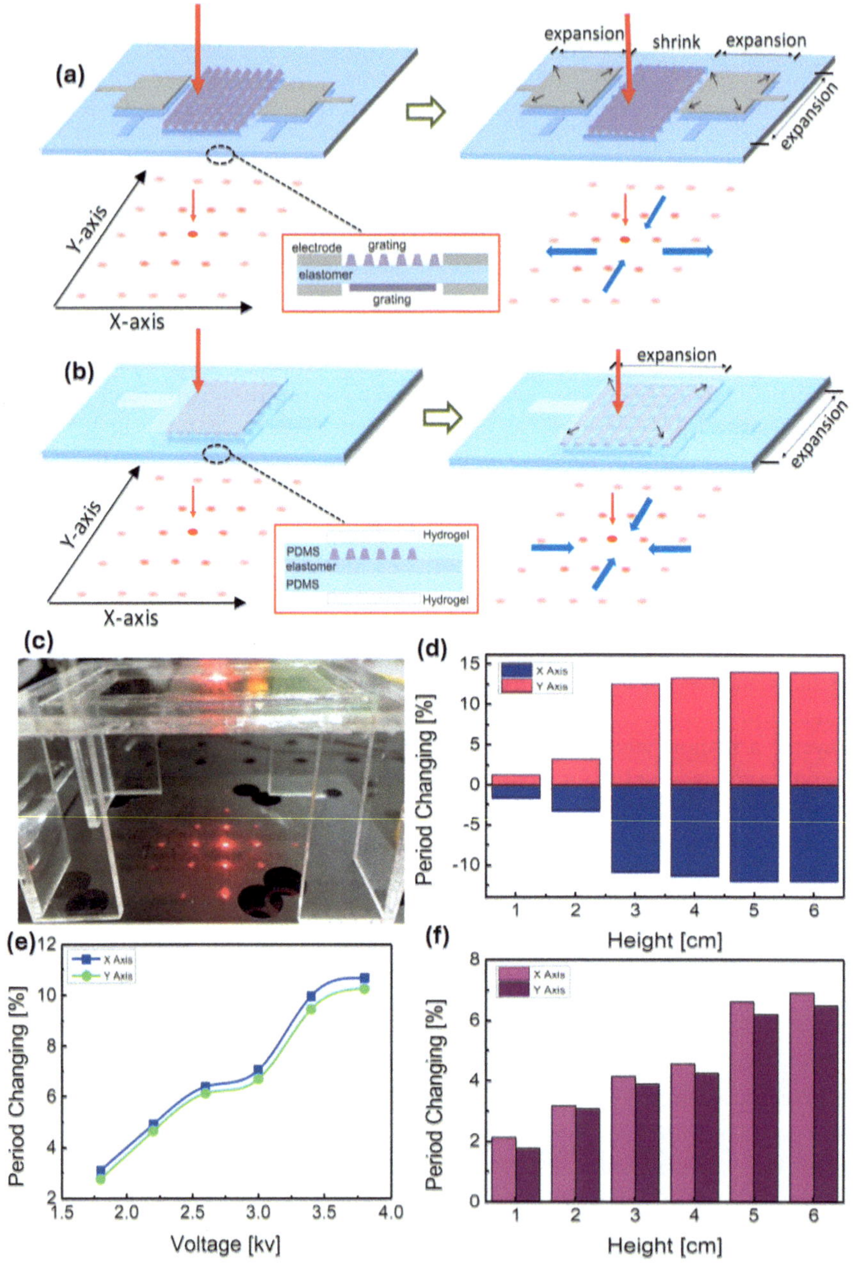

Fig. 5 Structural design, working principle, and performance analysis of the TENG-driven 2D tunable optical grating (TOG) system. **a** Schematic representation of the compression-mode TOG, featuring a double-layer Fe nanoparticle grating array and bidirectional laser diffraction modulation. **b** Expansion-mode TOG design with PDMS-sealed grating arrays and hydrogel electrodes for strain-induced optical tuning. **c** Photographic demonstration of the diffraction effect from the 2D TOG system. **d** Compression-mode TOG modulation dynamics under TENG actuation, showing directional grating period changes along the X- and Y-axes. **e** Expansion-mode TOG actuation using a high-voltage source, illustrating grating period variations. **f** Expansion-mode TOG response under TENG-driven operation, highlighting reduced actuation efficiency due to increased PDMS thickness and stiffness. Reproduced with permission from Chen et al. [63]

thereby enabling energy harvesting from ambient mechanical stimuli. This chapter comprehensively explores the advancements in PENG-powered optical devices, focusing on metal oxide-based hybrid composites. The discussion will include material synthesis, structural characterization, working mechanisms, and potential applications in next-generation optoelectronics.

4 Materials and Characterization

4.1 Structural and Optical Properties

The structural and optical characteristics of the synthesized metal oxide-based hybrid composites play a crucial role in determining their piezoelectric and optoelectronic efficiency. Scanning electron microscopy (SEM), X-ray diffraction (XRD), Raman spectroscopy, and Fourier-transform infrared (FTIR) spectroscopy were employed to analyze the morphology, phase composition, and molecular interactions of the composite materials. UV–Vis spectroscopy was utilized to investigate the optical absorption and bandgap properties, which directly influence the device's photo detection capabilities. Understanding these fundamental properties provides valuable insights into optimizing material performance for efficient energy conversion and self-powered applications.

Surface Morphology of SnS_2 Nanoparticles The structural characteristics of the synthesized SnS_2 nanoparticles were analyzed using scanning electron microscopy (SEM). As depicted in Fig. 6a, the SnS_2 nanoparticles exhibit a densely packed, aggregated texture with a granular morphology. The individual nanoparticles, with an average diameter of 150–180 nm, display small grain-like structures, whose formation is influenced by reaction temperature and growth kinetics. These morphological characteristics directly impact nucleation processes and crystal growth, which in turn affect optoelectronic and piezoelectric properties. To evaluate the optical properties of SnS_2, UV–Vis spectroscopy was employed. The absorption spectrum, shown in Fig. 6b, spans the 300-900 nm wavelength range, with a distinct absorption edge at 510 nm, indicative of a direct band transition. The bandgap energy (E_g) was determined using Tauc's plot of $(\alpha h\nu)^2$ versus ($h\nu$), as presented in Fig. 6c, yielding a value of ~2.4 eV, consistent with reported literature. This bandgap aligns well with the visible-light absorption range, making SnS_2 suitable for self-powered optoelectronic applications. The crystalline structure and phase composition of SnS_2, PVDF films, and SnS_2/PVDF composites were examined via X-ray diffraction (XRD), as illustrated in Fig. 6d(i-ii). The diffraction pattern of PVDF exhibits two dominant peaks at $2\theta = 20.4°$ and 19°, corresponding to the β-phase crystalline structure of PVDF. The XRD pattern of SnS_2 confirms its pure hexagonal phase, with diffraction peaks indexed according to JCPDS No. 89–3198. Notably, the XRD spectrum of the SnS_2/PVDF composite (Fig. 6d(iii)) reveals the coexistence of both the β-phase reflections of PVDF and hexagonal SnS_2 peaks, with an intensified β-phase peak at

20.4°, suggesting enhanced piezoelectricity due to SnS_2 incorporation. Raman spectroscopy was employed to further examine structural and phase composition changes. As illustrated in Fig. 6e(i), PVDF exhibits characteristic Raman peaks at 790, 840, and 870 cm^{-1}, corresponding to its α-phase, β-phase, and γ-phase, respectively. The Raman spectrum of SnS_2 (Fig. 6e(ii)) displays a dominant peak at 311 cm^{-1}, associated with the A_1g vibrational mode of hexagonal SnS_2, while a weaker E_g peak at 200 cm^{-1} is attributed to nanoscale effects. Upon incorporation into PVDF (Fig. 6e(iii)), the non-polar α-phase decreases, whereas the polar β-phase increases, confirming the piezoelectric enhancement induced by SnS_2. The increasing β-phase fraction (Fβ) with rising SnS_2 concentration indicates a strong interaction between SnS_2 nanofillers and the PVDF matrix, which improves charge transfer efficiency and piezoelectric properties. To investigate chemical bonding interactions, FTIR spectroscopy was performed, as shown in Fig. 6f. The FTIR spectrum reveals dominant PVDF absorption peaks, overshadowing a broad Sn-S stretching vibration peak at ~540 cm^{-1}, confirming the successful integration of SnS_2 into the PVDF matrix. This interaction is critical for stabilizing charge transfer at the composite interface, further enhancing piezoelectric and optoelectronic properties. Working Mechanism of the Piezoelectric-Photovoltaic Nanogenerator (PPENG) A schematic cross-section of the piezoelectric-photovoltaic nanogenerator (PPENG) is provided in Fig. 6g, illustrating its working principle under dark and illuminated conditions. In the absence of illumination (Fig. 6h(i)), the internal dipoles within the composite are randomly oriented. Upon applying mechanical strain (Fig. 6h(ii)), piezoelectric charges are generated at the SnS_2-PVDF and ITO electrode interfaces, leading to dipole realignment and the development of an electromotive force (EMF). This process results in alternating positive and negative voltage outputs during compression and relaxation cycles, primarily due to interfacial polarization between PVDF and SnS_2. Under illumination conditions, photogenerated charge carriers significantly influence the piezoelectric response of the composite (Fig. 6i(i)). Compared to the dark state, incident light enhances the piezoelectric potential, leading to an increased output voltage (Fig. 6i(ii&iii)). This effect is attributed to the photovoltaic properties of SnS_2, which contribute to improved charge separation and piezoelectric output enhancement. The photoresponse of the PPENG device was examined under green light illumination (510 nm, 10.2 mW/cm^2), as shown in Fig. 6j–l. The device exhibits a strong ON–OFF switching response during periodic light exposure (Fig. 6j), where the open-circuit voltage increases under illumination and returns to baseline when the light is turned off. This behavior is attributed to the excitation of electron–hole pairs, enhancing charge separation and photocurrent generation. The presence of SnS_2 nanofillers within the PVDF matrix facilitates efficient energy conversion and charge transport, further improving the conductivity and polarization effects. As illustrated in Fig. 6k, reducing the light intensity from 10.2 mW/cm^2 to dark conditions leads to a gradual decrease in output voltage, confirming the photodetection capabilities of the PPENG. The variation in open-circuit voltage (V_{oc}) with light intensity (Fig. 6l) indicates a linear increase from 24.4 V to 72.9 V, confirming enhanced photogenerated charge separation due to the piezoelectric effect in the SnS_2/PVDF hybrid composite.

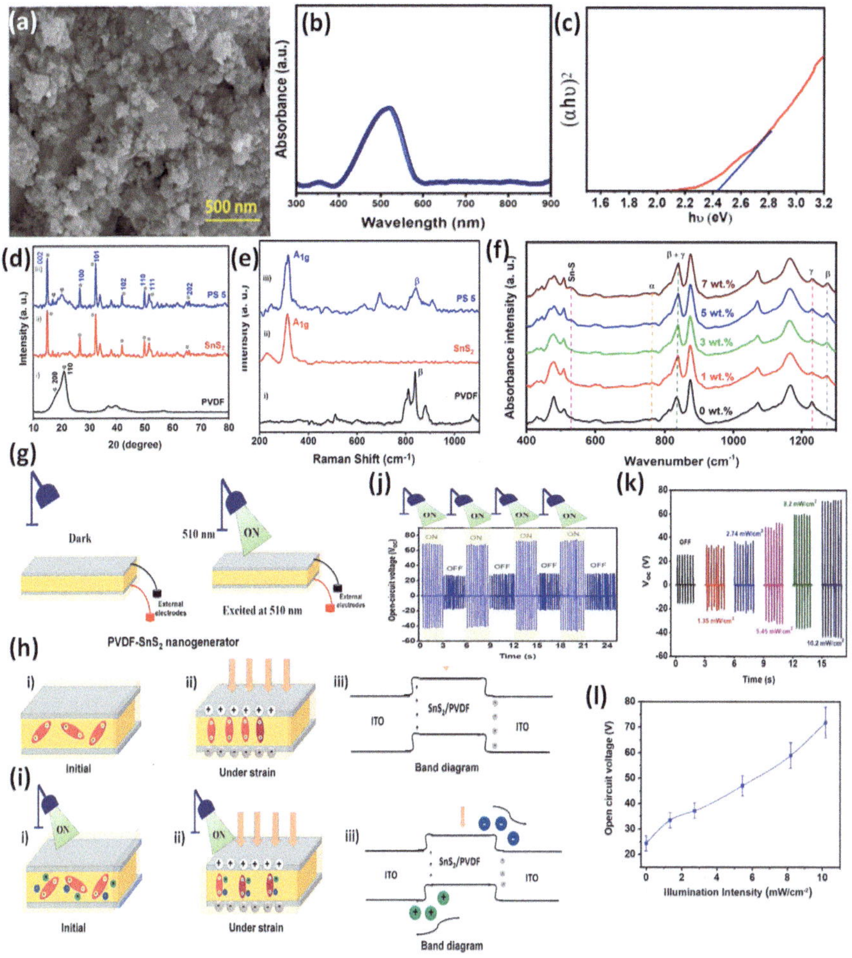

Fig. 6 Structural, optical, and piezoelectric characteristics of SnS_2 nanoparticles and their integration into PVDF. **a** SEM image of SnS_2 nanoparticles showing granular morphology. **b** UV–Vis absorption spectrum with a bandgap of ~2.4 eV. **c** Tauc's plot confirming direct band transition. **d** XRD patterns of PVDF, SnS_2, and SnS_2/PVDF composites indicating phase composition. **e** Raman spectra highlighting phase transformation in SnS_2/PVDF. **f** FTIR spectrum confirming SnS_2-PVDF interactions. **g** Schematic of PPENG structure. **h–i** Piezoelectric response under dark and illuminated conditions. **j–l** Photovoltaic-enhanced piezoelectric output under green light illumination. Reproduced with permission from Muduli et al. [64]

The integration of piezoelectric and photovoltaic mechanisms in hybrid materials has been extensively investigated, with SnS_2/PVDF composites demonstrating superior performance in self-powered photodetection, charge separation efficiency, and energy harvesting applications. The enhanced β-phase fraction of PVDF, combined with the optical absorption and charge transport properties of SnS_2, makes this composite an ideal candidate for next-generation optoelectronic and energy-harvesting applications. Despite these promising results, several challenges remain, particularly concerning long-term stability, large-scale fabrication, and material composition optimization to maximize device performance. Future research should focus on Interface engineering to improve charge transfer dynamics. Nanostructure optimization to achieve a higher β-phase fraction in PVDF. Alternative dopants or multi-component composites to enhance piezoelectric-photovoltaic coupling effects. By bridging the gap between energy harvesting and optoelectronics, SnS_2/PVDF-based piezoelectric-photovoltaic nanogenerators offer a promising route toward self-powered sensors, flexible photodetectors, and sustainable energy solutions [64].

4.2 Fabrication and Performance of the CdS:P_3HT Microwire Photodetector

The CdS:P_3HT microwire photodetector was fabricated following a standard method, as detailed in the experimental section. Figure 7a, b present the digital and optical microscopic images of the device, Fig. 7c illustrates the overall device structure and experimental setup. The device was designed to undergo tensile strain, with one end fixed on a rigid plate using a 3 cm long gasket, while the free end was stretched to apply controlled strain. Different tensile strain levels were introduced by varying the gasket thickness. The fabrication process involved depositing an ultralong (~0.1 cm) CdS nano/microwire at the center of a cleaned polystyrene (PS) substrate (dimensions: 6 cm × 0.6 cm × 0.1 cm), ensuring alignment along the long axis of the substrate. One end of the microwire was securely fixed using epoxy resin, while the device underwent oxygen plasma treatment for 5 min to enhance surface properties. A 30 mg/mL solution of P_3HT in chlorobenzene was then drop-cast onto the fixed end of the CdS microwire, forming a p–n junction, while the free end was electrically contacted using silver paste. To protect the microwire from mechanical damage and environmental contamination, the device was encapsulated with a thin layer of PDMS.To evaluate piezo-photodetection performance, a measurement system was established Fig. 7c. The device was mounted onto a rigid plate with varying gasket thicknesses to induce tensile strain. The [001] crystallographic direction of the CdS microwire was in direct contact with P_3HT, and tensile strain (ε) was quantified using the relation:

$$\varepsilon = \frac{h}{2R}$$

where h represents the substrate thickness, and R denotes the local bending radius. Since the CdS microwire was considerably shorter than the PS substrate, it experienced a pure tensile strain when the substrate was bent. The photodetection capability of the fabricated device was assessed under UV illumination (365 nm) across different light intensities (0.154 mW/cm^2 to 3.079 mW/cm^2) Fig. 7d–h. The current–voltage

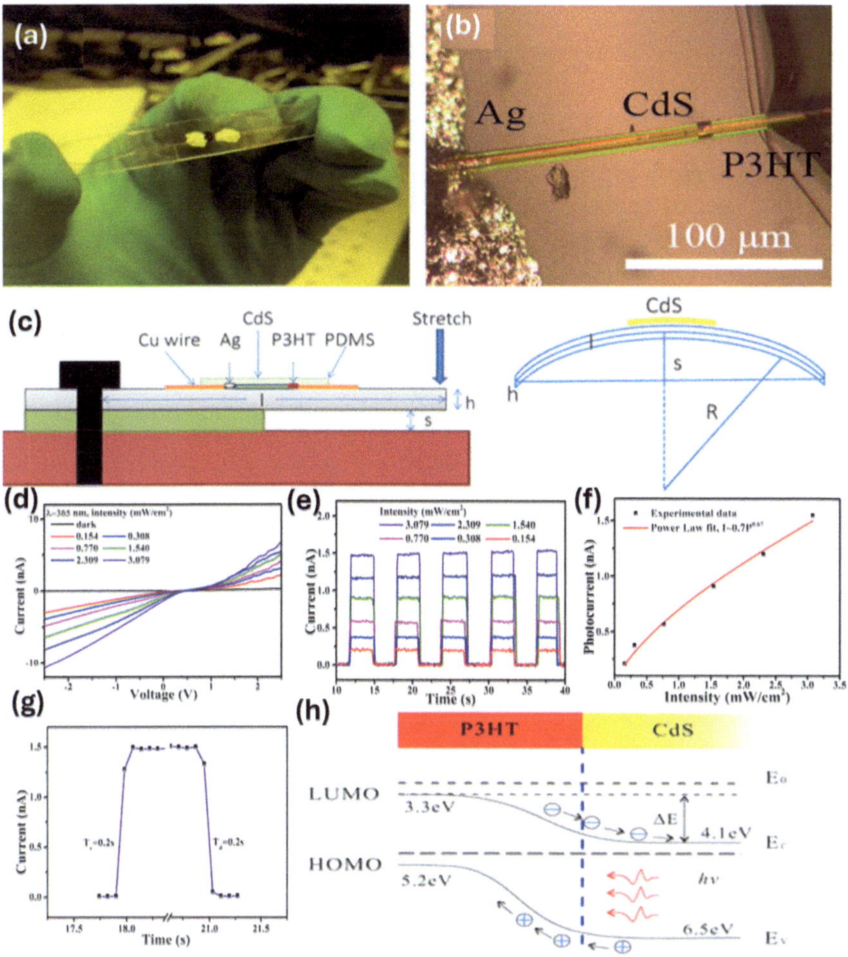

Fig. 7 Elaborates structural and optoelectronic characterization of the CdS:P3HT microwire photodetector. **a–b** Digital and optical microscopic images of the fabricated device. **c** Schematic illustration of the device structure and experimental setup for strain-induced measurements. **d** I-V characteristics under varying UV light intensities. **e** On–off current response demonstrating enhanced photocurrent. **f** Photocurrent versus light intensity showing a power-law dependence. **g** Fast response time (~0.2 s) for rise and decay. **h** Sensitivity comparison highlighting superior performance due to the CdS-P3HT heterojunction. Reproduced with permission from Yu et al. [65]

(I-V) characteristics in Fig. 7d indicate a clear photovoltaic response, with the V_{oc} and I_{sc} increasing proportionally with light intensity. Figure 7e illustrates the on–off current response at zero bias voltage, revealing a substantial increase in photocurrent, from approximately 1 pA (dark current) to 200 pA (0.154 mW/cm^2) and further rising to 1500 pA (3.079 mW/cm^2). The photocurrent versus light intensity curves in Fig. 7f demonstrate a continuous increase without saturation, following a power-law relationship:

$$I \sim 0.7P^{0.67}$$

where I represents the photocurrent and P denotes light intensity. This nonlinear relationship suggests efficient charge separation and transport mechanisms within the heterojunction. The response time of the device was measured to be less than 0.2 s for both rise and decay times, as shown in Fig. 7g, demonstrating its rapid switching capability. To determine the origin of the photovoltaic behavior, the I-V curves of a CdS microwire with both ends connected via silver paste were recorded. This control experiment exhibited Ohmic contact with no significant photovoltaic response, except for minor I-V asymmetry, which was attributed to imperfect contact at the CdS-silver interface. However, when the CdS microwire was integrated with P_3HT, a distinct rectifying behavior emerged, confirming that the p–n junction at the CdS-P_3HT interface is the primary source of the photovoltaic effect. Compared to conventional CdS-based photodetectors, the fabricated CdS:P_3HT microwire photodetector exhibits an exceptional sensitivity of 1500 under UV illumination (3.079 mW/cm^2), significantly surpassing previously reported CdS nanowire- and nanobelt-based photoconductive detectors. This enhancement is attributed to optimized device architecture and band alignment at the CdS-P_3HT heterojunction Fig. 7h. The device maintains a low dark current, owing to its zero applied bias voltage, which minimizes unwanted carrier injection. Furthermore, the P_3HT coating around the CdS microwire facilitates the formation of a type-II heterojunction, promoting spatial separation of photogenerated carriers and effectively suppressing electron–hole recombination. Similar heterojunction-based enhancements have been observed in other CdS nanowire heterostructures, further validating the approach. The combination of rapid response time, high sensitivity, and efficient charge carrier separation establishes the CdS:P_3HT microwire photodetector as a promising candidate for next-generation industrial photodetection applications, particularly in wearable optoelectronics, UV sensing, and self-powered detection systems. The combination of fast response time, high sensitivity, and efficient charge carrier separation establishes the CdS:P_3HT microwire photodetector as a promising candidate for next-generation industrial photodetection applications, particularly in: Wearable optoelectronics UV sensing and environmental monitoring Self-powered detection systems The integration of strain-tunable properties further extends its potential for flexible and stretchable photodetectors, making it ideal for biomedical sensing and smart textile applications. Despite its high performance, future improvements should focus on: Enhancing device stability and longevity through passivation layers or surface modifications. Optimizing the heterojunction interface to reduce charge recombination losses. Exploring alternative organic semiconductors to further improve absorption and charge transport properties. By addressing these challenges, CdS:P_3HT-based photodetectors can evolve into scalable and practical solutions for commercial optoelectronic devices [65].

4.3 Strain-Dependent Optoelectronic Properties of the CdS/P_3HT Heterojunction

Electrical Measurement System for Strain-Induced Modulation The effect of tensile strain on the electrical and optoelectronic properties of the CdS/P_3HT heterojunction was systematically investigated using a controlled bending setup, as illustrated in Fig. 8a. The CdS microwire/P_3HT heterojunction was subjected to external mechanical strain, enabling the study of strain-induced modulation of charge carrier transport. As the applied tensile strain increases, the piezoelectric effect alters the local band structure, significantly influencing carrier dynamics at the heterojunction interface.

Barrier Height Modulation Due to Tensile Strain To quantify the influence of strain on the electronic transport properties, the barrier height variation was analyzed using thermionic emission–diffusion theory, as depicted in Fig. 8b. The results indicate a progressive increase in barrier height with increasing strain, confirming that strain engineering effectively modulates charge carrier injection and recombination processes. This strain-induced change in barrier height is crucial for tuning optoelectronic performance in flexible photodetectors. Strain-Dependent Photodetection Response The strain-dependent photodetection capability of the device was evaluated under UV illumination (365 nm, 0.0115 mW/cm^2) at zero bias voltage, as shown in Fig. 8c. The results reveal a significant increase in photocurrent under UV exposure, demonstrating efficient optical-to-electrical signal conversion. A similar enhancement in photocurrent was observed under visible light illumination (Fig. 8d), confirming that applied tensile strain improves charge separation and transport efficiency in the heterojunction. However, under near-infrared (NIR) light (780 nm, 0.0225 mW/cm^2), the device exhibited minimal variation in photocurrent with applied strain (Fig. 8e). This is attributed to the weak absorption of CdS and P_3HT in the NIR region, where the incident photon energy is insufficient to induce band-to-band transitions. The negligible response in the NIR range further confirms that the strain-dependent enhancements are primarily due to piezoelectric-phototronic coupling, rather than thermal or defect-induced effects. Dark Current Variation Under Tensile Strain To further examine the underlying mechanisms, the dark current variation as a function of tensile strain was analyzed, as shown in Fig. 8f. The results indicate a gradual increase in dark current with increasing strain, suggesting that the strain-induced piezopotential modulates charge carrier transport at the heterojunction. This confirms that mechanical strain can effectively tune the electrical conductivity of the device. Barrier Height Extraction from I-V Characteristics A more detailed analysis of the barrier height variation was conducted using I-V characteristics under different strain conditions, as illustrated in Fig. 8g. A decrease in barrier height with increasing strain was observed, which correlates with the increase in dark current. This trend confirms that the piezoelectric effect plays a dominant role in tuning charge carrier behavior, leading to improved charge transport efficiency at the heterojunction Fig. 8h. Enhancement of Photocurrent and Sensitivity Under Strain The strain-induced enhancement in photocurrent and sensitivity was

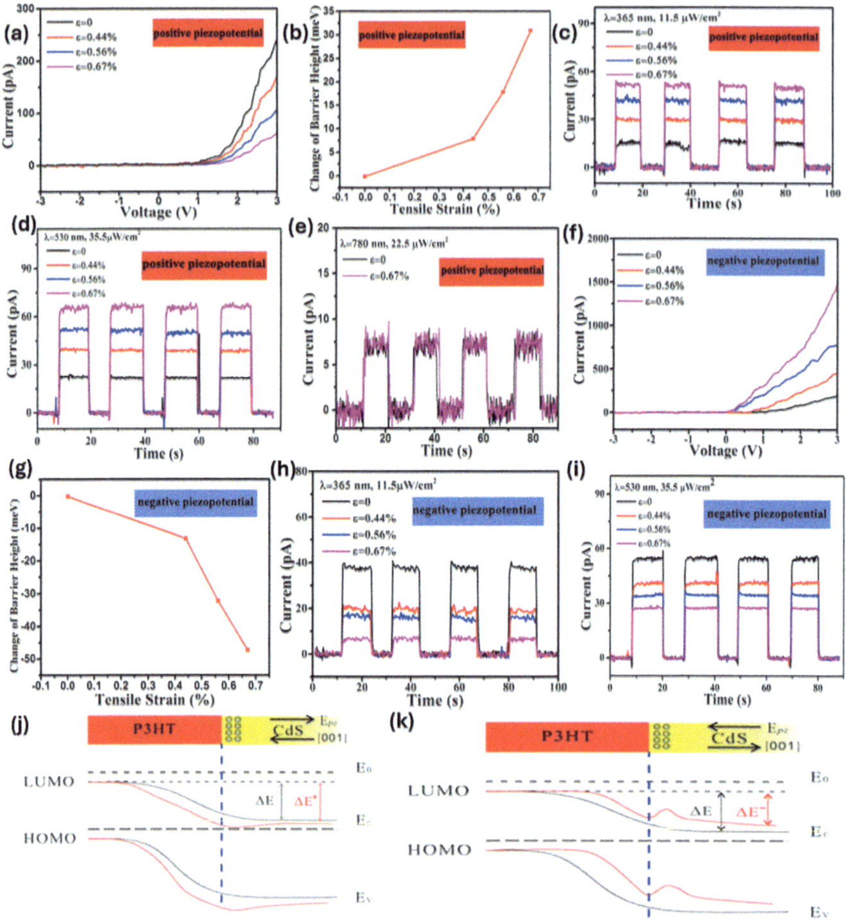

Fig. 8 Strain-dependent electrical and optoelectronic performance of the CdS:P3HT microwire photodetector. **a** Schematic of the electrical measurement setup for strain-induced modulation studies. **b** Barrier height variation with tensile strain based on thermionic emission–diffusion theory. **c–e** Photocurrent response under UV (365 nm), visible, and NIR (780 nm) illumination, highlighting enhanced charge separation under strain. **f** Dark current variation with tensile strain, showing strain-induced modulation of charge transport. **g** Barrier height reduction with increasing strain, confirming piezoelectric modulation. **h** Photodetector on–off response under different light conditions. **i** Sensitivity enhancement due to strain, showing a 230% increase in photocurrent at 0.67% strain. **j** Energy band diagram explaining strain-induced changes in conduction band alignment and charge transport efficiency. **k–l** Mechanism of piezopotential-driven carrier dynamics and its effect on photocurrent enhancement. Reproduced with permission from Yu et al. [65]

further quantified using on–off response measurements under different light conditions, as shown in Fig. 8i. Both photocurrent and sensitivity exhibited a significant increase with applied tensile strain, particularly under UV and visible light illumination at zero bias voltage. This enhancement is further illustrated in Fig. 8j, where a 230% increase in photocurrent and a 2000-fold improvement in sensitivity were observed at 0.67% tensile strain. Notably, despite these enhancements, response time and dark current remained largely unchanged, suggesting that strain primarily influences charge generation and transport rather than overall device stability. Energy Band Diagram Explaining Strain-Induced Modulation The fundamental mechanisms governing the strain-dependent optoelectronic response were analyzed using an energy band diagram, as depicted in Fig. 8k. When the [001] crystallographic direction of the CdS microwire is in contact with P_3HT, the piezopotential induced by tensile strain alters the local conduction band alignment. Under tensile strain, a negative piezopotential forms at the heterojunction interface, leading to a reduction in barrier height (E to E^-), thereby facilitating more efficient charge transport across the heterojunction. Conversely, a positive piezopotential increases the barrier height (E to E^+), reinforcing the built-in electric field, which enhances photoexciton separation and charge extraction efficiency. Influence of Strain on Photodetection Performance Across Wavelengths The influence of strain on photodetection efficiency across different wavelengths was further analyzed in Fig. 8l. The results demonstrate that the photocurrent and sensitivity increase under UV and visible light, whereas the response remains unchanged in the NIR region due to insufficient photon energy for band-to-band electronic transitions. This confirms that the strain-induced enhancements arise from piezoelectric-phototronic coupling rather than extrinsic photothermal effects. The results demonstrate that mechanical strain can effectively tune the optoelectronic properties of CdS/P_3HT heterojunction photodetectors through strain-induced piezoelectric-phototronic coupling. The ability to modulate charge carrier transport and barrier height using external mechanical strain opens new possibilities for flexible, tunable, and self-powered photodetectors. The key findings from this study highlight: Enhanced photocurrent and sensitivity under tensile strain, with a 230% increase in photocurrent and over 2000-fold sensitivity enhancement at 0.67% strain. Strain-induced barrier height modulation, facilitating more efficient charge carrier transport and reduced recombination losses. Selective enhancement under UV and visible light, while NIR response remains unchanged, confirming that the effect is intrinsic to piezoelectric-phototronic coupling. Future work should focus on: Exploring alternative organic semiconductors to further optimize heterojunction properties. Scaling up the device architecture for real-world flexible optoelectronic applications. Integrating strain-tunable photodetectors into wearable sensing platforms and adaptive optical systems. By leveraging strain engineering in piezoelectric-organic heterojunctions, this work provides a new pathway for next-generation flexible and tunable photodetection technologies, with potential applications in biomedical imaging, optical sensing, and smart optoelectronics [65].

4.4 Impedance Matching and Theoretical Modeling of S-TENG

The efficient operation of a Sliding-Mode Triboelectric Nanogenerator (S-TENG) is critically influenced by impedance matching with external resistance, a key parameter governing its electrical output and sensing capabilities. To elucidate this relationship, a theoretical model of the sensing electrode was developed, as illustrated in Fig. 9. The structural diagram and photographic image (Fig. 9a) depict the conventional S-TENG architecture, consisting of an ITO thin-film electrode, a PDMS triboelectric layer, and a flexible PET substrate. The device, with an effective area of 1 × 1 cm^2, exhibits high transparency, making it well-suited for applications demanding both optical clarity and mechanical flexibility. The electricity generation mechanism of the S-TENG follows a cyclic contact-separation process, as schematically represented in Fig. 9b. When an external force is applied, the triboelectric layers come into direct contact, leading to charge transfer via contact electrification (Step i). Upon force release, the layers separate, causing a redistribution of charges, which induces electron flow through the external circuit, generating a peak current (Step ii). As the separation distance increases, the potential difference between the ITO electrode and the ground reaches its maximum (Step iii). When the external force is reapplied, the layers reapproach, facilitating the return of free electrons to the electrode, thus completing the charge cycle (Step iv). A deeper analysis of the impedance matching conditions was conducted to evaluate the relationship between output voltage and external resistance in a series circuit, as shown in Fig. 9c. The results indicate that the output voltage inversely correlates with resistance, with a decline from ~3.2 V to ~0.87 V as the external load resistance (R_x) increases from 1 GΩ to 30 GΩ. Despite variations in operating frequency, the overall trend and working zones remain consistent, suggesting that external resistance plays a significant role in modulating S-TENG output. Replacing R_x with an environment-sensitive electrode significantly alters the voltage profile, demonstrating that changes in electrode resistance due to external stimuli impact impedance matching conditions and, consequently, the energy harvesting efficiency of the system. To provide a theoretical framework for this behavior, an equivalent circuit model for the sensing electrode was developed, as illustrated in Fig. 9d, e. This model incorporates key electrical components, including R_1, which represents sensor resistance variations, and C_1, which accounts for capacitance fluctuations associated with the sensing electrode. Additionally, the model considers C_t, the built-in capacitance formed between circuit nodes due to electrode material properties. Under equilibrium conditions, an electric potential difference (U) is established between the dielectric material and the ground, influencing charge transport dynamics. Traditional TENG models often approximate the system as a parallel-plate capacitor, assuming an ideal metallic electrode with negligible resistance fluctuations. However, in practical applications, the electrode material composition plays a crucial role in governing impedance characteristics. The refined model presented in Fig. 9e accounts for external UV light stimulation, which induces dynamic resistance variations in the photosensitive electrode material. These variations are modeled as an equivalent virtual external load resistance (R_x) in series with the conventional S-TENG, providing a more comprehensive representation of impedance matching effects under environmental stimuli. The incorporation of sensor resistance variations in the model reveals how external factors, such as UV exposure, influence the device's electrical performance. This refined circuit representation enables a deeper understanding of triboelectric charge transfer mechanisms, particularly in self-powered sensing applications, where real-time adaptability to environmental changes is essential. The findings highlight the potential for dynamic tuning of impedance in next-generation energy-harvesting and sensing devices, paving the way for intelligent, adaptive TENG-based technologies in wearable electronics, environmental monitoring, and self-sustained optoelectronic systems [66].

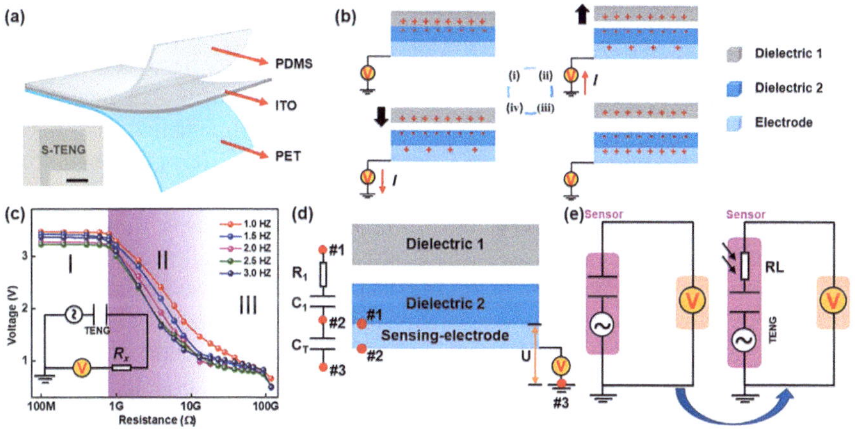

Fig. 9 Structural design, working mechanism, and impedance matching analysis of the Sliding-Mode Triboelectric Nanogenerator (S-TENG). **a** Structural diagram and photographic image of the S-TENG, featuring an ITO electrode, PDMS triboelectric layer, and flexible PET substrate. **b** Schematic representation of the cyclic contact-separation process governing charge transfer and electricity generation. **c** Voltage-output characteristics as a function of external resistance (R_x), highlighting the impact of impedance matching on energy harvesting efficiency. **d, e** Equivalent circuit model illustrating the role of sensor resistance (R_1) and capacitance (C_1) variations under environmental stimuli, such as UV exposure, in modulating S-TENG performance. Reproduced with permission from Peng et al. [66]

5 Device Applications

5.1 Self-Powered UV Photodetector System

The development of self-powered optoelectronic devices has gained significant attention due to their potential for sustainable and autonomous operation. In this context, a wearable self-powered UV photodetection system has been designed, integrating a strain-driven piezoelectric tribo-photodetector (SPTUD) on a flexible PET substrate. The system features two distinct functional zones: one half of the PET substrate is coated with an ITO thin film, which serves as both the electrode of a standard sliding-mode triboelectric nanogenerator (S-TENG) and the reference electrode for the system, while the other half is spin-coated with a ZnO nanoparticle (NP) thin film, acting as the UV-sensitive sensing electrode of the SPTUD. These electrodes are electrically connected in series to independent chip LEDs, where LED1, coupled with the S-TENG, confirms system operation by responding to mechanical motion, and LED2, connected to the SPTUD, detects UV light intensity based on variations in the conductivity of the ZnO NP layer Fig. 10a–c. To ensure stable performance and environmental protection, a PDMS layer is spin-coated on top, functioning as both a triboelectric charge-generating layer and a protective covering. The system's compact and flexible design enables seamless integration into wearable electronics for real-time UV exposure monitoring. For practical implementation, the self-powered UV photodetection system is embedded into a wearable bracelet, allowing continuous UV monitoring in outdoor environments Fig. 10d. The PDMS layer facilitates charge generation through skin contact and separation upon wrist movement, inducing triboelectric charges that drive the LED indicators. This integration enhances the user-friendliness of the system, making it a promising candidate for portable and on-the-go UV detection applications. To validate the system's functionality, a simulated test was conducted Fig. 10e, where the S-TENG converts mechanical energy into an electrical signal via triboelectrification and electrostatic induction, illuminating LED1 to indicate normal system operation Fig. 10f. In contrast, LED2 remains off under ambient conditions but activates under UV illumination due to a decrease in the resistance of the ZnO NP-based sensing electrode Fig. 10g. The system demonstrates high sensitivity, effectively detecting UV light intensities as low as 0.17 mW/cm^2, as shown in Fig. 10h, making it suitable for real-world applications. self-powered UV photodetection system offers a sustainable and efficient solution for real-time UV monitoring without reliance on external power sources. By providing instantaneous feedback on UV intensity, this system enables individuals to take timely protective measures against harmful UV radiation, reducing the risks associated with prolonged exposure. The wearable self-powered UV photodetection system offers a sustainable and efficient solution for real-time UV monitoring, eliminating the need for external power sources. By providing instantaneous feedback on UV intensity, this system enables individuals to take timely protective measures against prolonged UV exposure risks. Key advantages of this system include: Compact, flexible design for seamless integration into wearable electronics. Triboelectric and piezoelectric energy harvesting mechanisms, ensuring continuous self-powered operation. Real-time UV detection, enabling on-the-go monitoring and user safety. Future work should focus on: Optimizing ZnO NP thin-film sensitivity to expand detection capabilities across broader UV intensity ranges. Exploring additional hybrid nanomaterials for enhanced multi-wavelength sensing. Developing integrated wireless communication for real-time data transmission to mobile devices. By combining mechanical energy harvesting with optoelectronic sensing, this system paves the way for next-generation autonomous wearables, environmental monitoring tools, and self-sustained healthcare applications [66].

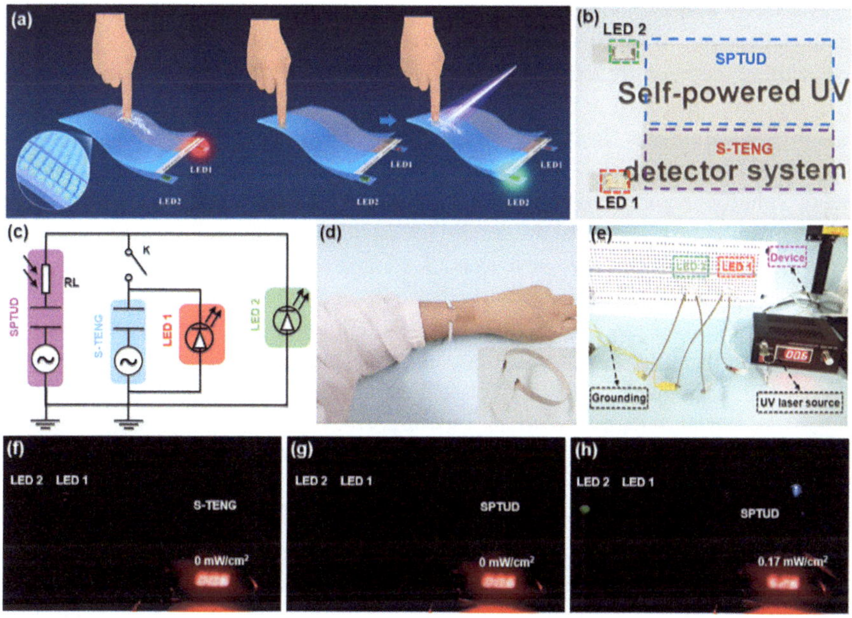

Fig. 10 Wearable self-powered UV photodetection system integrating a strain-driven piezoelectric tribo-photodetector with a TENG. **a–c** Schematic illustration of the device structure, showing the PET substrate with distinct functional zones: an ITO-coated electrode for triboelectric energy harvesting and a ZnO NP-coated sensing electrode for UV detection. **d** Wearable bracelet implementation for real-time UV monitoring. **e** Schematic of the working mechanism, where mechanical motion drives the triboelectric charge generation. **f** LED1 illumination confirms system operation via S-TENG output. **g** LED2 activation under UV exposure due to reduced resistance in the ZnO NP layer. **h** Sensitivity analysis demonstrating UV detection capability at intensities as low as 0.17 mW/cm^2. The system enables portable and autonomous UV monitoring for wearable electronics and healthcare applications. Reproduced with permission from Peng et al. [66]

5.2 *Flexible Self-Powered Photodetector for UV Sensing Applications*

The development of self-powered photo detectors capable of real-time UV monitoring has garnered significant attention for applications in wearable electronics, environmental sensing, and healthcare. Among various material strategies, hybrid nanocomposites integrating piezoelectric and semiconducting properties have demonstrated superior energy harvesting and optoelectronic performance. In this context, a Flexible Self-Powered Photodetector (F-SPPD) has been developed, utilizing Zinc Oxide Nanorods (F-ZnO NRs) grown on Zn^{2+}-seeded Polyvinylidene Fluoride (S-PVDF) film as the active layer. This configuration enables the simultaneous conversion of mechanical and optical energy, facilitating sustainable operation without the need for external power sources. The overall device architecture is schematically illustrated in Fig. 11a, where an optically transparent PMMA layer is incorporated to protect the nanorods from structural deformation during electrode deposition. The synthesis of the S-PVDF film yields a unique microsized pothole-like morphology, characterized by a contact angle (θC) of 85° Fig. 11b, inset. This porosity results from the rapid evaporation of acetone during film formation, leading to the creation of cavities that serve as nucleation sites for ZnO NR growth through a hydrothermal process. The ZnO nanorods exhibit a grassland-like structure, with an average length of ~2 μm and a diameter of ~100 nm Fig. 11c. Notably, the nanorod integration enhances the hydrophobicity of the surface, increasing the contact angle to 118° Fig. 11c, inset, which may contribute to improved environmental stability and charge carrier dynamics. Further structural analysis reveals that nanorod growth is predominantly localized within the porous regions of the S-PVDF film, forming a zigzag internal growth pattern Fig. 11d, e. A comparative study using pristine PVDF without Zn^{2+} seeding resulted in negligible ZnO NR growth, confirming the essential role of S-PVDF in enabling efficient nanorod formation. One of the primary challenges in designing flexible optoelectronic devices is maintaining structural integrity under mechanical deformation. The fabricated F-SPPD exhibits exceptional flexibility, as demonstrated in Fig. 11f, where it withstands various bending stresses without compromising structural integrity. The active layer thickness of ~60 μm ensures mechanical robustness while maintaining lightweight and flexible properties, making the device suitable for integration into wearable and flexible electronics. To evaluate the photodetection capabilities of the F-SPPD, the device was exposed to 365 nm UV light, and its current–voltage (I–V) characteristics were analyzed Fig. 11g. Upon UV illumination, the photocurrent exhibited a significant increase, attributed to the generation of photogenerated electron–hole pairs within the ZnO nanorods. The dark current (I_D) of 0.55 μA increased to 28, 39, and 50 μA under UV light intensities of 8, 16, and 21 mW/cm^2, respectively. The dependence of photocurrent (I_{Ph}) on light intensity follows a power-law relationship, given by $I_{Ph} = A\ P^{\theta}$, confirming the device's ideal photodetector behavior with consistent light-to-current conversion efficiency. Stability and reproducibility are critical parameters for practical photodetector applications. To assess these characteristics, ON–OFF switching tests were conducted under 8 mW/cm^2 UV illumination at a 10 V bias Fig. 11h. The device exhibited a rise time (τ_r) of 24 s and a decay time (τ_d) of 30 s, calculated using the exponential rise and decay factors ($\tau_r \approx 1-e^{-1}$, $\tau_d \approx e^{-1}$). These results highlight the device's stable and repeatable UV response, essential for long-term deployment in real-world applications. The working principle of the F-SPPD is further explained through an energy band diagram Fig. 11i. In the absence of UV illumination, the dark current arises due to the intrinsic charge carrier density in ZnO nanorods. Upon UV exposure, incident photons excite electrons from the valence band to the conduction band, increasing the charge carrier density and subsequently enhancing photocurrent. This mechanism ensures efficient photoelectric conversion, making the device highly suitable for self-powered UV sensing. To further investigate the spectral selectivity of the F-SPPD, the device was exposed to different visible light wavelengths Fig. 11j–l. The response was significantly reduced at 405 nm, while it became negligible at 535 nm, confirming the high selectivity of the device for UV light detection. The sensitivity analysis presented in Fig. 11l reveals a two-fold deviation in response across different spectral zones, ensuring accurate and reliable UV detection. The integration of Zn^{2+}-seeded S-PVDF with ZnO NRs enhances charge transport efficiency, leading to an optimized, self-powered photodetection system. The development of the F-SPPD highlights the potential of hybrid nanocomposite-based photodetectors for next-generation self-powered optoelectronic applications. By leveraging the synergistic effects of piezoelectric and semiconducting materials, this device achieves dual functionality as a mechanical energy harvester and a high-performance UV photodetector. The combination of fast response time, high spectral selectivity, and mechanical flexibility makes the F-SPPD an attractive candidate for wearable UV monitoring, environmental sensing, and real-time health monitoring applications. The ability to operate without external power sources further enhances its potential for integration into autonomous and sustainable electronic devices, paving the way for future advancements in flexible and self-sustained optoelectronic technologies [67].

5.3 *Piezoelectric-Phototronic Coupling in Self-Powered Photodetectors*

Piezoelectric Output Response of the F-SPPD The integration of piezoelectric and optoelectronic properties in nanogenerator-based photodetectors provides a promising strategy for self-powered sensing technologies. The realization of a self-powered photodetector (PD) is demonstrated by analyzing the response of a Flexible Self-Powered Photodetector (F-SPPD) under UV light (365 nm) irradiation, highlighting the strong coupling between piezoelectric energy harvesting and photodetection mechanisms. The observed piezoelectric output response of the F-SPPD is presented in Fig. 12a, where the device exhibits a stable piezoelectric output under

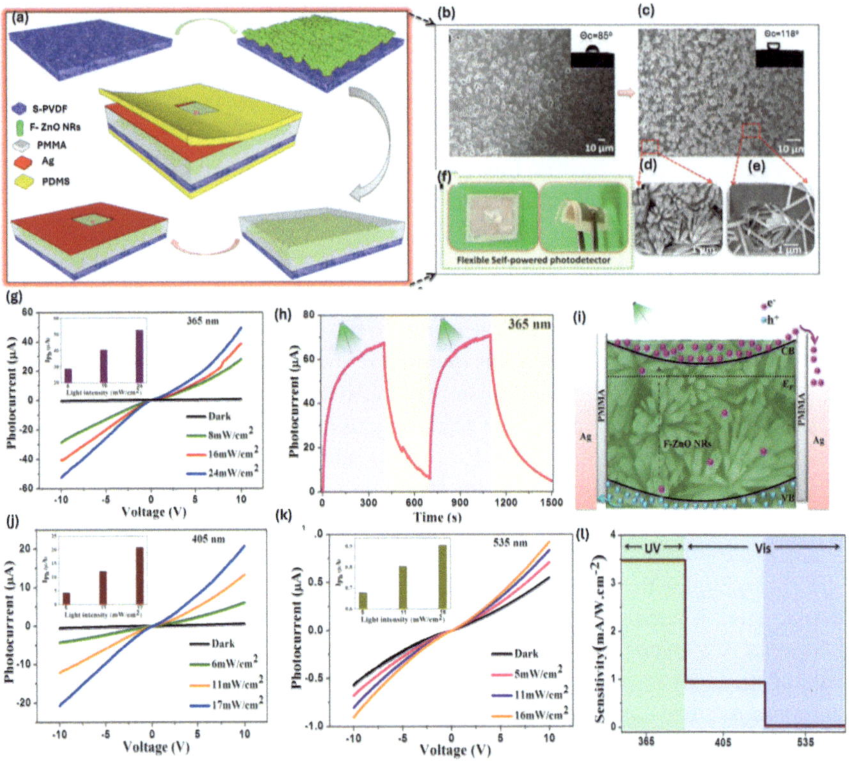

Fig. 11 Flexible self-powered photodetector (F-SPPD) based on Zn^{2+}-seeded PVDF and ZnO nanorods (NRs) for UV sensing. **a** Device architecture with a protective PMMA layer. **b**, **c** Morphology of S-PVDF and ZnO NRs, enhancing hydrophobicity. **d**, **e** Cross-sectional views showing nanorod growth in porous regions. **f** Device flexibility under mechanical deformation. **g** I–V characteristics under UV illumination. **h** ON–OFF switching response confirming stability. **i** Energy band diagram illustrating charge transport. **j–l** Spectral selectivity and sensitivity analysis, demonstrating strong UV response and high efficiency for self-powered optoelectronics. Reproduced with permission from Purusothaman et al. [67]

mechanical deformation. Working Principle of the F-SPPD A schematic representation of the working principle of the F-SPPD is provided in Fig. 12b. Under dark conditions, the F-SPPD maintains a stable piezoelectric output (~5 V) when subjected to mechanical bending forces. During tensile deformation, piezoelectric potentials are generated at the interface, inducing electron flow between electrodes to compensate for the dipole moment reorientation (Fig. 12b-UV–vis OFF). Once the strain-induced dipole alignment reaches equilibrium, the piezoelectric output gradually decays to zero, maintaining a steady-state potential if the strain is applied. Upon release, the piezoelectric potential collapses, causing a backflow of electrons, which manifests a negative peak in the output cycle. This characteristic piezoelectric behavior confirms the efficient energy conversion capability of the F-SPPD in mechanical energy harvesting applications. Experimental Setup for Performance Evaluation The experimental setup used to evaluate the dual functionality of the F-SPPD is illustrated in Fig. 12c, d. These figures provide insight into the operational conditions under which the device exhibits both energies harvesting and photodetection capabilities. Photodetection Response Under UV Illumination Upon UV illumination, the photodetection response of the device is significantly altered due to the generation of photoinduced electron–hole (e^{-}–h^{+}) pairs within the ZnO nanorods of the device (Fig. 12b—UV–vis ON). The increased charge carrier density leads to a reduction in the internal resistance of the device, which, in turn, diminishes the piezoelectric output. This effect is attributed to the screening behavior of photogenerated carriers, which neutralize the piezoelectric potential (σ^{+}) at the electrode interfaces, thereby reducing the net polarization-induced voltage. The drop in piezoelectric output under UV exposure further validates the strong coupling between mechanical and optical excitation in the F-SPPD. This piezoelectric-phototronic effect allows for tunable optoelectronic performance, where the piezoelectric response is modulated based on incident light intensity. Synergistic Piezoelectric and Photodetection Mechanisms The demonstrated synergistic interplay between piezoelectric and semiconducting properties makes the F-SPPD a highly efficient self-powered photodetector, with a tunable response to UV exposure. This hybrid approach enhances the versatility of the device, enabling dual-mode energy harvesting and sensing without requiring external power sources. The F-SPPD effectively combines mechanical energy harvesting with optical sensing, making it an ideal platform for next-generation wearable electronics, smart sensing systems, and autonomous environmental monitoring applications. The key advantages of this system include: Simultaneous mechanical energy harvesting and UV photodetection in a single device. Tunable optoelectronic performance due to piezoelectric-phototronic coupling. Compact, flexible design for wearable and real-time sensing applications. Future directions for enhancing the F-SPPD include Optimizing the ZnO nanorod morphology to further improve charge separation efficiency. Integrating wireless data transmission capabilities for real-time environmental monitoring. Expanding spectral sensitivity to detect a broader range of visible and UV wavelengths. By demonstrating multifunctional energy conversion in a compact, flexible configuration, the F-SPPD paves the way for future self-sustained optoelectronic applications, such as radiation monitoring, bioelectronics, and intelligent sensor networks [67].

6 Summary

The integration of TENGs and PENGs into optoelectronic devices has paved the way for self-powered, highly efficient optical systems with enhanced tunability and energy sustainability. This chapter has provided a detailed overview of metal oxide-based hybrid composites, emphasizing their role in triboelectric and piezoelectric energy harvesting, optical modulation, and photodetection applications. By harnessing the piezoelectric-phototronic coupling and charge transfer dynamics at material interfaces, these devices demonstrate superior performance in real-time optical tuning,

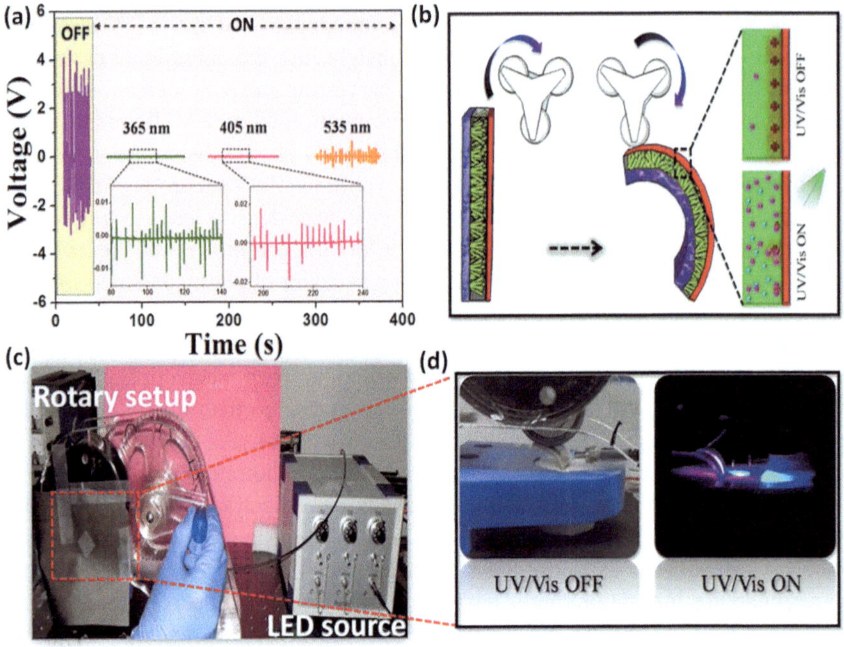

Fig. 12 Piezoelectric and optoelectronic coupling in the F-SPPD under UV illumination. **a** Piezoelectric output response. **b** Schematic of the working principle under mechanical strain with and without UV exposure. **c, d** Experimental setup for performance evaluation. The piezoelectric potential (~5 V) is generated under bending stress, while UV exposure induces photogenerated charge carriers, reducing the piezoelectric output due to screening effects. This tunable piezoelectric-phototronic interaction enables self-powered sensing and energy harvesting, making the F-SPPD ideal for wearable electronics, real-time UV monitoring, and autonomous sensing applications. Reproduced with permission from Purusothaman et al. [67]

UV sensing, and adaptive photonic circuits. Despite the remarkable advancements, challenges remain in optimizing material compositions, improving charge transport mechanisms, and scaling up fabrication processes for practical implementation. Future research should focus on enhancing the structural stability, exploring novel hybrid materials, and integrating advanced signal processing techniques to further refine self-powered optical systems. The development of multifunctional and flexible optoelectronic platforms holds great potential for next-generation wearable sensors, biomedical imaging, smart display technologies, and autonomous environmental monitoring systems. By bridging the gap between energy harvesting and optoelectronics, TENG/PENG-powered devices are set to revolutionize sustainable photonic applications and redefine the future of self-sufficient optical technologies.

Acknowledgements VV thank KLEF for the support through Internal Funding Project KLEF/IFP/2022-2023/ECE/007, and the Royal Society of Chemistry for the Researcher Grant R22-7381975972. The work is also supported by Department of Science and Technology, India through PURSE Scheme (SR/PURSE/2023/196).

References

1. M. Cui, P. Zhao, C. Jiang, B. Tian, C. Luo, H. Lin, H. Peng, C.-G. Duan, Self-Powered optoelectronic device with depression-related synaptic functions. ACS Appl. Electron. Mater. **5**, 3403–3409 (2023). https://doi.org/10.1021/acsaelm.3c00407
2. T. Das, P. Biswas, A. Dev, J. Mallick, M. Kar, Multi-functional piezoelectric nanogenerator based on relaxor ferroelectric materials (BSTO) and conductive fillers (MWCNTs) for self-powered memristor and optoelectronic devices. Chem. Eng. J. **479**, 147900 (2024). https://doi.org/10.1016/j.cej.2023.147900
3. W. Huang, H. Zhang, J. Tang, Z. Lin, T. Guo, Y. Zhou, S. Jiang, P. Hang, M. Jiao, C. Zhu, L. Wang, D. Yang, X. Yu, X. Li, Self-Powered optoelectronic synaptic devices for neuromorphic computing with the lowest energy consumption density. ACS Photonics **11**, 3095–3104 (2024). https://doi.org/10.1021/acsphotonics.4c00379
4. S. Kim, M. Kim, H. Kim, Self-powered photodetectors based on two-dimensional van der Waals semiconductors. Nano Energy **127**, 109725 (2024). https://doi.org/10.1016/j.nanoen.2024.109725
5. L. Qi, W. Tang, X. Weng, K. Wu, Y. Cen, Y. Sun, S. Zhou, Z. Li, X. Wu, C. Kang, D. Zhao, S. Dai, Y. Xie, H. Liang, W. Zhang, Y.-J. Zeng, S. Ruan, Flexoelectricity-enhanced self-powered photodetection in 2D van der waals heterojunctions with large curvatures. Adv. Funct. Mater. **34**, 2315991 (2024). https://doi.org/10.1002/adfm.202315991
6. Y. Feng, Y. Zhang, Y. Zhang, X. Miao, Y. Lin, T. Min, Y. Pan, Self-powered broadband photodetectors based on large-scale continuous films of weyl semimetal WTe2. ACS Appl. Opt. Mater. **2**, 333–340 (2024). https://doi.org/10.1021/acsaom.3c00450
7. Z. Yin, Y. Zeng, D. Yang, Y. Jiao, J. Song, P. Hu, H. Fan, F. Teng, Multifunctional optoelectronic device based on CuO/ZnO heterojunction structure. J. Lumin. **257**, 119762 (2023). https://doi.org/10.1016/j.jlumin.2023.119762
8. P. Vashishtha, P. Goswami, P. Prajapat, A. Kumar Gangwar, P. Singh, G. Gupta, Highly efficient, self-powered, and air-stable broadband photodetector based on SnSe thin film. Mater. Sci. Eng. B **297** 116808 (2023). https://doi.org/10.1016/j.mseb.2023.116808
9. X. Zhang, Z. Li, T. Yan, L. Su, X. Fang, Phase-Modulated multidimensional perovskites for high-sensitivity self-powered UV photodetectors. Small **19**, 2206310 (2023). https://doi.org/10.1002/smll.202206310
10. J. Zhang, X. Feng, L. Mei, W. Yu, H. Song, N. Cui, T. Yun, H. Mu, S. Lin, A MAPbBr3/PdSe2 Schottky junction-based optoelectronic sensor with self-powered and switchable photocurrents. Mater. Des. **234**, 112368 (2023). https://doi.org/10.1016/j.matdes.2023.112368
11. M.K. Sharma, Y.-J. Lin, C.-K. Lin, B.K. Nahak, C.-J. Huang, C.C. Chen, Y.-L. Wang, Z.-H. Lin, Self-powered AlGaN/GaN HEMT-based sensor integrated with rotational TENG for comprehensive water quality analysis. Nano Energy, **135**, 110637 (2025). https://doi.org/10.1016/j.nanoen.2024.110637
12. X. Qu, X. Liu, Y. Yue, Y. Tang, P. Miao, Triboelectric nanogenerator-enabled self-powered strategies for sensing applications. TrAC Trends Anal. Chem. **185**, 118191 (2025). https://doi.org/10.1016/j.trac.2025.118191
13. W. Zhou, Z. Liu, Y. Sun, H. Teng, W. Wang, Bayanheshig, W. Li, Bidirectional Littrow double grating interferometry for quadruple optical interpolation. Opt. Laser Technol. **175**, 110751 (2024). https://doi.org/10.1016/j.optlastec.2024.110751
14. X. Shu, M. Kang, Y. Zhou, X. Cui, J. Nie, Z. Zhao, Y. Zhang, Triboelectric nanogenerators based self-powered visualized weight sensor with high accuracy and real time. Nano Energy **135**, 110624 (2025). https://doi.org/10.1016/j.nanoen.2024.110624
15. O. Cakir, D. Doganay, M. Cugunlular, M.O. Cicek, O. Demircioglu, S. Coskun, H.E. Unalan, Post-treatment optimization for silver nanowire networks in transparent droplet-based TENG sensors. Nano Energy **128**, 109940 (2024). https://doi.org/10.1016/j.nanoen.2024.109940
16. G. Wu, L. Wu, H. Zhang, X. Wang, M. Xiang, Y. Teng, Z. Xu, F. Lv, Z. Huang, Y. Lin, Y. Zhao, H. Zhao, C. Xue, L. Gao, Research progress of screen-printed flexible pressure sensor. Sens. Actuat. A Phys. **374**, 115512 (2024). https://doi.org/10.1016/j.sna.2024.115512

17. Y. Teng, B. Zhang, Z. Qiu, X. Hu, Z. He, High-sensitivity dual-FBG acceleration sensor for low frequency vibration measurement. J. Opt. **53**, 2361–2373 (2024). https://doi.org/10.1007/s12596-023-01477-3
18. Z. Qiu, M. Mai, Y. Teng, X. Wang, J. Ma, Bearing-based FBG acceleration sensor for low-frequency vibration measurement. Measurement **228**, 114322 (2024). https://doi.org/10.1016/j.measurement.2024.114322
19. H. Fu, S. Peng, P. Li, C. Teng, C. Caucheteur, H. Qu, X. Hu, Highly sensitive fiber force sensor based on cascaded Fabry-Perot cavities and Vernier effect. Opt. Laser Technol. **175**, 110825 (2024). https://doi.org/10.1016/j.optlastec.2024.110825
20. M.V. Paranjape, P. Manchi, A. Kurakula, V.S. Kavarthapu, J.K. Lee, S.A. Graham, J.S. Yu, Generalized utilization of energy harvesting ability of TENG for concurrent energy storage and motion sensing application with effective external circuitry. Nano Energy **129**, 109983 (2024). https://doi.org/10.1016/j.nanoen.2024.109983
21. Y. Wu, S. Wang, R. Zhang, T. Yu, M. Xu, X. Li, M. Pu, X. Ma, Y. Guo, Y. Su, H. Tai, X. Luo, A novel multifunctional photonic film for colored passive daytime radiative cooling and energy harvesting. Small **20**, 2308661 (2024). https://doi.org/10.1002/smll.202308661
22. A. Mirsepah, L. Shooshtari, R. Mohammadpour, A. Esfandiar, A. Irajizad, Wearable broadband MoS2 photodetector for dual heart rate and UV detection powered by PDMS-MXene TENG. Chem. Eng. J. **499**, 155953 (2024). https://doi.org/10.1016/j.cej.2024.155953
23. D. Teng, R. Hu, Y. Tang, K. Wang, A sodium-based hybrid plasmonic waveguide with ultra-deep subwavelength confinement and low cross-talk. Opt. Mater. (Amst). **147**, 114729 (2024). https://doi.org/10.1016/j.optmat.2023.114729
24. B. Li, M. Xu, B. An, W. Sun, R. Teng, S. Luo, C. Ma, Z. Chen, J. Li, W. Li, S. Liu, Mechanical and thermal responsive chiral photonic cellulose hydrogels for dynamic anti-counterfeiting and optical skin. Mater. Horizons (2025). https://doi.org/10.1039/D4MH01646G
25. Y. Qin, M. Liu, H. Teng, N. Chen, C. Wu, C. Jiang, Z. Xue, H. Zhu, J. Gui, X. Liu, Y. Xiao, H. Hu, Tunable hybridized plasmons–phonons in a graphene/mica-nanofilm heterostructure. Nanoscale **16**, 20522–20531 (2024). https://doi.org/10.1039/D4NR00942H
26. Z. Man, Z. Chen, W. Tao, Z. Lv, F. Teng, Q. Cui, H. Fu, Q. Liao, A. Tang, Host-Guest systems enable efficient organic afterglow via a facile spraying method. Adv. Funct. Mater. **35**, 2412652 (2025). https://doi.org/10.1002/adfm.202412652
27. G. Li, G. Li, M. Liao, W. Liu, H. Zhang, S. Huang, T. Huang, S. Zhang, Z. Li, H. Peng, Unlocking mixed-metal oxides active centers via acidity regulation for K&SO2 poisoning resistance: self-detoxification mechanism of zeolite-confined deNOx catalysts. Environ. Sci. Technol. **58**, 10388–10397 (2024). https://doi.org/10.1021/acs.est.4c03060
28. F. Liu, Y. Yan, G. Chen, D. Wang, Recent advances in ambient electrochemical methane conversion to oxygenates using metal oxide electrocatalysts. Green Chem. **26**, 655–677 (2024). https://doi.org/10.1039/D3GC03513A
29. Y. Ding, J. Jiang, Y. Wu, Y. Zhang, J. Zhou, Y. Zhang, Q. Huang, Z. Zheng, Porous conductive textiles for wearable electronics. Chem. Rev. **124**, 1535–1648 (2024). https://doi.org/10.1021/acs.chemrev.3c00507
30. Y. Lu, J. Bai, B. Sun, N. Li, Z. Yang, H. Yu, C. Wang, C. Gu, H. Liu, P. Tang, Q. Wang, Enhancing the capacitive energy storage ability of Ti_3C_2Tx MXene and PVA-derived carbon composite aerogels through structural disorder. Electrochim. Acta **521**, 145910 (2025). https://doi.org/10.1016/j.electacta.2025.145910
31. P. Tang, J. Zhang, X. Li, F. Yang, Q. Zhao, J. Ma, Z. Hu, H. Sun, X.-B. Wang, Z. Sun, Y. Yang, Cryogenic photoelectron spectroscopic and theoretical study of the electronic and geometric structures of undercoordinated osmium chloride anions OsCln– (n = 3–5). J. Phys. Chem. A **128**, 5500–5507 (2024). https://doi.org/10.1021/acs.jpca.4c01713
32. Y.-L. Wu, C.-K. Jiao, F.-F. Wang, P.-F. Tang, Y.-T. Yan, Y.-Y. Song, C. Fu, X.-H. He, G.-P. Yang, Y. Wang, High electrocatalytic oxygen evolution reaction and hydrogen evolution reaction of N-doped Co@CN materials derived from two cobalt-based metal-organic frameworks (MOFs). J. Mol. Struct. **1316**, 138945 (2024). https://doi.org/10.1016/j.molstruc.2024.138945

33. J. Teng, H. Yu, Z. Liu, L. Bai, Z. Liu, Fabrication of novel metal oxide nanosheets-decorated carbon nanofibers for highly efficient removal of ultra-small nanoplastics. J. Environ. Chem. Eng. **12**, 114094 (2024). https://doi.org/10.1016/j.jece.2024.114094
34. A. Singh, S. Singh, B.C. Yadav, In_2O_3 nanocubes and $ZnWO_4$ nanorod-based triboelectric nanogenerator for self-powered humidity sensors. Sens. Actuat. B Chem. **398**, 134721 (2024). https://doi.org/10.1016/j.snb.2023.134721
35. Z. Peng, Y. Huang, A.G. Bannov, S. Li, L. Tang, L. Tan, Y. Chen, Regulating oxygen vacancies and coordination environment of manganese dioxide for enhanced high-mass-loading energy storage. Energy Environ. Sci. **17**, 3384–3395 (2024). https://doi.org/10.1039/D4EE00949E
36. Y. Wu, Q. Qi, T. Peng, J. Yu, X. Ma, Y. Sun, Y. Wang, X. Hu, Y. Yuan, H. Qin, In situ flash synthesis of ultra-high-performance metal oxide anode through shunting current-based electrothermal shock. ACS Appl. Mater. Interfaces **16**, 16152–16163 (2024). https://doi.org/10.1021/acsami.3c19174
37. J. Zhang, X. Mao, Q. Jia, R. Nie, Y. Gao, K. Tao, H. Chang, P. Li, W. Huang, Body-worn and self-powered flexible optoelectronic device for metronomic photodynamic therapy. Npj Flex. Electron. **8**, 60 (2024). https://doi.org/10.1038/s41528-024-00345-9
38. H. Hu, C. Deng, H. Gao, T. Han, S. Xue, Y. Tang, M. Zhang, M. Li, H. Liu, L. Deng, W. Xiong, 3D nanoprinting of heterogeneous metal oxides with high shape fidelity. Adv. Mater. **36**, 2405053 (2024). https://doi.org/10.1002/adma.202405053
39. S. Anbalagan, K. Manojkumar, M. Muthuramalingam, S. Hajra, S. Panda, R. Sahu, H. Joon Kim, A. Sundaramoorthy, N. Nithyavathy, V. Vivekananthan, Progress and recent advances in self-powered gas sensing based on triboelectric and piezoelectric nanogenerators. Chem. Eng. J. **497**, 154740 (2024). https://doi.org/10.1016/j.cej.2024.154740
40. M. Xu, M. Peng, H. Tang, W. Zhou, B. Qiao, D. Ma, Renaissance of strong metal-support interactions. J. Am. Chem. Soc. **146**, 2290–2307 (2024). https://doi.org/10.1021/jacs.3c09102
41. J. Yao, Q. Wang, Y. Zhang, Y. Teng, J. Li, P. Zhao, C. Zhao, Z. Hu, Z. Shen, L. Liu, D. Tian, S. Qiu, Z. Wang, L. Kang, Q. Li, Ultra-low power carbon nanotube/porphyrin synaptic arrays for persistent photoconductivity and neuromorphic computing. Nat. Commun. **15**, 6147 (2024). https://doi.org/10.1038/s41467-024-50490-y
42. Y. Liu, Y. Wu, R. Duan, J. Fu, M. Ovesen, S.C.E. Lai, T.-E. Yeo, J.Y. Chee, Y. Chen, S.L. Teo, H.R. Tan, W. Zhang, J.K.W. Yang, K.S. Thygesen, Z. Liu, Y.-W. Zhang, J. Teng, Linear electro-optic effect in 2D ferroelectric for electrically tunable metalens. Adv. Mater. **36**, 2401838 (2024). https://doi.org/10.1002/adma.202401838
43. Y. Liu, S. Yue, Z. Tian, Z. Zhu, Y. Li, X. Chen, Z.L. Wang, Z.-Z. Yu, D. Yang, Self-Powered and self-healable extraocular-muscle-like actuator based on dielectric elastomer actuator and triboelectric nanogenerator. Adv. Mater. **36**, 2309893 (2024). https://doi.org/10.1002/adma.202309893
44. A. Roszkiewicz, M. Garlińska, A. Pregowska, Advancements in piezoelectric-enabled devices for optical communication. Phys. Status Solidi **222**, 2400298 (2025). https://doi.org/10.1002/pssa.202400298
45. F. Dong, M. Zhu, Y. Wang, Z. Chen, Y. Dai, Z. Xi, T. Du, M. Xu, AI-enabled rolling triboelectric nanogenerator for bearing wear diagnosis aiming at digital twin application. Nano Energy **134**, 110550 (2025). https://doi.org/10.1016/j.nanoen.2024.110550
46. Y. Wang, G. Liu, Q. Meng, X. Jiang, X. Li, H. Liu, Z. Li, Bending the straight into curved: a tree-ring-inspired fully printed omnidirectional triboelectric nanogenerator with ring-nested structure for all-in-one wearable self-powered systems and IoT smart packaging. Nano Energy **135**, 110631 (2025). https://doi.org/10.1016/j.nanoen.2024.110631
47. N.K. Das, M. Ravipati, S. Ray, S. Badhulika, Electrospun Fe/Co metal-organic framework-PVDF composite nanofiber-based triboelectric nanogenerator as a self-powered sterilizing mask. ACS Appl. Energy Mater. **8**, 2308–2317 (2025). https://doi.org/10.1021/acsaem.4c02880
48. L. Wang, Y. Zhang, X. Zhang, X. Cheng, S. Zhai, X. Bi, H. Li, Z. Wang, T. Cheng, Harvesting broadband breeze vibration energy via elastic bistable triboelectric nanogenerator for in situ, self-powered monitoring of power transmission lines. Adv. Energy Mater. **15**, 2403318 (2025). https://doi.org/10.1002/aenm.202403318

49. N. Xu, Z. Wang, H. Askari, B.H.G. Barbosa, A. Khajepour, Intelligent mobility: unveiling triboelectric nanogenerator-powered intelligent tire realization. Appl. Mater. Today **42,** 102506 (2025). https://doi.org/10.1016/j.apmt.2024.102506
50. G. Yin, X. Liang, Y. Zhang, J. Li, S. Wei, Fountain-inspired triboelectric nanogenerator as rotary energy harvester and self-powered intelligent sensor. Nano Energy **137**, 110779 (2025). https://doi.org/10.1016/j.nanoen.2025.110779
51. K. Munirathinam, A. Shanmugasundaram, Y.J. Jeong, J.Y. Kim, D.W. Lee, High-performance triboelectric nanogenerator powered by flowing water for self-powered wireless sensor Platform. Chem. Eng. J. **505**, 159106 (2025). https://doi.org/10.1016/j.cej.2024.159106
52. H. Wang, M. Gao, L. Zhang, Z. Su, C. Chen, W. Yang, P. Xu, D. Niu, P. Ma, P. Ma, Advanced low-temperature self-healable bio-polyurethanes with double-alkane-tailed ringing units for applications in self-powered flexible control panels. Chem. Eng. J. **506**, 160019 (2025). https://doi.org/10.1016/j.cej.2025.160019
53. T. Yang, J. Xie, Z. Huang, J. Liu, H. Luo, X. Jing, Bio-inspired vibration isolator with triboelectric nanogenerator for self-powered monitoring. Mech. Syst. Signal Process. **223**, 111854 (2025). https://doi.org/10.1016/j.ymssp.2024.111854
54. S. An, G. Li, X. Zhou, H. Pu, J. Wang, Y. Cheng, S. Liu, T. Zhou, Y. Zhou, X. Pu, Tin can telephone-inspired self-powered mechanical wave communication integrated with self-charge excitation triboelectric nanogenerator. Nano Energy **133**, 110470 (2025). https://doi.org/10.1016/j.nanoen.2024.110470
55. Z. Shen, Y. Fang, B. Yuan, Chitosan-based films with excellent flame retardancy and highly sensitive fire response for application in self-powered dual fire-alarm systems. Int. J. Biol. Macromol. **299**, 140131 (2025). https://doi.org/10.1016/j.ijbiomac.2025.140131
56. L. Wang, Y. Gao, J. Luo, W. Xu, C. Zhang, H. Yuan, Y. Zhang, Y. Pang, High-performance triboelectric nanogenerator based on NH_2-MXene/TiO_2@sodium alginate composite film for self-powered cathodic protection. Chem. Eng. J. **506**, 159837 (2025). https://doi.org/10.1016/j.cej.2025.159837
57. K.-Y. Hsu, S.-M. Huang, B.T. Murti, C.-C. Chen, Y.-C. Chao, I.-C. Ha, C.-C. Tsai, C.-Y. Chen, M.-L. Tsai, P.-K. Yang, A flexible, sustainable, and deep learning-assisted triboelectric patch for self-powered interactive sensing and wound healing applications. Nano Energy **134**, 110501 (2025). https://doi.org/10.1016/j.nanoen.2024.110501
58. X. Zeng, L. Teng, X. Wang, T. Lu, W. Leng, X. Wu, D. Li, Y. Zhong, X. Sun, S. Zhu, Y. Dong, P. Tan, Z. Zeng, Z. Hu, Z. Li, Q. Zheng, Efficient multi-physical crosslinked nanocomposite hydrogel for a conformal strain and self-powered tactile sensor. Nano Energy **135**, 110669 (2025). https://doi.org/10.1016/j.nanoen.2025.110669
59. Y. Sun, Z. Qian, Y. Wang, Y. Li, Y. Zheng, Y. Liu, A flexible humidity-resistant nanofiber-based triboelectric nanogenerator with high electrical output stability as self-powered sensors for motion monitoring. Chem. Eng. J. **506**, 159845 (2025). https://doi.org/10.1016/j.cej.2025.159845
60. J. Wu, X. Zhou, J. Luo, J. Zhou, Z. Lu, Z. Bai, Y. Fan, X. Chen, B. Zheng, Z. Wang, L. Wei, Q. Zhang, Stretchable and self-powered mechanoluminescent triboelectric nanogenerator fibers toward wearable amphibious electro-optical sensor textiles. Adv. Sci. **2401109**, 1–10 (2024). https://doi.org/10.1002/advs.202401109
61. J. Huang, X. Yang, J. Yu, J. Han, C. Jia, M. Ding, J. Sun, X. Cao, Q. Sun, Z.L. Wang, A universal and arbitrary tactile interactive system based on self-powered optical communication. Nano Energy **69**, 104419 (2020). https://doi.org/10.1016/j.nanoen.2019.104419
62. J. Wang, H. Wang, X. Li, Y. Zi, Self-powered electrowetting optical switch driven by a triboelectric nanogenerator for wireless sensing. Nano Energy **66**, 104140 (2019). https://doi.org/10.1016/j.nanoen.2019.104140
63. X. Chen, Y. Wu, A. Yu, L. Xu, L. Zheng, Y. Liu, H. Li, Z. Lin Wang, Self-powered modulation of elastomeric optical grating by using triboelectric nanogenerator. Nano Energy **38**, 91–100 (2017). https://doi.org/10.1016/j.nanoen.2017.05.039
64. S.P. Muduli, S. Veeralingam, N.K. Das, S. Badhulika, Lead-free polymer composite film for photon energy influenced piezoelectric nanogenerator based on photoactive SnS_2. Mater. Res. Bull. **178**, 112923 (2024). https://doi.org/10.1016/j.materresbull.2024.112923

65. X.X. Yu, H. Yin, H.X. Li, W. Zhang, H. Zhao, C. Li, M.Q. Zhu, Piezo-phototronic effect modulated self-powered UV/visible/near-infrared photodetectors based on CdS:P_3HT microwires. Nano Energy **34**, 155–163 (2017). https://doi.org/10.1016/j.nanoen.2017.02.033
66. M. Peng, H. Ji, S. Liu, P. Lv, Y. Kuang, Y. Liu, Y. Liu, X. Sun, Z. Wen, A self-powered triboelectric UV photodetector based on coupling impedance matching and photoresistive effect by sensing-electrode model. Nano Energy **109**, 108294 (2023). https://doi.org/10.1016/j.nanoen.2023.108294
67. Y. Purusothaman, N.R. Alluri, A. Chandrasekhar, V. Vivekananthan, S.J. Kim, Direct in situ hybridized interfacial quantification to stimulate highly flexile self-powered photodetector. J. Phys. Chem. C **122**, 12177–12184 (2018). https://doi.org/10.1021/acs.jpcc.8b02604

Detection of Nitrous Oxide Gas Using 2D Molybdenum Diselenide Monolayer

Neha Mishra, Bramha P. Pandey, and Santosh Kumar

Abstract This chapter presents a comprehensive review of the interaction between nitrous oxide (N_2O) gas and a two-dimensional molybdenum diselenide ($MoSe_2$) monolayer (ML), with a focus on its electronic, magnetic, and optical properties. N_2O is a highly toxic greenhouse gas, significantly more potent than CO_2, and its detection is vital due to its detrimental environmental and health effects. Its accumulation in the atmosphere impacts air quality and is linked to neurotoxic effects in humans. Given the pressing need for sensitive gas detection technologies, this chapter explores the potential of $MoSe_2$ ML as an active sensing material. Supported by both experimental and computational insights, it is shown that N_2O adsorption significantly alters the material's electronic structure—evident through changes in the band structure, density of states, adsorption energy, and charge transfer. These effects highlight the suitability of $MoSe_2$ ML for gas sensing applications. Remarkably, N_2O adsorption induces magnetism in the otherwise non-magnetic $MoSe_2$ ML, transitioning it into a magnetically dilute semiconductor (MDS) with anti-ferromagnetic characteristics, making it relevant for spintronic devices. Additionally, the optical properties of the material—such as the absorption coefficient, dielectric function, and refractive index—also exhibit substantial shifts upon gas interaction and transition metal doping (e.g., Fe, Mn). These modifications position $MoSe_2$ ML as a versatile material for optoelectronic sensors and photodetectors. In the context of this book, the chapter contributes to the broader narrative of advanced optical sensing by showcasing how 2D materials—especially transition metal dichalcogenides—can complement noble

N. Mishra · B. P. Pandey (✉)
Department of Electronics and Communication Engineering, Madan Mohan Malaviya University of Technology, Gorakhpur, India
e-mail: bppece@mmmut.ac.in

N. Mishra
e-mail: nmece@mmmut.ac.in.com

S. Kumar
Department of Electronics and Communication Engineering, Koneru Lakshmaiah Education Foundation, Vaddeswaram, Andhra Pradesh, India
e-mail: santosh@kluniversity.in

B. Ramachandran et al. (eds.), *Advanced Optical Sensors*, Progress in Optical Science and Photonics 38, https://doi.org/10.1007/978-981-95-1687-2_7

metal-based hybrids in gas detection platforms. The findings support the development of sensitive, tunable, and multifunctional sensor systems in emerging optical sensor technologies.

Keywords Optical · Magnetic · Transition metal dichalcogenide · N_2O · Adsorption

1 Introduction

Graphene, a novel two-dimensional (2D) material with excellent properties, is being studied widely, but its applications are limited due to its zero-bandgap property [1]. The two-dimensional 2D nanomaterials have become an efficient alternative to graphene or few-layer materials due to their ubiquitous properties [2–4]. The advantage is a high surface-to-volume ratio, atomically thin size, and tunable bandgap [5]. The shift in bandgap from indirect and direct semiconductor when size reduces to a monolayer makes them appropriate for optoelectronic applications [6]. Different synthesis methods accurately detach the monolayer from bulk transition metal dichalcogenides (TMD), such as mechanical exfoliation [7], chemical vapor deposition [8], and atomic layer deposition [9].

Researchers have long awaited a class of materials that combine high sensitivity with low cost [6]. The gas sensors designed using (transition metal dichalcogenides) TMD materials hold an advantage over graphene sensors in terms of sensitivity, electron mobility, and, lastly, intense chemical activity for gas interface [10, 11]. One of the applications of these materials is to detect toxic gases that are a threat to both the surrounding environment and human health. Thus, these gases should be detected accurately at sub-ppm (parts per million) levels to know the actual level of damage caused by them. Some experimental methods are brought in application to detect the gases at ppm level, such as gas chromatography-mass spectrometry (GCMS) [12], non-dispersive infrared analysis (NDIR) [13], and infrared spectrophotometry (IRSP) [14]. Still, the results obtained are unreliable and time-consuming as the process is complex. The complexity issues include lower absorption coefficient, overlapping stretches developed between interacting atoms of analyte and adsorbent for chemoresistive gas sensors [15].

Among the various gases, nitrous oxide (N_2O) is a highly toxic gas with severe impacts on agriculture, like soil denitrification and nitrification. Human health results in cerebral flow imbalance, cell death, and brain attack. Thus, all these suggestions for the detection of N_2O gas present in the surroundings. Presently, the experimental techniques involved in the N_2O detection include gas chromatography [12], chamber method [13], infrared-based optical method [14] and laser absorption spectroscopy [16] respectively. Also, in this direction, a low-cost, high-precision NDIR spectrometer is designed for detection of N_2O gas detection from the soil region and the atmosphere [17]. The reported prototype of N_2O gas detector is based on silicone material.

The silicone is a polymer of silicon containing carbon, hydrogen and oxygen separately. Infra-red spectrometers based on silicon-on-insulator are designed to detect the N_2O gas that uses infra-red sources and photodetectors for fast response [18]. However, no two-dimensional material (other than $MoSe_2$) is reported to study the detection of N_2O gas molecule specifically. Therefore, the detection of N_2O gas on molybdenum diselenide ($MoSe_2$) monolayer (ML) is reviewed in this chapter, and its impact on electronic, magnetic, and optical properties are discussed in depth with comparison to other NO_x (NO, NO_2) gases.

2 Impact of N_2O Gas on the Human Body

N_2O has been used as a source of anesthesia for a more extended period [19]. It is given to the patients in a ratio of 50:50 proportions with oxygen to provide analgesia during surgeries. However, the analgesic property of N_2O is questioned at present due to harmful consequences administered by the medical fraternity. The reason is the cardiovascular problems, and neurotoxicity reported post analgesic treatment of this gas [19]. Furthermore, it claims that N_2O neurotoxicity is more pronounced during the perinatal brain period and aged brain.

2.1 Mechanism of Neurotoxicity

In the human body, several mechanisms contribute to neurotoxicity [20]. It is classified as (i) NMDA antagonism, (ii) enzyme inhibition, and (iii) modulation in the flow of cerebral blood [21]. Furthermore, it shows different vulnerabilities in different brain conditions as the neonatal brains are more prone to NMDA antagonism. Again, while patients with vitamin B_{12} deficiency are sensitive to homocysteine problems and the patients with existing brain problems are more affected by changes in cerebral blood flow [21]. Thus, it is concluded that the level of toxicity differs with brain conditions in patients, and similarly, the risks also vary from one to another. The different neurotoxicity mechanisms as discussed in the following sections are therefore necessary for study.

2.2 N-Methyl-D-Aspartate Antagonism

The NMDA are excitatory receptors that respond to endogenous glutamate in the human body [22]. As a result, depending on how they are activated, this antagonist has both useful and harmful properties. To begin with, glutamate is an excitatory neuromodulator whose excessive release causes excite toxicity and a higher flux of Ca^{2+} ions in the neurons of the human brain [23]. As a result, it protects the brain from harmful toxins. Second, N_2O can inactivate tissue plasminogen activator and exacerbate blood–brain barrier disorder [20].

2.3 Enzyme Imbalance

Another property of N_2O that demonstrates its toxic nature is its indirect destruction of methionine synthase, which indirectly shows its toxic nature by inactivating methionine synthase [24]. Homocysteine is converted to methionine by this enzyme. N_2O fixes to the cobalt (Co) atom in Vitamin B_{12}, leading to oxidation of enzyme and inactivation of Vitamin B_{12} [25]. Vitamin B_{12} is a factor in methionine synthase. Thus, inactivation causes a deficit of enzymes. In the process cycle, methionine converts to homocysteine through *S*-adenosyl-methionine and *S*-adenosyl-homocysteine [26]. Homocysteine, a sulfur-containing amino acid, is known to have many negative effects in the body, so it can be dangerous in either form. It has been linked to high-level cardiac problems [27, 28]. Figure 1 depicts the above-mentioned information.

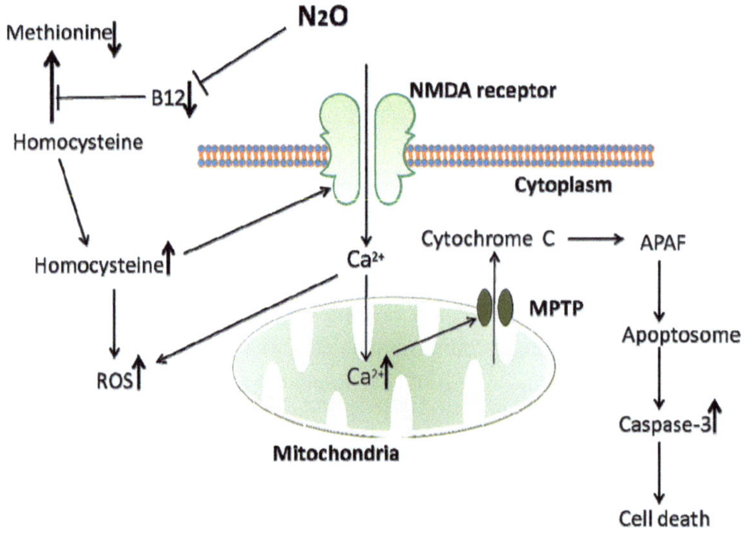

Fig. 1 A diagram of a homocysteine-facilitated cell death pathway induced by N_2O revelation [20]. N_2O hinders the action of vitamin B_{12}, a cofactor for conversion of homocysteine to methionine. This hindrance of B_{12} leads to growth of homocysteine-a toxic amino acid. Homocysteine increases reactive oxygen species (ROS) leading to eventual apoptotic cell death, and NMDA receptor activation. NMDA receptor activation can lead to an increase in ROS due to an influx of calcium into the cell

2.4 Modulated Cerebral Blood Flow

It improves cerebral blood flow and metabolic rate. It is especially noticeable when cerebral auto-regulation is disrupted, resulting in a significant reduction in cerebral perfusion [29]. It is most commonly seen in patients who have undergone neurosurgical procedures.

2.5 Analgesic Actions

It works by way of opioid receptors in the midbrain's periaqueductal grey. It also affects nociceptive pathways by releasing norepinephrine and a2 adrenoceptors in the dorsal horn of the spinal cord, which modulates them. As a result, it is concluded that N_2O's analgesic properties cause opioid peptide depletion [30].

2.6 Impact of N_2O on Environment and Climate

The impact of N_2O gas on the environment and climate is discussed in this section. It focuses on the contribution of N_2O in increasing the greenhouse effect. It is known to be more toxic than a CO gas molecule. Nearly, 70% of N_2O gas emissions are contributed by agricultural soil by processes of nitrification and denitrification. The various factors affecting the pathway of N_2O production are therefore dealt with.

2.6.1 N_2O as a Significant Greenhouse Pollutant

Human activities have resulted in environmental deprivation over the discharge of greenhouse gases and pollution. These activities led to the paradigm of economic development also [25]. N_2O, carbon monoxide (CO), and methane (CH_4) are major greenhouse pollutants. Specifically, N_2O is present in our surroundings since longe is the ozone-depleting substance of the stratosphere. Its lifetime is estimated to be 116 $\pm$ 9 years [31]. The unexpected increase in the concentration level of atmospheric N_2O over the last five decades from 270 to 331 ppb is a growing concern today [32]. The two processes, namely, nitrification and denitrification, used to control N_2O emissions in the terrestrial region regulate multiple environmental and biological factors.

The rate of rising N_2O emissions continues with the same trend due to the rising demands of food, fiber, energy, and waste generation from industries [33]. This anthropogenic N_2O emission is more noticeable in developed countries globally. It is reported that N_2O losses due to photolysis and reaction with atomic oxygen [O] is 13.5 $TgNyr^{-1}$. Due to this reason, it requires writing the biannual emission of this gas under Paris Agreement. Therefore, a global N_2O budget covers different natural and anthropogenic sources and sinks [33]. It is reported that N_2O losses due to photolysis and reaction with atomic oxygen [O] is 13.5 $TgNyr^{-1}$. The soil and oceans contribute to 57% of total N_2O oxide emissions. Also, the anthropogenic sources account for 43% of total N_2O oxide emissions directly and indirectly.

2.6.2 *N_2O Emission from Agriculture*

Agriculture soil is a significant contributor to N_2O oxide emissions, nearly more than 70%. The N_2O emissions from the soil are produced through a microbial process of nitrification and denitrification. The nitrogen fertilizers compute 42% of the N_2O emissions with the use of urea in the soil. Zhao et al. in 2020 estimated that agro-ecosystems would establish themselves as leading sources of N_2O emissions [34]. Signor and Cerri in 2013 mentioned the role of microbial processes in increasing N_2O concentrations. The different microbial processes include nitrification, denitrification, nitrifier denitrification, and ammonification, to name a few. It is noticed that 60% is due to the byproducts of nitrification and denitrification. In this process, N_2O behaves as an electron acceptor for microbial respiration [35].

2.6.3 *Pathway to N_2O Production*

The nitrogen cycle (or N cycle) is a complex pathway that includes two processes for N_2O production, namely, nitrification and denitrification. The Nitrification process takes place aerobically, while the denitrification process is anaerobic. The diagrammatic N_2O output path through the N cycle is shown in Fig. 2. Firstly, nitrification is an essential process of agriculture –ecosystems. It governs the flow of nitrogen in the system. It is a two-stage process of conversion of NH_3 to NO_2^- and NO_2^- to NO_3^- respectively [36]. The first stage of conversion occurs through NH_2OH by ammonia-fixing bacteria and ammonia-oxidizing archaea [37]. The oxidizing bacteria are *Nitrosomonas* and *Nitrosococcus,* and oxidization of NH_3 is carried out both by oxidizing bacteria and archaea, respectively. The NH_2OH oxidization and denitrifier denitrification through nitrification organisms are the two processes for producing N_2O. In the oxidation of NH_2OH, conversion of NH_2OH to NOH takes place rather than NO_2^- and hydrolysis of NO_2^- forms N_2O [37].

Secondly, the denitrification process is the conversion of NO_3^-. This process occurs in anoxic conditions deprived of dissolved oxygen and denitrifies transit from aerobic to anaerobic respiration to molecular N_2 in which several anaerobic bacteria participate [28]. It is a process opposite to nitrification, with more than one enzymatic pathway governed by heterotrophic bacteria and pseudomonas [29]. This process occurs in anoxic conditions deprived of dissolved oxygen and denitrifies transit from aerobic to anaerobic respiration. Thus, complete denitrification gives N_2, while incomplete denitrification gives N_2O as its end product.

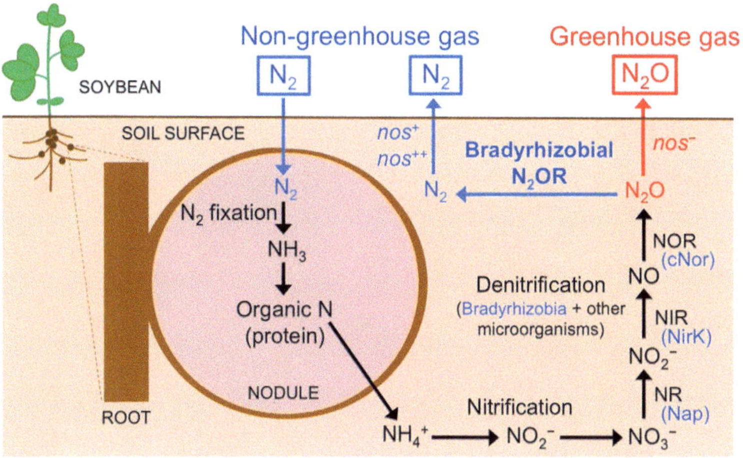

Fig. 2 An overview of the pathway for N_2O production with **a** microbial processes of nitrification and denitrification and **b** N_2O conversion taking place in the soil where red lines denote uploading of gaseous N-compounds from the soil. Organic nitrogen is mineralized into NH_4^+, transformed into N_2O through nitrification and denitrification processes. Soybean bradyrhizobia (blue) and other microorganisms contribute to the denitrification process. The N_2O formed is either emitted into the atmosphere (red) or further reduced to N_2 exclusively by soybean bradyrhizobia that produce N_2O reductase (N_2OR; blue). Nase, nitrogenase; NR, dissimilatory nitrate reductase; Nap, periplasmic nitrate reductase; NIR, dissimilatory nitrite reductase; NirK, Cu-containing nitrite reductase; NOR, nitric oxide reductase; cNor, c-type nitric oxide reductase [38]

Apart from different microbial processes, other factors such as moisture regime, soil aeration, WFPS, soil texture, fertilizer application, C/N ratio, oxygen and pH, and temperature show a vibrant role in N_2O production. The diagrammatic representation of the above factors is shown in Fig. 3 [39].

3 $MoSe_2$ ML Crystal Structure and Composition

TMDs are a large family of crystals that exist in a variety of lattice structures and exhibit a variety of physical properties, as shown in Fig. 4a [40]. The chemical abbreviation of these atoms is MX2, with M as transition metal atom and X as chalcogenides. The MX2 crystal structure is a three-folded arrangement of atoms in which occupies the center position bounded by two chalcogen atoms upwards and downwards, as shown in Fig. 4b. The bulk $MoSe_2$ unit cell is studied to exist in three polymorphs: 1T, 2H, and 3R. Here, the letter T denotes trigonal, H is for hexagonal, and R is rhombohedral. These variations originate due to the three-layer arrangement of atoms, as shown in Fig. 4c.

For monolayer transition metal dichalcogenides (TMDs) there are two widely reported structures: 2H and IT. The 2H means trigonal prismatic, and IT means

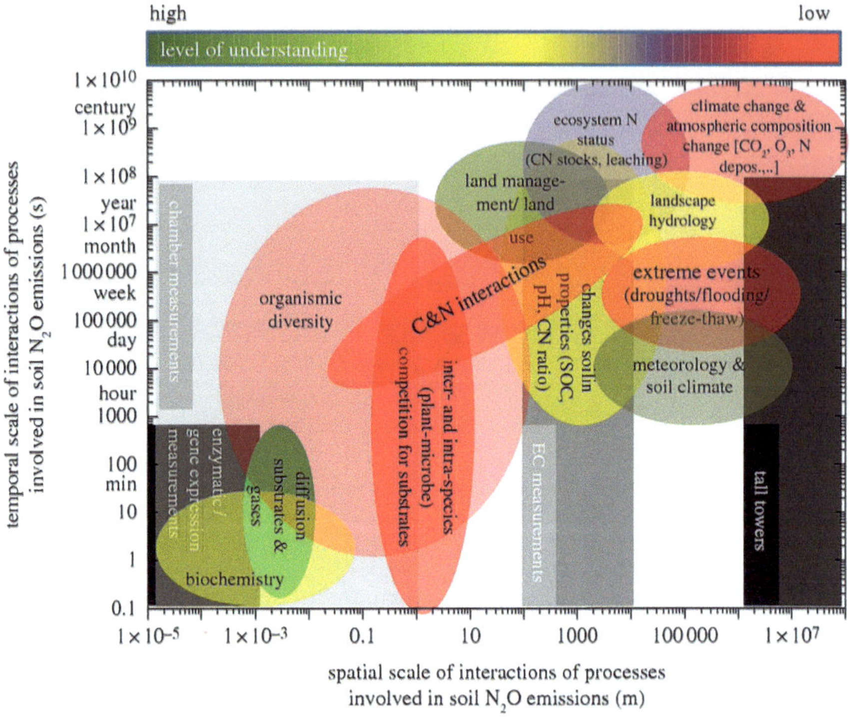

Fig. 3 Different factors are affecting soil microbial population in N_2O production. Different colors represent the level of understanding. Various measuring techniques such as enzymatic, chamber, eddy are marked by grey colors, commonly used for identifying the processes of N_2O production and consumption and soil surface fluxes [39]

octahedral separately. The 2H belongs to the D3h space group, while D3d defines the space group for 1T polymorphs [41]. These structural variations are also due to M and X's atomic planes stacking, forming MX2 crystals. A-B-A stacking is present in 2H, and A-B-C stacking is present in 1T of monolayer TMDs [42]. In A-B-A stacking, the X atoms in different atomic planes reside in the same position at the top in a direction perpendicular to the layer, as shown in Fig. 5 [43].

Further, apart from the two stable polymorphs (2H and 1T), metastable states are arising due to M–M bond formation [44]. The bonds are a result of distortions in the structure. A typical example of such distortions is the 1T metastable state formed due to dimerization of IT polymorph structure [44]. The stable states possess basic properties like 2H is semiconducting, and 1T is metallic.

The symmetry occupied by the structural phases is based on the electron count of the *d*-orbitals. The group 6 TMDs are mostly semiconducting with two electrons in $d_z{}^2$-orbitals, and group 4 TMDs are metallic with one electron in the $d_z{}^2$-orbitals.

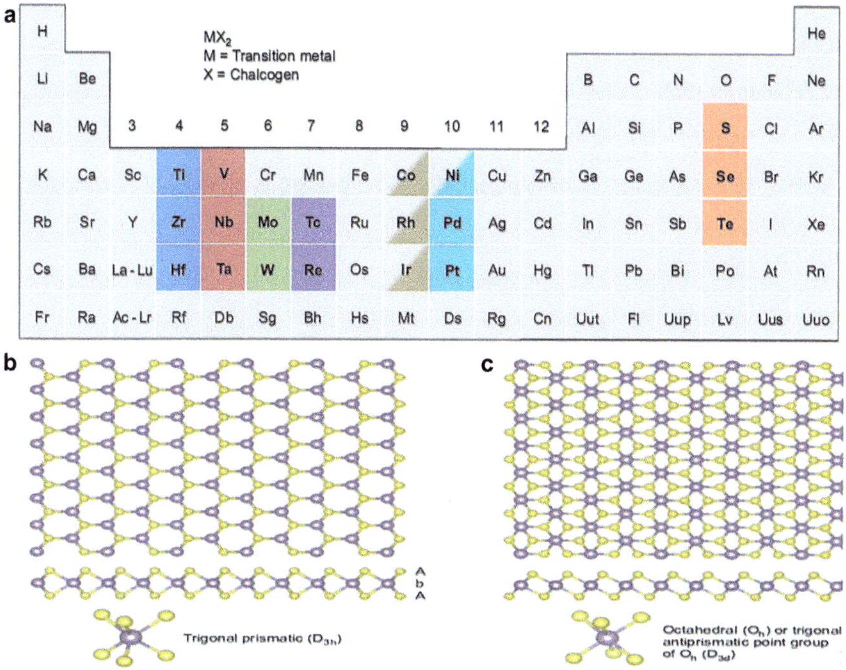

Fig. 4 **a** Representation of TMDs in (**a**) the periodic table **b** trigonal prismatic **c** octahedral or trigonal antiprismatic structure of $MoSe_2$ ML. Reprinted with permission from Nature Chemistry, Copyright 2019, Springer Nature [40]

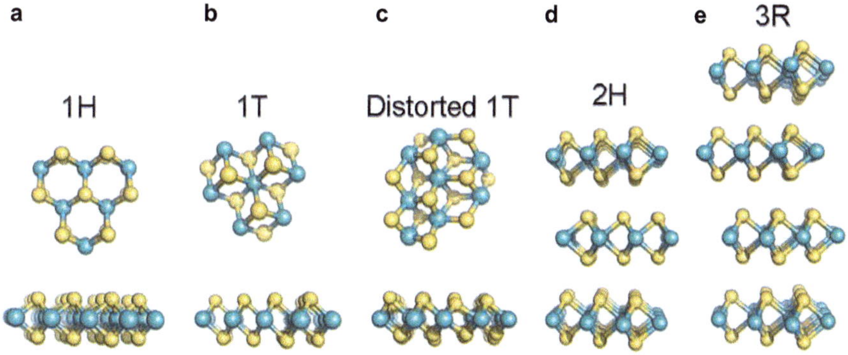

Fig. 5 **a–e** Various polytypes of 1H, 1T, 2H, and 3R phases in $MoSe_2$ ML. 1H,1T is used to present semi-metallic, 2H represents the semiconducting nature present in monolayer $MoSe_2$ [41]

4 Splitting of *d*-Orbitals in $MoSe_2$ ML

The 2H and 1T phases have trigonal prismatic and octahedral structures, as shown in Fig. 6a. The ligand field theory explains the splitting of *d*-orbitals [45]. According to this theory, the metallic and semiconducting nature shown by 1T and 2H polymorphs of $MoSe_2$ ML is attributed to the part and filled d-electrons of Mo and X atoms. The 't_{2g}' and 'E_g' splitting is present in 1T $MoSe_2$ ML. In the metallic 1T phase, the splitting of *d*-orbitals into two degenerate levels, namely, E_g constituted by $d_{x^2-y^2}$ and d_{z^2} and d_{xy}, d_{yz}, d_{xz} as t_{2g}, respectively. The E_g orbital partially occupies while the t_{2g} orbitals are empty, giving a metallic nature to the 1T phase [46]. Similarly, in the 2H phase, there are three degenerate orbitals e_1 ($d_{x^2-y^2}$, d_{xy}), e_2 (d_{yz}, d_{xz}), and a_1 (d_{z^2}).

The a_1 orbitals fully occupy, and e_1 and e_2 are either partially occupied or unoccupied, giving it a semiconducting nature. There is an apparent gap of 1 eV in the two orbitals a_1, e_1 as revealed in Fig. 6b. Also, it is observed that for 2H and 1T phases, the nonbonding d-bands are positioned between the bonding and antibonding bands of transition metal (Mo) and chalcogen (X) atom Fig. 7. The phase adopted by the TMDs depends upon the number of d-electron count of their orbitals. As with group VI, TMDs have two electrons in their d_{z^2}-orbitals, showing semiconducting nature, while group IV aligns into an octahedral structure with d^{10} electrons of the central atom.

Group V TMDs have trigonal and octahedral phases with one electron in d_{z^2} orbitals of the central atom. Group VII TMDs are distorted octahedral structures with three electrons in the d_{z^2} orbitals, and group 10 TMDs are always octahedral will fill the d-orbital of the central atom [46]. Figure 8 Shows the scanning transmission electron microscopy (STEM) image for 1H and IT with M-M distortion.

5 Synthesis Methods

The different synthesis methods used in extracting atomically thin monolayers are mechanical exfoliation, chemical vapor deposition, wet chemical method, and hydrothermal method. The commonly used plan is discussed in detail in this section.

5.1 Mechanical Exfoliation Technique

The monolayer samples prepared by mechanical exfoliation utilizing the scotch tape method are of utmost quality and more suitable for high-act devices and condensed matter phenomena [42, 45, 47–55]. On the other hand, liquid exfoliation methods are suited for those solicitations where many materials are utilized, like energy storage and catalysis or fillers for composites. Furthermore, the liquid exfoliation methods

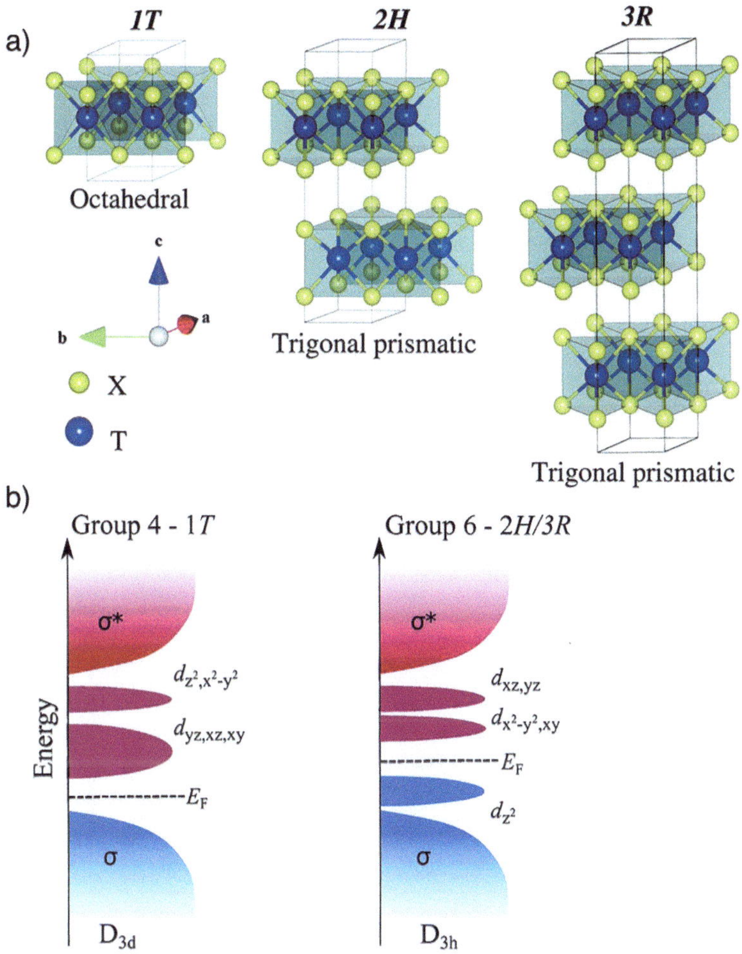

Fig. 6 **a** Trigonal-prismatic structure of 2H phase and octahedral structure of 1T phase **b** metallic nature possessed by dband filling in 2H and 1T phase of $MoSe_2$ ML. 2H phase shows semiconducting nature and 1T shows the metallic nature [43]

by direct sonification in some common solvents are castoff to dissolve graphene. Thus, this method fabricates single-layer structures such as MoS_2, WS_2, and $MoSe_2$.

The connection between chemical structure and its stability in other solvents of mechanically exfoliated single layers is investigated by dispersing graphene. The dispersion of monolayers by sonication into an aqueous solution of sodium cholate prevents the re-aggregation of monolayers. Furthermore, the sonication techniques depend on the solvent to overcome the cohesive energy of the layers placed adjacent to them. So, selected solvents should match the surface energies of the material. The challenge in these methods is to counteract the surrounding layer's surface energies and keep the dimension of the exfoliated monolayers unchanged. The drawback

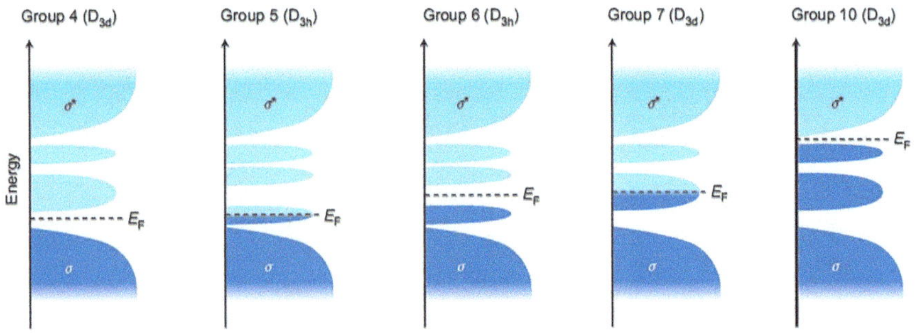

Fig. 7 The bands filling for different space groups of TMDs. Reprinted with permission from Nature Chemistry, Copyright 2013, Springer Nature. Group 6 shows the semiconducting nature represented by 2H phase whereas Group 7 shows the metallic phase given by 1T phase [40]

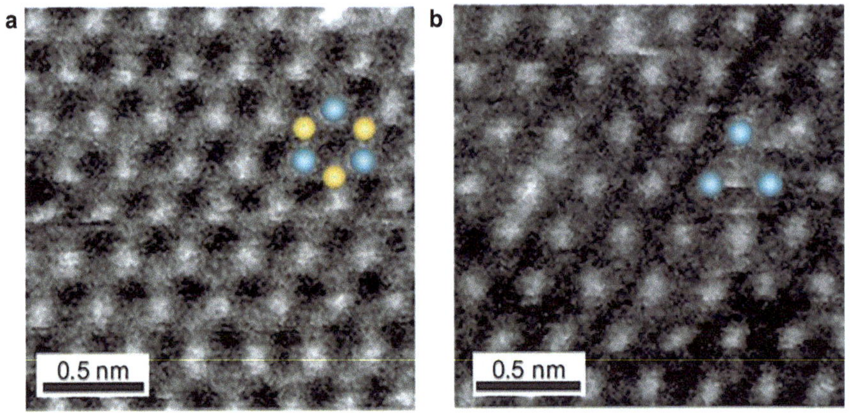

Fig. 8 Dark-field scanning transmission electron microscopy image of single-layer MoS_2 showing the contrast variation of 1H (a) and 1T (b) phases. Blue and yellow balls indicate Mo and S atoms, respectively [40]

with the sonication technique is that it breaks the lateral dimensions of exfoliated single-layer [42].

The most efficient method of exfoliating single layers with field production is ultrasound stimulated hydration of lithium intercalated compounds. MoS_2 monolayer is prepared with n-butyl merged in hexane as an intercalating agent. The outcome of this method is very high. The only reason behind the Li intercalation method is to entree the metallic 1T-phase tempted by charge transfer from Li to the TMDs. A bilayer of water passivates the 1T obtained in solid form and residual charges. Although this method has better yield over others but with has certain limitations.

The first limitation is the high-temperature requirement followed by long-duration hours. Careful and controlled intercalation is required to prevent the formation of metal nanoparticles. Figure 9 shows the exfoliation of single-layer TMDs. The gain

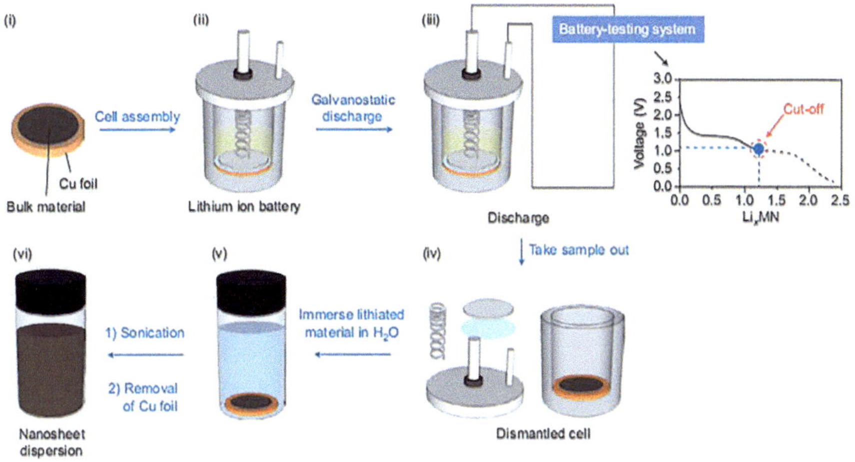

Fig. 9 Chemical exfoliation method for preparation of single MoS_2, Reprinted with permission from Nature Chemistry, Copyright 2013, Springer Nature [40]

of this process is that lithium intercalation can be scrutinized and controlled at the desired level to prevent the deposition of lithium-intercalated compounds. Thus, single layers of high–quality are possible to prepare with this method [46].

5.2 Chemical Vapor Deposition Method

In the beginning, graphene is synthesized with the CVD method and other approaches [56, 57]. The successful extraction of graphene nanosheets has prevailed the researchers to apply this method to MoS_2 and other TMDs. Recently, this method prepares the large area ultrathin MoS_2 sheets [58, 59]. The thermolysis process presents the preparation of multilayered MoS_2 in Fig. 10a. It follows by dip-coating multi-sheets in ammonium thiomolybdates ($(NH_4)_2MoS_4$) and converting them of multi-sheets in ($(NH_4)_2MoS_4$) and converting them back into the desired MoS_2 by annealing and sulphurization back into the desired MoS_2 at 500 °C and 1000 °C respectively.

The chemical reaction involved in the process is $(NH_4)_2MoS_4 + H_2 = MoS_2 + 2H_2S + 2\,NH_3$. Next, it follows second annealing at 1000 °C to improve the domains of extracted MoS_2 layer. It is to observe photoluminescence and gate dependency in FET devices of MoS_2 sheets. The MoS_2 based FETs pretend n-type behavior with a high ON/OFF ratio up to 10^5 and mobility of 6 $cm^2V^{-1}\ s^{-1}$ [60]. This precursor was opted to deposit the MoS_2 layer on graphene, as shown in Fig. 10b [61].

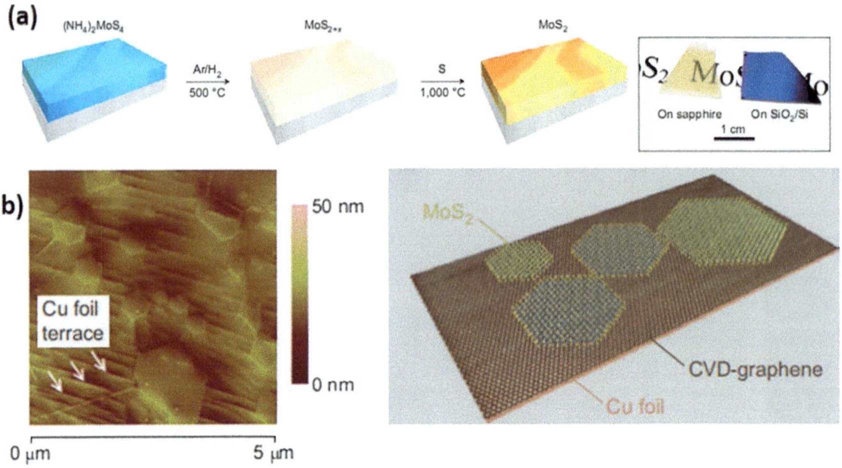

Fig. 10 **a** The thermolysis method of preparing MoS_2 monolayer **b** MoS_2 single layer deposited on graphene using van der Waal epitaxy method, Reprinted with permission from Nature Chemistry, Copyright 2013, Springer Nature [40]

6 Surface Dynamics of Synthesized Single-Layer TMDs

Other limitations arise at the surface of single-layer TMDs post-synthesis. The experimental methods, basal planes, and prismatic edges form with the exfoliation of the MX_2 single layer. Their termination is necessary and controllable either by M or X atom, relying on the capability of the growth environment. One of the drawbacks of MX_2 exfoliation is the lack of *s-p* orbital interaction with adjacent layers that widen the material's bandgap. Also, changes in orbital hybridization cause monolayer MX_2 sheets to change their behavior from indirect to direct semiconductors. Thus, it enhanced photonic properties over the bulk semiconductor [62].

The other way to control surface chemistry is by containing Mo and X atoms in the MX_2 monolayer. When the size of the multi-layer MX_2 sheet reduces to monolayer, the formation of edges, corner atoms that constitute surplus local chemical effects minimize. The open sites at the surface of MX_2 sheets form metallic state edges, which have an essential role in catalysis [63]. Figure 11a shows the sulfur vacancy created in the MoS_2 sheet and are active sites for hydrodesulfurization. The refractory of the sulfur molecule, which is a crucial contaminant in fuel, can be removed by dibenzothiophene (DBZ) is used as a probe to test the catalytic activities of MoS_2 nanosheets, as shown in Fig. 11b.

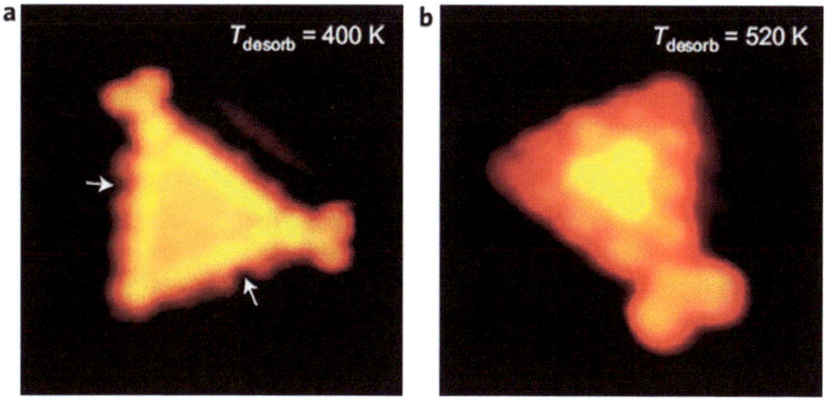

Fig. 11 **a** STM image of MoS_2 cluster with Mo terminated edge after dosing of DBT at 300 K with sulphur vacancies **b** S terminated edge of MoS_2 cluster after dosing DBT at 300 K. Reprinted with permission from Nature Chemistry, Copyright 2013, Springer Nature [40]

7 Interaction of N_2O with $MoSe_2$ ML

7.1 Electronic Properties

The electronic properties analyze the change in properties of $MoSe_2$ ML with the approaching gas molecule in this section. It discusses the modulation in the properties of monolayer such as adsorption energy, band structure, and density of states below.

7.1.1 Modulation in Gas Adsorption Energy (E_{ads})

Adsorption energy (E_{ads}) is the term that relates the interaction of the target gas molecule (N_2O) with the absorbent ($MoSe_2$ ML). It is expressed in eV. The word is so-called because when the gas molecule approaches the monolayer, it perturbs the electronic field of the monolayer, and loss or gain of electrons between them occurs. It is through this transfer or increase of electrons by the gas molecule said as adsorption. The adsorption is of two types: Chemical (or attractive) and Physical (or repulsive). A chemical bond forms in the chemisorption between the adsorbent ($MoSe_2$ ML) and analyte (N_2O). The adsorption energy is negative during chemisorption, and the adsorption distance is less than 3 Å [64].

In physisorption, the interaction occurs through weak Van der Waal forces without any bond formation. The adsorption energy is positive, and the adsorption distance is higher than 3 Å [64]. It is governed by the following equation [65]:

$$E_{ads} = E_{gas+monolayer} - E_{gas} - E_{monolayer} \quad (1)$$

$E_{gas+monolayer}$ is the system's total energy in the above equation, including adsorbent and analyte; E_{gas} signifies the energy for the isolated gas molecule (N_2O), and Emonolayer denotes the significant energy of the individual monolayer ($MoSe_2$ ML) respectively. The adsorption energy for N_2O on $MoSe_2$ ML for different configurations (namely Top Mo, Top Se) estimates to be − 1.05, 1.38 eV at an adsorption distance of 2 Å.

At 1.5 Å, it is 3.00 eV for Top Se and 1.08 eV for Top Mo separately as shown in Fig. 12 [66]. In comparison with other NO_x gases (NO, NO_2) it is noticed as − 1.33 eV for NO and − 1.15 eV for NO_2 gas molecule on $MoSe_2$ ML consecutively [67]. It is estimated for Rh-doped $MoSe_2$ ML to be − 2.00 and − 2.56 eV for NO and NO_2 gas molecules [68]. For MoS_2 as adsorbent, it is − 0.04 eV for NO gas molecule [69]. For Co, Ni, Ag, Au, Rh, Pd, Pt., Ir doped MoS_2 the adsorption energies are 2.73, 1.69, 1.41, 0.87, 1.12, 2.53, 0.93, 1.24, 3.93 eV subsequently [70]. The charge transfer is associated with the adsorption energy in direct proportion as shown in Table 1 [66].

The Chemoresistive gas sensor follows the charge transfer mechanism to interact with the gas molecule [66]. The charge transfer is extracted through the following equation [66]:

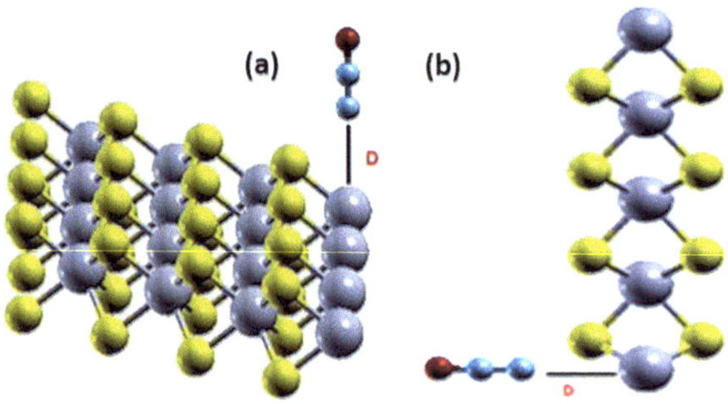

Fig. 12 Graphical presentation of N_2O adsorption on $MoSe_2$ ML with varying distance (D). The impact of distance 'D' is shown on electronic properties such as adsorption energy and charge transfer. Reprinted with permission from IEEE Sensors Journal, Copyright 2021, IEEE [66]

Table 1 Adsorption Energies (E_{ads}) and Charge Transfer (Q_T) for N_2O and NO_x gas molecules on $MoSe_2$ ML

S. No	Configuration	Charge transfer (Q_T)(e)	Adsorption energy (E_{ads})(eV)
1.	$N_2O/MoSe_2$ (Top Mo)	1.07	− 1.05
2.	$N_2O/MoSe_2$ (Top Se)	1.11	1.38
3.	NO/Pd-$MoSe_2$	− 0.10	− 1.33
4.	NO_2/Pd-$MoSe_2$	− 0.39	− 1.15
5.	NO/Rh-$MoSe_2$	0.32	− 2.56
6.	NO_2/Rh-$MoSe_2$	− 0.36	− 2.00
7.	NO/$MoSe_2$	0.035	0.167
8.	$NO_2/MoSe_2$	0.149	0.199

$$Q_{total} = Q_{ads} - Q_{iso} \tag{2}$$

7.1.2 Band Structure of Adsorbed N_2O on $MoSe_2$ ML

The band structure calculation gives the graphical representation of the different energy levels presents inside the material. The group of energy levels constitutes energy bands. For plotting of band structure of monolayer materials, the preferred high symmetry points are Γ-M-K-Γ. These materials have a tunable bandgap of 1–2 eV. The bandgap of the $MoSe_2$ ML is reported from 1.44 eV [66] to 1.90 eV [71] using different exchange functional such as GGA, LDA, and HSE06 under the first principle calculations. The commonly used function is GGA, but it has the disadvantage that it underestimates the material's bandgap. So, GGA + U is a correction parameter used for the accurate scheming of the band structure.

The value of U parameter varies from 2.5 to 4 eV [72]. Figure 13a shows the bandgap of the $MoSe_2$ monolayer to be 1.44 eV without N_2O adsorption. In Fig. 13b, with N_2O adsorption in Top Mo configuration, a rise in energy states close to the Fermi level (here, shifted to zero) with a bandgap value of 1.39 eV occurs. Similarly, for the Top Se configuration, the increase in energy states is maintained with a bandgap value of 1.25 eV, as shown in Fig. 13c.

Another approach for analyzing the impact of N_2O gas adsorption is through doping of transition metal atom (Mo) with other 3d series atoms. The 3d series atoms have unoccupied *d*-orbitals, which, when doped on the Mo atom, modulates the behavior of the $MoSe_2$ monolayer. The maximum and minimum values of the valence (conduction) band (VBM) (CBM) are along Z-D for metallic nature and D-B for semiconducting nature. Also, for Pd-doped Mo atom on $MoSe_2$ ML, it is found to have semiconducting nature with a bandgap value of 1.469 eV, as shown in Fig. 14 [67].

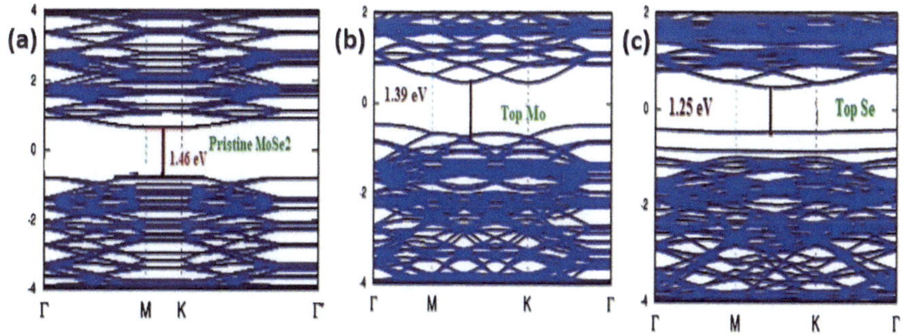

Fig. 13 Band structure of **a** Pristine $MoSe_2$ **b** Top Mo with N_2O adsorption **c** Top Se with N_2O adsorption. The impact of gas adsorption is shown from 1.46 to 1.25 eV from pristine to Top Mo and Top Se separately. Reprinted with permission from IEEE Sensors Journal, Copyright 2021, IEEE [66]

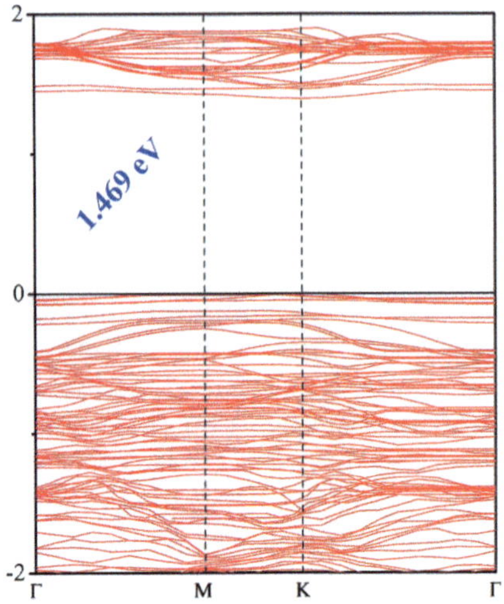

Fig. 14 Band structure for Pd-doped $MoSe_2$ monolayer, Reprinted with permission from Physics Letters Copyright 2019, Elsevier [61]

The Pd-atom is showing the electron-accepting nature by withdrawing electrons from the Mo atom in the $MoSe_2$ ML. As a result, the Fermi level shifts upwards or towards the conduction band, displaying the n-type behavior. The narrowing of the bandgap from the original $MoSe_2$ ML increases the conductivity of the monolayer. The dangling bonds present at the surface of Pd-doped $MoSe_2$ constitute the conductive band formation of the Pd-$MoSe_2$ monolayer.

7.1.3 Band Structure of NO_x Adsorbed $MoSe_2$ ML

In Fig. 15a and b, NO and NO_2 adsorbed on Pd-$MoSe_2$ ML are displayed. The NO-adsorbed Pd-$MoSe_2$ ML is semiconducting with a bandgap value of 1.208 eV [67]. In addition, there occur some novel energy levels close to Fermi level with the VBM (CBM) located at K-point. For the NO_2 system, the semiconducting bandgap is 1.053 eV, as shown in Fig. 15b. Also, the novel states are observed nearby to the Fermi level.

The adsorption of NO_2 gas on the Pd-$MoSe_2$ surface shows narrowed bandgap with increased electron concentration constituting higher conductivity.

7.1.4 The Density of States (DOS)

Figure 16 presents the TDOS of N_2O adsorption on Top Mo, Top Se, and Pristine $MoSe_2$ ML. The pristine $MoSe_2$ ML is non-magnetic as both the (spin-up and -down) channels are symmetric to each other. The flat bands near the Fermi level indicate the semiconducting nature of the monolayer. It verifies from the DOS that flat bands are in the range of 1.90 eV on the energy axis, equal to the bandgap value. On the

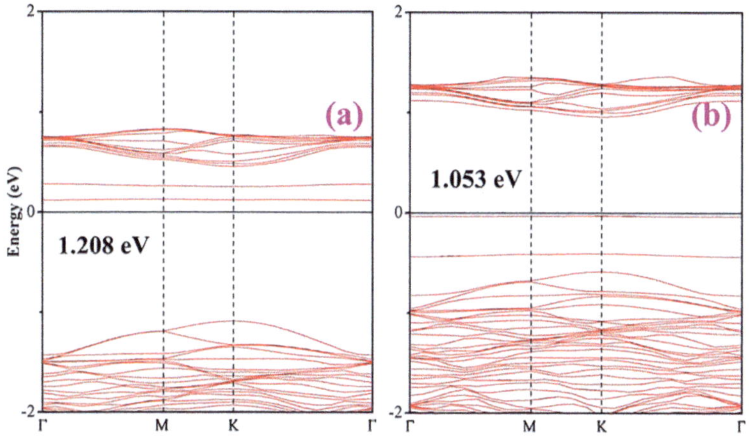

Fig. 15 Comparative Band structure of **a** NO and **b** NO_2 adsorbed on Pd-doped $MoSe_2$ ML, Reprinted with permission from Physics Letters Copyright 2019, Elsevier [67]

other hand, with the N_2O adsorption on the two configurations: Top Mo and Top Se, the flat regions near the Fermi level disappear.

The induced states as a result of adsorption show that Top Se is more favorable than Top Mo. Also, both the configurations have asymmetric nature as the spin (up and down) channels are not symmetric to each other. Higher peaks introduced in

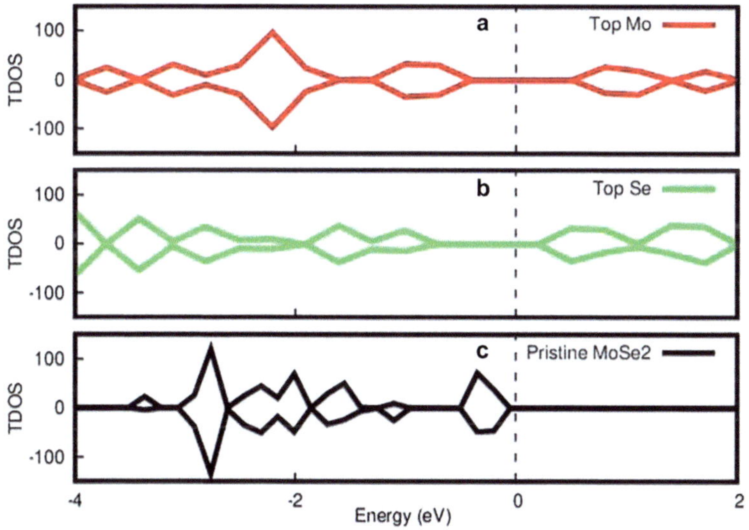

Fig. 16 The TDOS presentation of N_2O adsorption on Top Mo and Top Se $MoSe_2$ ML. Impurities are incorporated due to gas adsorption in both the Top Mo and Top Se configurations, shown by the asymmetric nature. Reprinted with permission from IEEE Sensors Journal, Copyright 2021, IEEE [66] (TDOS: Total density of states)

the doped system near the Fermi level, maintaining the magnetic nature. It can also attribute more electrons in the valence configuration of the Fe atom relative to the Mn atom. Later, with the adsorption of N_2O gas molecule, increased peaks with reduced magnitude are obtained near the Fermi level. The spin-down channel shows increased participation with higher elevations than the spin-up channel, as shown in the figure. In this case, too, the electron-withdrawing nature of the N_2O gas molecule is maintained.

7.1.5 *Density of States of NO_x Adsorbed Gas Molecules*

The TDOS representation of NO and NO_2 adsorbed gas molecules on Pd-doped $MoSe_2$ ML is shown in Fig. 17a [67]. For NO adsorption, there are spots of induced states at the Fermi level. The magnetic nature maintains asymmetry in spin channels, the DOS plot shifts towards the left after adsorption. The NO gas molecule shows an electron-donating nature as the peak minimizes post- gas adsorption. In the isolated state, the NO gas molecule shows to possess magnetic nature. The increased area of overlapping between N-2*p* orbitals and P*d*-4*d* orbitals defines the level of interaction in the system.

Similarly, the NO_2 gas molecule's more width is occupied by the TDOS relative to the NO system. Likewise, NO molecule, NO_2 molecule also carries a magnetic character. Interestingly, the TDOS plot for NO and NO_2 adsorbed Rh-doped $MoSe_2$ ML, as shown in Fig. 17b, c, represents similar behavior. The NO_2 molecule is more overlapped than the NO molecule with the Rh atom, as shown above. Also, the NO_2 molecule splits into several peaks post adsorption. Unlike the NO molecule, the amplitude of peaks is reduced post adsorption for the NO_2 molecule. The magnetic nature maintains for both NO and NO_2 adsorbed molecules on Rh-doped $MoSe_2$ ML, respectively.

7.1.6 *Density of States for Doped $MoSe_2$ ML with 3d Metal Atoms*

The DOS plot displays the electronic behavior of Mn, Co, and Ni-doped $MoSe_2$ ML in Fig. 18a–d. The orbital DOS is shown adjacent to the TDOS in Fig. 19a–d [73]. The Fermi level shifts towards the conduction band in the doped system. A magnetic asymmetry is observed in all four systems, respectively indicating the presence of magnetic states. The flat band states near the Fermi level represent the semiconducting nature for Mn, Co, and Ni $MoSe_2$ ML, respectively. The Fe-doped $MoSe_2$ ML owns the half-metallic nature, as shown in Fig. 18b [73].

The orbital PDOS shows the hybridization of doped atoms and Se atoms to locate each other with the same energy level. It also suggests the more robust interaction present between them. Thus, from the above analysis, it infers that doping of 3d metal atoms imparts magnetic character in the doped $MoSe_2$ ML.

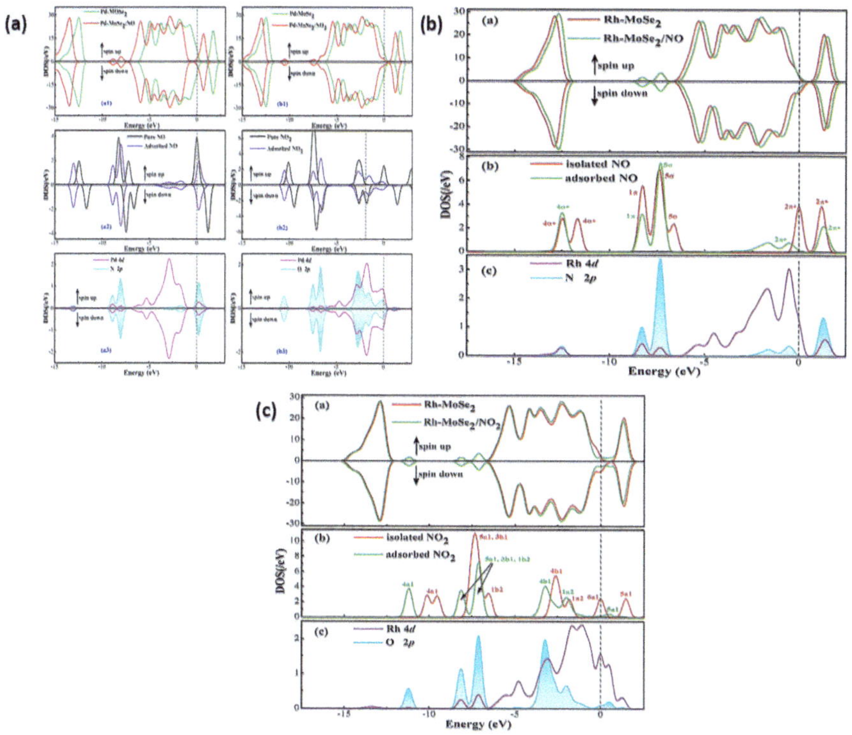

Fig. 17 Partial and total DOS for **a** Pd-doped NO and NO_2 **b** Rh-doped NO **c** Rh-doped NO_2 on $MoSe_2$ ML. Reprinted with permission from Physics Letters A, Copyright 2019, Elsevier [67, 68]

7.2 *Optical Properties*

7.2.1 *Optical Properties of NO_x Adsorbed on $MoSe_2$ ML*

This section studies the fundamental optical properties of $MoSe_2$ with (without) NO_x (NO, NO_2) gas adsorption. The actual optical properties studied for monolayer are dielectric constant (real and imaginary), absorption index, and refraction index, respectively [74]. The most important property is the imaginary part ($\varepsilon_2(\omega)$) of the dielectric constant to which other optical properties are dependent [75]. In general, the $\varepsilon_2(\omega)$ defines the material's ability to absorb light. The $\varepsilon_2(\omega)$ is extracted through Kramer-Kronig relation [75], and real part $\varepsilon_1(\omega)$ derives from $\varepsilon_2(\omega)$. The equations involved in taking out of $\varepsilon_2(\omega)$ is given in (3)–(4):

$$\varepsilon(\omega) = \varepsilon_1(\omega) + i\varepsilon_2(\omega) \tag{3}$$

$$\varepsilon_2(\omega) = \frac{e^2 h}{\pi m^2 \omega^2} \sum_{V.C} \int_{BZ} |M_{CV}(k)|^2 \delta[\omega_{CV}(k) - \omega] \mathrm{d}^3 k \tag{4}$$

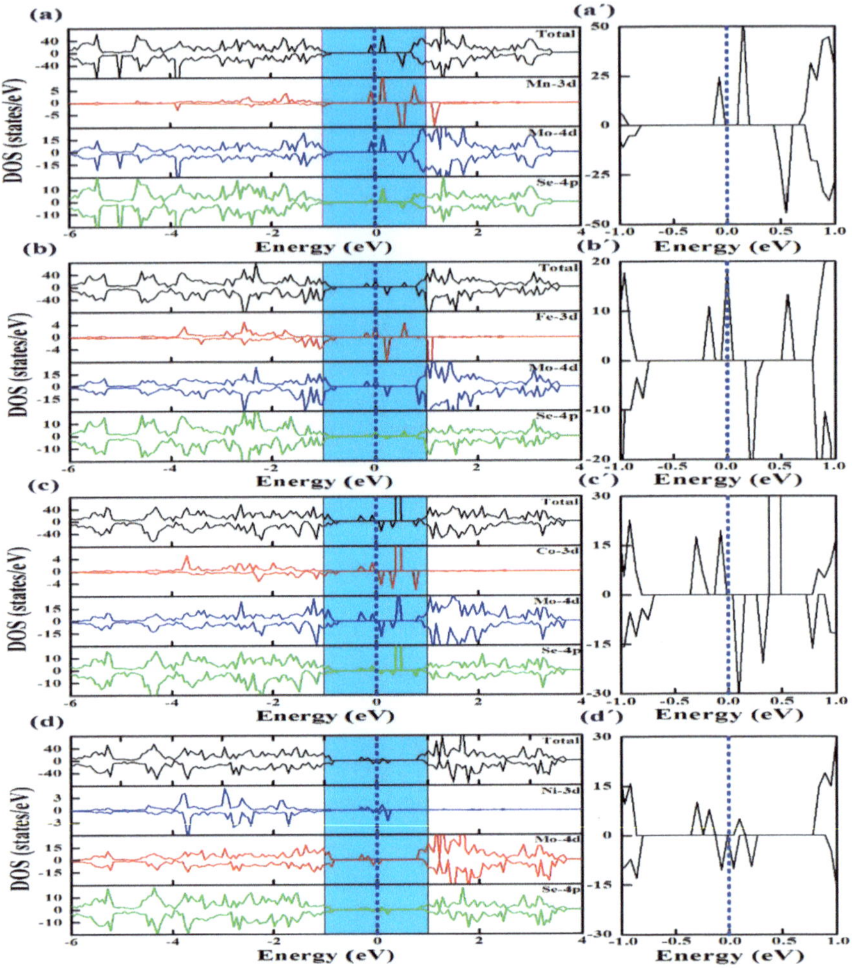

Fig. 18 The total and partial DOS plot for **a** Mn **b** Fe **c** Co **d** Ni-doped $MoSe_2$ ML. The asymmetry is shown for the doped configurations. The impurities are shown near the fermi level (shifted to zero). Reprinted with permission from Physica E, Copyright 2019, and Elsevier [73]

Also, the optical properties of the $\varepsilon_2(\omega)$ are connected to the bandgap value as electronic fragments densely [75]. It is reported through the Penn model that bandgap and static dielectric constant $\varepsilon_1(\omega)$ are inversely related to each other [76, 77]. An extensive bandgap value results in smaller $\varepsilon_1(\omega)$ and vice versa [77]. The relation between $\varepsilon_2(\omega)$ and other optical properties are given by the Eqs. (5–7):

$$\varepsilon_1(\omega) = 1 + \frac{2}{\pi} \int_0^{\infty} \frac{\omega \prime \varepsilon_2(\omega')d(\omega')}{\omega'^2 - \omega^2} \tag{5}$$

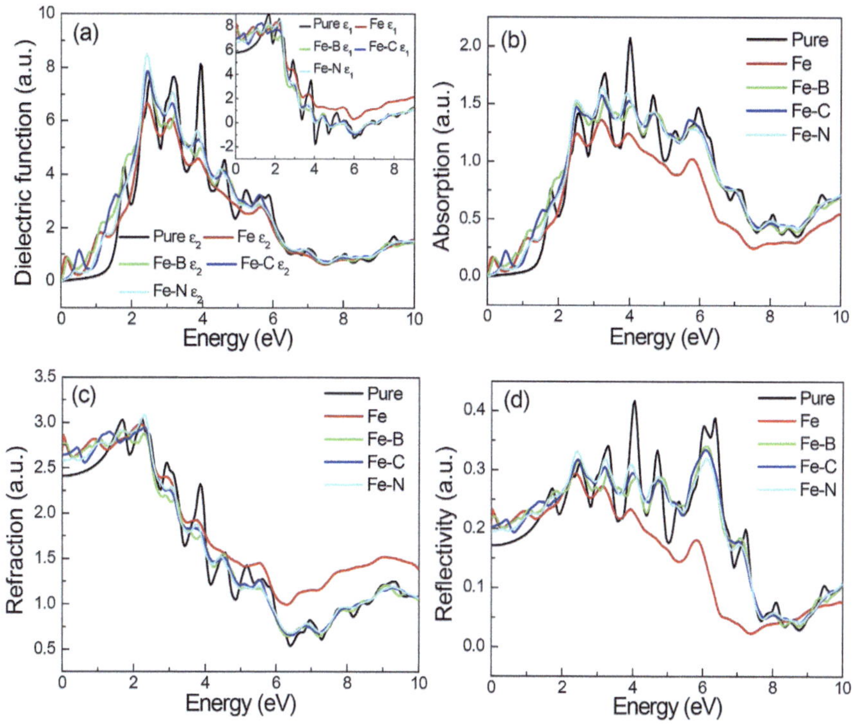

Fig. 19 Optical attributes **a** imaginary part of dielectric constant **b** real part of dielectric constant **c** absorption index and **d** refractive index of Pure, Fe-doped and Fe-co-doped (B, C, N) $MoSe_2$ ML. The peaks in the plots are lying in the range of 2–4 eV on energy scale. Reprinted with permission from Journal of Alloys and Compounds, Copyright 2019, Elsevier [81]

$$\alpha(\omega) = \frac{4k\pi}{\lambda} = \frac{\omega}{nc}\varepsilon_2(\omega) \tag{6}$$

$$n(w) = \frac{1}{\sqrt{2}}\left[(\varepsilon_1^2 + \varepsilon_2^2)^{\frac{1}{2}} + \varepsilon_1\right]^{\frac{1}{2}} \tag{7}$$

The imaginary part of the dielectric constant ($\varepsilon_2(0)$) is analyzed To investigate the effect of gas adsorption on the optical properties of $MoSe_2$ ML [78]. The dielectric constant has a higher value for the pristine $MoSe_2$ monolayer than other semiconducting TMDs. The pure monolayer has a value of 2.3 eV while it decreases to 0.47 eV with NO_2 adsorption to no change or zero value for NO adsorption as given in Table 2 [78]. The dielectric constant has a higher value for the pristine $MoSe_2$ monolayer than other semi-conducting TMDs. Thus, these optical properties can be simulated for N_2O adsorbed $MoSe_2$ ML and experimentally verified using various spectrometric techniques.

Table 2 The imaginary part of the dielectric constant of pristine and NO_x adsorbed $MoSe_2$ ML

Structure	Imaginary part of dielectric constant ($\varepsilon_2(0)$) (eV)
$MoSe_2$ ML	2.3
$NO/MoSe_2$ ML	0
$NO_2/MoSe_2$ ML	0.47

7.2.2 *Optical Properties Present in Doped and Co-doped $MoSe_2$ ML*

This section discusses the optical properties owned by Fe-doped (Fe-co-doped (B, C, N)). Figure 19a shows the real and imaginary part of the dielectric constant for the above system. It is observed that Fe-doped has higher $\varepsilon_1(\omega)$ than co-doped configurations. It is increasing in the energy range of 0–0.7 eV, describing its half-metallic character [79]. The dielectric constant further decreases and becomes negative with the increasing energy range of 4–6.5 eV. The negative value shows that it does not allow energy to propagate through it quickly [80]. The imaginary part $\varepsilon_2(\omega)$ first increases then decreases downwards.

There is a relation between the density of states and $\varepsilon_2(\omega)$, as there are shared singularities among them showing the excited electrons of different wave vector positions. The high steep in the small energy region for Fe-doped (Fe-B co-doped) indicates the inter-and intra-band transitions from the half-metallic and semiconducting nature [81]. The optical absorption is shown in Fig. 19b is a derived quantity from the imaginary part ($\varepsilon_2(\omega)$) of the dielectric function. It is observed that pristine $MoSe_2$ ML has higher absorption than doped and co-doped in the energy range of 3–6 eV. The absorption that occurred here is the infra-red absorption. The red-shift phenomenon is also noticed in the energy range of 0–1.5 eV.

With the increase in doping, the intensity of infra-red absorption increases, s increased showing their ability to absorb infrared light.

Specific peaks were reported in the 0–2 eV energy scale of 0–2 eV due to the impurity doping in the Fe-doped and Fe-NM co-doped configurations [81]. The electrons and holes absorb photons and are excited to conduction band from valence bands with increased impurity. It creates absorption peaks with a smaller energy range and insufficient excitation. Also, absorption within a semiconductor occurs only when a photon absorbs light to a value higher than the energy bandgap, the maximum absorption intensity at 2.7 eV.

The Fe-co-doped configurations show absorption in the visible region in contrast to the Fe-doped and pure $MoSe_2$ monolayer. Doped configurations dominate over the pure $MoSe_2$ ML in reflectivity (r(0)). It is the same as the trend with the real part of dielectric function ($\varepsilon_1(\omega)$) as shown in Fig. 19c. As shown in Fig. 19d, the refractivity pattern illustrates that doped and co-doped configurations have high n(0) than pure $MoSe_2$ ML. Another reason for n(0) different values is the dispersion of light when it penetrates abysmal into monolayer material and interacts with the material's atoms [81]. Again, it maintains a trend similar to the real part of the dielectric part of $\varepsilon_1(\omega)$

Table 3 Optical properties of pristine and co-doped (B, C, N) $MoSe_2$ ML

S. No	Configuration	Static dielectric constant ($\varepsilon_1(0)$)	Absorption index ($\alpha = 0$)	Refraction index ($n = 0$)
1	Pristine $MoSe_2$	5.81	1.14	2.4
2	Fe–B co-doped $MoSe_2$	7.668	0.000	2.77
3	Fe–C co-doped $MoSe_2$	6.995	0.220	2.64
4	Fe–N co-doped $MoSe_2$	6.669	0.110	2.54

displayed below. The optical properties for pristine, Fe-doped and Fe-co-doped (B, C, N) $MoSe_2$ monolayer is shown in Table 3 [81].

7.3 *Magnetic Properties*

7.3.1 *Magnetic Properties of NO_x Adsorbed on $MoSe_2$ ML*

The section focuses on the magnetic properties of pristine $MoSe_2$ and NO_x adsorbed $MoSe_2$ ML. It is studied that spin-polarization favors non-spin polarization for magnetic properties in $MoSe_2$ ML. Although, $MoSe_2$ in its bare state is non-magnetic, as shown in Table 4. There are no unpaired electrons present in the pristine ML to induce magnetism. The magnetic moment both total and local, are the parameters mentioned in this section for the comparative study of magnetic behavior in pristine and NO_x (NO, NO_2) adsorbed $MoSe_2$ ML, respectively. The magnetic moment for pristine $MoSe_2$ is a zero value portraying its non-magnetic nature.

The adsorption of NO and NO_2 molecules on $MoSe_2$ ML induces magnetic moment, as shown in Table 4. The NO gas molecule has donating nature and increases the unpaired electrons upon adsorption on $MoSe_2$ ML. Due to this reason, it shows magnetic behavior.

In the same way, the NO_2 serves as a magnetic host when adsorbed on $MoSe_2$ ML and gives a magnetic attribute to the whole system. When applied to both the configurations mentioned above, the external electric field alters the magnetic capacity of

Table 4 Variation of magnetic anisotropic energy, Magnetic Moment (MM) (μ_B) (total) of (1) pristine (2) $NO/MoSe_2$ and (3) $NO_2/MoSe_2$ ML

Structure	Total MM (μ_B)
$MoSe_2$	0
$NO/MoSe_2$	1
$NO_2/MoSe_2$	0.99

Table 5 Magnetic Moment (MM) (μ_B) (local and total) variation of (1) pristine (2) NO/$MoSe_2$ and (3) NO_2/$MoSe_2$ ML due to external electric field of 1 V/Å

Structure	Due to external electric field (= 1 V/Å) (μ_B)
$MoSe_2$	0
NO/$MoSe_2$	1.08
NO_2/$MoSe_2$	0.74

the system, as shown in Table 5. It increases the magnetic moment to 1.08 μ_B for NO on the one hand and decreases it to 0.74 μ_B for NO_2 on the other [82]. Thus, in the same way, the magnetic nature of N_2O gas molecules with and without adsorption on $MoSe_2$ ML can be simulated, and its experimental verification is awaited.

7.3.2 *Magnetic Properties of Doped and Co-doped Atoms on $MoSe_2$ ML*

The magnetic property induced upon doping of metal atoms on $MoSe_2$ ML is studied in this section. The concept of doping is closely attached to the magnetic stability (ΔE) is extracted by the difference between ferromagnetic (FM) orderings and anti-ferromagnetic (AFM) orderings ($\Delta E = E_{FM} - E_{AFM}$). A negative value predicts FM states are more stable, whereas the positive values indicate AFM is more desired. So, it concludes that ΔE is negative for the doped (Fe-, Mn-, Co-, Ni-doped $MoSe_2$ ML) configurations, as shown in Table 6. Thus, all the arrangements (doped and co-doped) are Ferro-magnetically stable and agree with the reported work as displayed in Table 7 [73].

Table 6 Total magnetic moment (MM) and magnetic stability (**ΔE**) for different 3d metal doped $MoSe_2$ ML

Structure	Total MM (μ_B)	Magnetic stability (ΔE) (eV)
Fe-$MoSe_2$ ML	1.963	– 0.14
Mn-$MoSe_2$ ML	1.074	– 0.17
Co-$MoSe_2$ ML	2.760	– 0.21
Ni-$MoSe_2$ ML	1.765	– 0.07

Table 7 Total magnetic moment of Fe-doped and Fe-codoped (B, C, N) $MoSe_2$ ML

Structure	Total MM (μ_B)	Local MM (μ_B)
Fe-$MoSe_2$	2.00	1.371
Fe–N-$MoSe_2$	0.990	– 0.750
Fe–B-$MoSe_2$	0.997	0.770
Fe–C-$MoSe_2$	0	0

Table 8 Total magnetic moment of different non-metals doped on $MoSe_2$ ML

Structure	Local MM (μ_B)	Total MM (μ_B)
H-$MoSe_2$	− 0.004	1
B-$MoSe_2$	0.029	1
C-$MoSe_2$	0	0
Si-$MoSe_2$	0	0
N-$MoSe_2$	0.185	1
P-$MoSe_2$	0.144	1
As-$MoSe_2$	0	0
O-$MoSe_2$	0	0
S-$MoSe_2$	0	0
Te-$MoSe_2$	0	0
Cl-$MoSe_2$	0.038	1
Br-$MoSe_2$	0.046	1
I-$MoSe_2$	0.053	1

The nature of Fe-atom doped $MoSe_2$ ML, both spin-polarized and non-spin polarized, is required for ground-state calculations. Therefore, it results in weak antiferromagnetic coupling per neighboring (Mo, Se) atom and is in sequence with the stated work [78, 83] as presented in Table 6. Similarly, the total and local MM varies with non-metal doping, as shown in Table 8 [84–86].

8 Summary

The chapter concludes by highlighting the environmental and health implications of nitrous oxide (N_2O), a potent greenhouse gas with a global warming potential far exceeding that of carbon monoxide (CO). Research identifies N_2O as a major pollutant, with over 70% of its atmospheric emissions originating from agricultural activities. It is produced through both aerobic (nitrification) and anaerobic (denitrification) pathways within the nitrogen cycle. Additionally, the chapter underscores how N_2O adsorption significantly modifies the electronic, optical, and magnetic properties of $MoSe_2$ monolayers, reinforcing their potential use in advanced sensing applications. Microbial respiration uses this gas as an acceptor. Later, the biological aspect of N_2O gas is associated with negative health effects. In the medical field, N_2O gas is primarily used as an analgesic or anaesthesia during surgeries. The increased neurotoxicity and cardiovascular problems caused by N_2O gas are a growing source of concern. It's also the leading cause of high homocysteine levels, which leads to cell death in the human body. These harmful effects of N_2O gas require depth study, an attempt to study the impact of N_2O gas on $MoSe_2$ ML, a transition metal dichalcogenide monolayer. There are different phases of $MoSe_2$ ML such as 2H, 1T, and 3R, of which 2H is and shows semiconducting nature and more stability.

Thus, 2H $MoSe_2$ ML is opted for adsorption of N_2O gas and realizes its impact on modulation in electronic, magnetic, and optical properties. The N_2O gas molecule adsorbs on the surface of the monolayer at a variable distance of 2–1.5 Å. There is modulation in properties of N_2O gas is studied relative to other NO_x gases (NO, NO_2). The changes observed in electronic properties are mainly in the adsorption energy (E_{ads}), Charge Transfer (Q_T), (total and partial) DOS, and Band structure. Firstly, the adsorption is more preferred by the top Se configuration with a maximum adsorption energy of 1.38 eV for pristine $MoSe_2$. The chemisorption occurs in this case. It is relatively greater than NO_x gases for the same configuration. Secondly, there is a linear relation of adsorption energy with the charge transfer maintained for pristine and doped cases. Third, the DOS plot predicts the change from non-magnetic to induced magnetic nature in doped and N_2O adsorbed $MoSe_2$ ML. It observes that there are shifts in the peak values close to the Fermi level due to gas adsorption. Many mid-gap states introduced near the Fermi level evidence magnetic conditions in the doped and adsorbed ML, respectively. The modulation trend is noticed in the band structure from pristine to adsorbed state via doping of Mo atom. The change is observed mainly in the Top Se configuration with N_2O adsorption. Further, the magnetic behavior is also well pronounced with NO_x adsorption on $MoSe_2$ ML. The magnetic array forms with doping and gas molecule adsorption verified through the magnetic moment. Lastly, the behavior of the $MoSe_2$ ML towards visible light is explored through optical properties. The four optical properties, namely, absorption index, reflection index, the real and imaginary part of dielectric constant separately. The peaks (before and after) adsorption locates in the energy series of 2–4 eV, which belongs to the visible range in the electromagnetic spectrum. Also, in the Fe-doped (co-doped) configuration, red end phenomena are noticed. Therefore, this indicates that the $MoSe_2$ ML is more suitable for optoelectronic applications such as ultra-fast photo-detectors, high ON/OFF ratio $> 10^5$, display devices, and diagnosing diseases in the human body. The practical techniques available for synthesis of pristine $MoSe_2$ monolayers are chemical vapor deposition, meachanical exfoliation, wet chemical and hydrothermal method. The practically used method is the chemical vapor deposition as it extracts defect-free monolayer. The issues like defects, edges and basal planes are assoiciated with other methods and this increases the roughness in the extracted monolayer. Recently, chemical vapor deposition is adopted for dpositing a $MoSe_2$ monolayer on SiO_2 substrate in two-dimensional FETs. Finally, it is concluded that the impact of N_2O gas adsorption is greater than that of its relative NO_x gases, having a significant impact on the $MoSe_2$ ML's various properties.

References

1. A.K. Geim, K.S. Novoselov, The rise of graphene. Nat. J. **6**, 183–191 (2007). https://doi.org/10.1038/nmat1849
2. K.S. Novoselov, D. Jiang, F. Schedin, T.J. Booth, V.V. Khotkevich, S.V. Morozov, A.K. Geim, Two-dimensional atomic crystals. Proc. Natl. Acad. Sci. **102**, 10451–10453 (2005). https://doi.org/10.1073/pnas.0502848102
3. C. Lee, Q. Li, W. Kalb, X.-Z. Liu, H. Berger, R.W. Carpick, J. Hone, Frictional characteristics of atomically thin sheets. Science **328**, 76–80 (2010). https://doi.org/10.1126/science.1184167
4. L. Ci et al., Atomic layers of hybridized boron nitride and graphene domains. Nat. Mate. **9**, 430–435 (2010). https://doi.org/10.1038/nmat2711
5. Y.H. Zhang, Y.B. Chen, K.G. Zhou, C.H. Liu, J. Zeng, H.L. Zhang, Y. Peng, Improving gas sensing properties of graphene by introducing dopants and defects: a first-principles study. Nanotechnology **20**, 1–23 (2009). https://doi.org/10.1088/0957-4484/20/18/185504
6. H. Wang, C. Li, P. Fang, Z. Zhang, J.Z. Zhang, Synthesis, properties, and optoelectronic applications of two-dimensional MoS_2 and MoS_2-based heterostructures. Chem. Soc. Rev. **47**, 6101–6127 (2018). https://doi.org/10.1039/C8CS00314A
7. J.K. Huang, J. Pu, C.L. Hsu, M.H. Chiu, Z.Y. Juang, Y.H. Chang, W.H. Chang, Y. Iwasa, T. Takenobu, L.J. Li, Large-area synthesis of highly crystalline WSe_2 monolayers devices and applications. ASC Nano. **8**, 923 (2014). https://doi.org/10.1021/nn405719x
8. V. Nicolosi, M. Chhowalla, M.G. Kanatzidis, M.S. Strano, J.N. Coleman, Liquid exfoliation of layered materials. Science **340**, 568 (2013). https://doi.org/10.1126/science.1226419
9. S.V. Morozov, K.S. Novoselov, M.I. Katsnelson, F. Schedin, D.C. Elias, J.A. Jaszczak, A.K. Geim, Giant intrinsic carrier mobilities in graphene and its bilayer. Phys. Rev. Lett. **100**, 016602–016606 (2008). https://doi.org/10.1103/PhysRevLett.100.016602
10. Y.H. Zhang, Y.B. Chen, K.G. Zhou, C.H. Liu, J. Zeng, H.L. Zhang, Y. Peng, Improving gas sensing properties of graphene by introducing dopants and defects: a first-principles study. Nanotechnology **20**, 185504–185511 (2009). https://doi.org/10.1088/0957-4484/20/18/185504
11. H. Namduri, S. Nasrazadani, Quantitative analysis of iron oxides using Fourier transform infrared spectrophotometry. Corros. Sci. **50**, 2493–2497 (2008). https://doi.org/10.1016/j.corsci.2008.06.034
12. M.T. Silva, E. Nouli, N.P. Xekoukoulotakis, D. Mantzavinos, Effect of key operating parameters on phenols degradation during H_2O_2-assisted TiO_2 photocatalytic treatment of simulated and actual olive mill wastewaters, Appl. Catal. B, Environ. **73**, 11–22 (2007). https://doi.org/10.1016/j.apcatb.2006.12.007
13. S.H. Brewer, A.M. Allen, S.E. Lappi, T.L. Chasse, K.A. Briggman, C.B. Gorman, F. Stefan, Infrared detection of phenylboronic acid terminated alkanethiol monolayer on gold surfaces. Langmuir **20**, 5512–5520 (2004). https://doi.org/10.1021/la035037m
14. M. Brotman, S.C. Cullen, Supplementation with Demerol during nitrous oxide anesthesia. Anesthesiology **10**, 696–705 (1949). https://doi.org/10.1097/00000542-194911000-00005
15. W.T. Honeycutt, M.T. Ley, N.F. Materer, Precision and limits of detection for selected commercially available, low-cost carbon dioxide and methane gas. Sensors **19**, 1–15 (2019). https://doi.org/10.3390/S19143157
16. D. Trevor, T.D. Rapson, D. Helen, Analytical techniques for measuring nitrous oxide, TrAC. Trends Anal. Chem. **54**, 65–74 (2014). https://doi.org/10.1016/j.trac.2013.11.004
17. K.M.T.S. Bandara, K. Sakai, T. Nakandakari, K. Yuge, A low-cost NDIR-based N_2O gas detection device for agricultural soils: assembly, calibration model validation, and laboratory testing. Sensors **21**, 1189–11105 (2021). https://doi.org/10.3390/s21041189
18. M.S. Yazici, B. Dong, D. Hasan, F. Sun, C. Lee, Integration of MEMS IR detectors with MIR waveguides for sensing applications. Opt. Express **28**, 11524–11538 (2020). https://doi.org/10.1364/OE.381279

19. K. Leslie, P.S. Myles, M.T. Chan, A. Forbes, M.J. Paech, P. Peyton, B.S. Silbert, E. Williamson, Nitrous oxide and long-term morbidity and mortality in the enigma trial. Anesth. Analg. **112**, 387–393 (2011).https://doi.org/10.1213/ANE.0b013e3181f7e2c4
20. S. Savage, D. Ma, The neurotoxicity of nitrous oxide: the facts and ""Putative"" mechanisms. Brain Sci. **4**, 73–90 (2014). https://doi.org/10.1038/nm0498-460
21. V. Jevtovic-Todorovic, S.M. Todorovic, S. Mennerick, S. Powell, K. Dikranian, N. Benshoff, C.F. Zorumski, J.W. Olney, Nitrous oxide (laughing gas) is an NMDA antagonist, neuroprotectant and neurotoxin. Nat. Med. **4**, 460–463 (1998). https://doi.org/10.1038/nm0498-460
22. J.H. Abraini, H.N. David, O. Nicole, E.T. MacKenzie, A. Buisson, M. Lemaire, Neuroprotection by nitrous oxide and xenon and its relation to minimum alveolar concentration. Anesthesiology **101**, 260–261 (2004). https://doi.org/10.1097/00000542-200407000-00049
23. J.F. Nunn, Clinical aspects of the interaction between nitrous oxide and vitamin B12. Br. J. Anaesth. **59**, 3–13 (1987). https://doi.org/10.1093/bja/59.1.3
24. E.L. Mayer, D.W. Jacobsen, K. Robinson, Homocysteine and coronary atherosclerosis. J. Am. Coll. Cardiol. **27**, 517–527 (1996). https://doi.org/10.1093/eurheartj/ehn515
25. J. Blacher, K. Demuth, A.P. Guerin, C. Vadez, N. Moatti, M.E. Safar, G.M. London, Association between plasma homocysteine concentrations and cardiac hypertrophy in end-stage renal disease. J. Nephrol. **12**, 248–255 (1999). PMID: 10493568
26. N.H. Badner, K. Drader, D. Freeman, J.D. Spence, The use of intraoperative nitrous oxide leads to postoperative increases in plasma homocysteine. Anesth. Analg. **87**, 711–713 (1998). https://doi.org/10.1097/00000539-199809000-00041
27. P. Nagele, D. Tallchief, J. Blood, A. Sharma, E.D. Kharasch, Nitrous oxide anesthesia and plasma homocysteine in adolescents. Anesth. Analg. **113**, 843–848 (2011). https://doi.org/10.1213/ANE.0b013e31822402f5
28. F. Nappo, N. de Rosa, R. Marfella, D. de Lucia, D. Ingrosso, A.F. Perna, B. Farzati, D. Giugliano, Impairment of endothelial functions by acute hyperhomocysteinemia and reversal by antioxidant vitamins. J. Am. Med. Assoc. **281**, 2113–2118 (1999). https://doi.org/10.1001/jama.281.22.2113
29. A.A. Ermens, H. Refsum, J. Ruprecht, L.J. Spijkers, A.B. Guttormsen, J. Lindemans, P.M. Ueland, J. Abels, Monitoring cobalamin inactivation during nitrous oxide anesthesia by determination of homocysteine and folate in plasma and urine. Clin. Pharmacol. Ther. **49**, 385–393 (1991). https://doi.org/10.1038/clpt.1991.45
30. A.M. Yusuf, A.B. Abubakar, S.O. Mamman, Relationship between greenhouse gas emission, energy consumption, and economic growth: evidence from some selected oil-producing African countries. Environ. Sci. Pollut. Res. **27**, 15815–15823 (2020). https://doi.org/10.1007/s11356-020-08065-z
31. B. Hall, G. Dutton, J. Elkins, The NOAA nitrous oxide standard scale for atmospheric observations. J. Geophys. Res. D **112**, D09305 (2007). https://doi.org/10.1029/2006JD007954
32. R.G. Prinn et al., History of chemically and radiatively important atmospheric gases from the Advanced Global Atmospheric Gases Experiment (AGAGE). Earth Syst. Sci. Data **10**, 985–1018 (2018). https://doi.org/10.5194/essd-10-985-2018
33. S. Zhao, J. Zhou, D. Yuan et al., NirS-type N_2O-producers and nosZ II-type N_2O-reducers determine the N_2O emission potential in farmland rhizosphere soils. J. Soils Sediments **20**, 461–471 (2020). https://doi.org/10.1007/s11368-019-02395-3
34. S. Ishii, H. Ohno, M. Tsuboi et al., Identification and isolation of active N_2O reducers in rice paddy soil. Int. Soc. Microb. Ecol. J. **5**, 1936–1945 (2011). https://doi.org/10.1038/ismej.2011.69
35. A. Kumar, K. Medhi, R.K. Fagodiya, G. Subrahmanyam, R. Mondal, P. Raja, S.K. Malyan, D.K. Gupta, C.K. Gupta, H. Pathak, Molecular and ecological perspectives of nitrous oxide producing microbial communities in agro-ecosystems. Rev. Environ. Sci. Biotechnol. **19**, 717–750 (2020). https://doi.org/10.1007/s11157-020-09554-w
36. L. Guo, X. Wang, T. Diao et al., N_2O emission contributions by different pathways and associated microbial community dynamics in a typical calcareous vegetable soil. Environ. Pollut. **242**, 2005–2013 (2018). https://doi.org/10.1016/j.envpol.2018.07.028

37. F. Sabba, C. Picioreanu, J. Perez, R. Nerenberg, Hydroxylamine diffusion can enhance N_2O emissions in nitrifying biofilms: a modeling study. Environ. Sci. Technol. **49**, 1486–1494 (2015). https://doi.org/10.1021/es5046919
38. C. Sánchez, K. Minamisawa, Nitrogen cycling in soybean rhizosphere: sources and sinks of nitrous oxide (N_2O). Front. Microbiol. **10**, 1–7 (2019). https://doi.org/10.3389/fmicb.2019.01943
39. K. Butterbach-Bahl, E.M. Baggs, M. Dannenmann, R. Kiese, S. Zechmeister-Boltenstern, Nitrous oxide emissions from soils: how well do we understand the processes and their controls? Phil. Trans. R. Soc. B **368**, 1–13 (2013). https://doi.org/10.1098/rstb.2013.0122
40. M. Chhowalla, H.S. Shin, G. Eda, L.J. Li, K.P. Loh, H. Zhang, The chemistry of two-dimensional layered transition metal dichalcogenide nanosheets. Nat. Chem. **5**, 263–275 (2013). https://doi.org/10.1038/nchem.1589
41. D. Veoiry, A. Mohite, M. Chhowalla, Phase engineering of transition metal dichalcogenides. Chem. Soc. Rev. **44**, 2702–2712 (2015). https://doi.org/10.1039/C5CS00151J
42. E. Benavente, M.A. Santa Ana, F. Mendizabal, G. Gonzalez, Intercalation chemistry of molybdenum disulfide. Coord. Chem. Rev. **224**, 87–109. https://doi.org/10.1016/S0010-8545(01)00392-7
43. A. Kuc, T. Heine, The electronic structure calculations of two-dimensional transition-metal dichalcogenide in the presence of external electric and magnetic fields. Chem. Soc. Rev. **44**, 2603–2614 (2015). https://doi.org/10.1039/C4CS00276H
44. A.N. Enyashin, G. Seifert, Electronic properties of MoS_2 monolayer and related structures. Nanosys.: Phys. Chem. Math. **5**, 517–539 (2014)
45. J.N. Coleman et al., Two-dimensional nanosheets produced by liquid exfoliation of layered materials. Science **331**, 568–571 (2011). https://doi.org/10.1126/science.1194975
46. P. Joensen, R.F. Frindt, S.R. Morrison, Single-layer MoS_2. Mater. Res. Bull. **21**, 457–461 (1986). https://doi.org/10.1016/0025-5408(86)90011-5
47. M.B. Dines, Lithium intercalation via n-butyllithium of layered transition-metal dichalcogenides. Mater. Res. Bull. **10**, 287–291 (1975). https://doi.org/10.1016/0025-5408(75)90115-4
48. S. Golub, Y.V. Zubavichus, Y.L. Slovokhotov, Y.N. Novikov, Single-layer dispersions of transition metal dichalcogenides in the synthesis of intercalation compounds, Russian. Chem. Rev. **72**, 123–141 (2003). https://doi.org/10.1070/RC2003v072n02ABEH000789
49. Z.Y. Zeng et al., Single-layer semiconducting nanosheets: high-yield preparation and device fabrication. Angew. Chem. Int. Ed. **50**, 11093–11097 (2011). https://doi.org/10.1002/anie.201106004
50. K.-G. Zhou, N.-N. Mao, H.-X. Wang, Y. Peng, H.L. Zhang, A mixed-solvent strategy for efficient exfoliation of inorganic graphene analogues. Angew. Chem. Int. Ed. **50**, 10839–10842 (2011). https://doi.org/10.1002/anie.201105364
51. G. Cunningham et al., Solvent exfoliation of transition metal dichalcogenides: dispersibility of exfoliated nanosheets varies only weakly between compounds. ACS Nano **6**, 3468–3480 (2012). https://doi.org/10.1021/nn300503e
52. R.J. Smith et al., Large-scale exfoliation of inorganic layered compounds in aqueous surfactant solutions. Adv. Mater. **23**, 3944–3948 (2011). https://doi.org/10.1002/adma.201102584
53. P. May, U. Khan, J.M. Hughes, J.N. Coleman, Role of solubility parameters in understanding the steric stabilization of exfoliated two-dimensional nanosheets by adsorbed polymers. J. Phys. Chem. C **116**, 11393–11400 (2012). https://doi.org/10.1021/jp302365w
54. Z. Zeng et al., An effective method for the fabrication of few-layer-thick inorganic nanosheets. Angew. Chem. Int. Ed. **51**, 9052–9056 (2012). https://doi.org/10.1002/anie.201204208
55. X. Li et al., Large-area synthesis of high-quality and uniform graphene films on copper foils. Science **324**, 1312–1314 (2009)
56. S. Bae et al., Roll-to-roll production of 30-inch graphene films for transparent electrodes. Nature Nanotech. **5**, 574–578 (2010). https://doi.org/10.1038/nnano.2010.132
57. K.-K. Liu et al., Growth of large-area and highly crystalline MoS_2 thin layers on insulating substrates. Nano Lett. **12**, 1538–1544 (2012). https://doi.org/10.1021/nl2043612

58. H.S. Lee et al., MoS_2 nanosheet phototransistors with thickness-modulated optical energy gap. Nano Lett. **12**, 3695–3700 (2012). https://doi.org/10.1021/nl301485q
59. Y. Shi et al., Van der Waals epitaxy of MoS_2 layers using graphene as growth templates. Nano Lett. **12**, 2784–2791 (2012). https://doi.org/10.1021/nl204562j
60. Y. Zhan, Z. Liu, S. Najmaei, P.M. Ajayan, J. Lou, Large-area vapor-phase growth and characterization of MoS_2 atomic layers on a SiO_2 substrate. Small **8**, 966–971 (2012). https://doi.org/10.1002/smll.201102654
61. Y.H. Lee et al., Synthesis of large-area MoS_2 atomic layers with chemical vapor deposition. Adv. Mater. **24**, 2320–2325 (2012). https://doi.org/10.1002/adma.201104798
62. K.F. Mak, C. Lee, J. Hone, J. Shan, T.F. Heinz, Atomically thin MoS_2: a new direct-gap semiconductor. Phys. Rev. Lett. **105**, 136805–136809 (2010). https://doi.org/10.1103/PhysRevLett.105.136805
63. V.E.C. Padilla, M.T.R. de la Cruz, Y.E.Á. Alvarado, R.G. Díaz, C.E.R. García, G.H. Cocoletzi, Studies of hydrogen sulfide and ammonia adsorption on P- and Si-doped graphene: density functional theory calculations. J. Mol. Model. **25**, 1–12 (2019). https://doi.org/10.1007/s00894-019-3974-y
64. A. Shokri, N. Salami, Gas sensor based on MoS_2 monolayer. Sens. Act. B **236**, 378–385 (2016). https://doi.org/10.1016/j.snb.2016.06.033
65. Y. Wang, B. Wang, R. Huang, B. Gao, F. Kong, Q. Zhang, First-principles study of transition-metal atoms adsorption on MoS_2 monolayer. Physica E: Low-Dimen. Sys. Nano. **63**, 276–282 (2014). https://doi.org/10.1016/j.physe.2014.06.017
66. N. Mishra, B.P. Pandey, S. Kumar, Impact of N_2O Gas adsorption upon electronic properties of 2D $MoSe_2$ monolayer: a DFT approach. IEEE Sens. J. **21**, 9756–9762 (2021). https://doi.org/10.1109/JSEN.2021.3057375
67. S. Ma, L. Su, L. Jin, J. Su, Y. Jin, A first-principles insight into Pd-doped $MoSe_2$ monolayer: a toxic gas scavenger. Phys. Lett. A **383**, 1–9 (2019). https://doi.org/10.1016/j.physleta.2019.125868
68. H. Cui, G. Zhang, X. Zhang, J. Tang, Rh-doped $MoSe_2$ as a toxic gas scavenger: a first-principles study. Nanoscale Adv. **1**, 772–780 (2019). https://doi.org/10.1039/C8NA00233A
69. F. Li, C. Shi, Effect of surface dangling bonds on transport properties of phosphorous doped SiC nanowires. Appl. Surf. Sci. **434**, 294–306 (2018). https://doi.org/10.1016/j.physe.2018.08.001
70. Y. Fan, J. Zhang, Y. Qiu, J. Zhu, Y. Zhang, G. Hu, A DFT study of transition metal (Fe, Co, Ni, Cu, Ag, Au, Rh, Pd, Pt, and Ir)- embedded monolayer MoS_2 for gas adsorption. Comput. Mater. Sci. **138**, 255–266 (2017). https://doi.org/10.1016/j.commatsci.2017.06.029
71. A.A. Koós, P. Vancsó, M. Szendro, G. Dobrik, D.A. Silva, Z.I. Popov, P.B. Sorokin, L. Henrard, C. Hwang, L.P. Biro, L. Tapasztó, Influence of native defects on the electronic and magnetic properties of CVD grown $MoSe_2$ single layers. J. Phys. Chem. C **123**, 24855–24864 (2019). https://doi.org/10.1021/acs.jpcc.9b05921
72. J. He, G. Ding, C. Zhong, S. Li, D. Li, G. Zhang, Cr_2TiC_2-based double MXenes: novel 2D bipolar anti-ferromagnetic semiconductor with a gate-controllable spin orientation toward anti-ferromagnetic spintronics. Nanoscale **11**, 356–364 (2019). https://doi.org/10.1039/C8NR07692H
73. Y. Tian, Z. Zhu, Z. Ge, A. Sun, Q. Zhang, S. Huang, H. Li, J. Meng, Electronic and magnetic properties of 3d transition metal-doped $MoSe_2$ monolayer. Phys. E: Low-Dimen. Sys. Nano. **116**, 113745 (2019). https://doi.org/10.1016/j.physe.2019.113745
74. L. Feng J. Su, Z. Liu, Effect of vacancies on structural, electronic and optical properties of monolayer MoS_2: a first-principles study. J. Alloys Compd. **613**, 122–127 (2014). https://doi.org/10.1016/j.jallcom.2014.06.018
75. N. Korozlu, K. Colakoglu, E. Deligoz, Y.O. Cifci, The structural, electronic and optical properties of CdxZn1-xSe ternary alloys. Opt. Commun. **284**, 1863–1867 (2011). https://doi.org/10.1016/j.optcom.2010.11.032
76. D.R. Penn, Wave-number-dependent dielectric function of semiconductors. Phys. Rev. **128**, 2093–2097 (1962). https://doi.org/10.1103/PhysRev.128.2093

77. F. Lahourpour, A. Boochani, S.S. Parhizgar, S.M. Elahi, Structural, electronic and optical properties of graphene-like nano-layers MoX_2(X:S, Se, Te): DFT study. J. Theor. Appl. Phys. **13**, 191–201 (2019). https://doi.org/10.1007/s40094-019-0333-4
78. K. Ding, Y. Lin, M. Huang, The enhancement of NO detection by doping strategies on monolayer MoS_2. Vacuum **130**, 146–153 (2016). https://doi.org/10.3390/nano8100851
79. V. Nagarajan, R. Chandiramouli, Adsorption studies of alcohol molecules on monolayer MoS_2 nanosheet—a first-principles insights. Appl. Surf. Sci. **413**, 109–117. https://doi.org/10.1016/j.apsusc.2017.04.008
80. S. Coca, D. Aravena, R. Morales, E. Ruiz, Large magnetic anisotropy in mononuclear metal complexes. Coord. Chem. Rev. **289**, 379–392 (2015). https://doi.org/10.1016/j.ccr.2015.01.021
81. Z. Lin, J. Ye, Y. An, Tunable electronic structures, half-metallicity and optical properties in Fe-NM (NM=B, C, N) co-doped monolayer 2H-$MoSe_2$. J. All. Comp. **805**, 578–584 (2019). https://doi.org/10.1016/j.jallcom.2019.07.096
82. W. Ai, L. Kou, X. Hu, Y. Wang, A.V. Krasheninnikov, L. Sun, X. Shen, Enhanced sensitivity of $MoSe_2$ monolayer for gas adsorption induced by electric field. J. Phys.:Condens. Matter **31**, 1–23.https://doi.org/10.1088/1361-648X/ab29d8
83. A. Peyghan, N.L. Hadipour, Z. Bagheri, Effects of Al doping and double antisite defect on the adsorption of HCN on a BC_2N nanotube: density functional theory studies. J. Phys. Chem. C **117**, 2427–2432 (2013). https://doi.org/10.1021/jp312503h
84. Y. Zhao, W. Wang, C. Li, Y. Xu, L. He, Tuning the magnetic properties of the monolayer $MoSe_2$ by nonmetal doping: first-principles study. Solid State Commun. **281**, 6–11 (2018). https://doi.org/10.1016/j.ssc.2018.06.010
85. M.J. Prather et al., Measuring and modeling the lifetime of a nitrous oxide including its variability. J. Geophys. Res. D **120**, 5693–5705 (2015). https://doi.org/10.1002/2015JD023267
86. S.N. Shirodkar, U.V. Waghmare, Emergence of ferroelectricity at a metal-semiconductor transition in a 1T monolayer of MoS_2. Phys. Rev. Lett. **112**, 157601–157606 (2014). https://doi.org/10.1103/PhysRevLett.112.157601

Improvement of Solar Cell Performance Using Plasmonics Nanoparticles and Metamaterials

Arijit Bardhan Roy and Moumita Saha

Abstract Plasmonics is the study of the interaction between electromagnetic field and free electrons in a metal, dielectric and semiconductor nanoparticle. Due to such interaction, the plasmonic particles can increase the path length of incident light. Optical path length into a solar cell is one of the most significant parameters for photon collection and conversion of them to short circuit current in the device. A major portion of light gets transmitted through the thin silicon substrate (< 30 μm) which is the main drawback of using such material in PV applications. For this reason, optically thick wafers (180–300 μm) are more suitable for solar cell fabrication. On the other hand, in case of thick silicon substrates, the carriers which are generated far away from the p–n junction, are unable to contribute to photocurrent generation due to bulk recombination process and material cost will be more compare to the thin substrate. However through this chapter authors going to established the fact that, by applying the plasmonic nanoparticles and metamaterials with optimized diameter and distribution, the path length of light in both thin and thick Si solar cells can be increased, which in turn gives better carrier generation and short circuit current.

Keywords Plasmonics · Metal · Dielectric · Semiconductor nanoparticle · Optical path · Thin silicon · Solar cell

1 Introduction

Plasmonics is the study of how free electrons in a metal or dielectric nanoparticle interact with an electromagnetic field. This interaction results in localized surface plasmon resonances (LSPRs) or propagating surface plasmon polaritons (SPPs), which have significant applications in various optical and electronic systems. One

A. B. Roy (✉)
Department of Electronics and Communication Engineering, SRM Institute of Science and Technology, Kattankulathur, Chengalpattu, Tamil Nadu, India

M. Saha
Department of Botany, Shri Gnanambica Degree College, Madanapalle, Andhra Pradesh, India

B. Ramachandran et al. (eds.), *Advanced Optical Sensors*, Progress in Optical Science and Photonics 38, https://doi.org/10.1007/978-981-95-1687-2_8

of the most compelling applications of plasmonics is in enhancing the performance of photovoltaic (PV) devices, particularly in silicon-based solar cells.

Plasmonic particles, typically composed of noble metals such as gold (Au) and silver (Ag), have the ability to manipulate light at the nanoscale. When incident light interacts with these nanoparticles, it can experience scattering, absorption, or near-field enhancement, which can effectively modify the optical path length of photons. This modification is particularly valuable in solar cells, where efficient light trapping and absorption are crucial for high energy conversion efficiency. By extending the optical path length of light within the active layer, plasmonic nanoparticles contribute to increased photon absorption, thereby enhancing the generation of electron–hole pairs.

One of the most important factors for photon collection and conversion to short-circuit current in a solar cell is the optical path length. The primary disadvantage of employing thin silicon substrates (less than 30 μm) in photovoltaic applications is that a significant amount of light passes through them without being absorbed. To counter this issue, wafers that are optically thick (180–300 μm) have traditionally been used for the production of solar cells. Thick substrates enhance absorption simply by increasing the physical thickness of the material, but this approach has its own limitations.

The major drawback of thick silicon substrates is the bulk recombination process. In these substrates, photogenerated carriers that are produced far from the p–n junction may recombine before they can contribute to the photocurrent, reducing the overall efficiency of the solar cell. Additionally, thick silicon wafers are more expensive and require more material, making them less attractive for cost-effective large-scale PV production. As a result, there has been a growing interest in developing strategies that improve light absorption in thin-film solar cells while minimizing material usage and recombination losses.

The application of plasmonic nanoparticles provides an effective solution to these challenges. By carefully selecting the size, shape, and dispersion of these nanoparticles, researchers can optimize their optical properties to maximize light trapping in both thin and thick Si solar cells. For instance, spherical Ag nanoparticles with diameters ranging from 50 to 150 nm have been shown to significantly enhance light absorption in silicon due to their strong scattering properties. Similarly, Au nanoparticles can provide broadband light absorption enhancement by supporting multiple plasmonic resonances.

Moreover, the integration of plasmonic structures into solar cells can be achieved through various approaches, such as embedding nanoparticles within the active layer, coating them on the surface, or incorporating them into a back-reflector configuration. Each method offers distinct advantages in terms of absorption enhancement, cost-effectiveness, and ease of fabrication. For instance, embedding nanoparticles within the silicon matrix enhances near-field effects, leading to localized absorption, while placing them on the rear surface can improve backscattering and total internal reflection within the cell.

Beyond simple nanoparticle incorporation, advanced plasmonic architectures such as nanorods, core–shell structures, and metasurfaces are being explored for next-generation solar cells. These designs offer tunable optical responses that can be tailored to specific spectral regions, further optimizing light management in PV devices. Additionally, hybrid approaches combining plasmonics with other light-trapping mechanisms, such as photonic crystals and textured surfaces, are proving to be highly effective in pushing the efficiency limits of solar cells.

In conclusion, plasmonics presents a promising avenue for enhancing the performance of silicon-based solar cells by improving light absorption and carrier generation. By strategically integrating plasmonic nanoparticles, both thin and thick silicon substrates can benefit from increased optical path lengths and improved short-circuit currents. As research in this field progresses, further advancements in nanoparticle engineering, fabrication techniques, and hybrid light-trapping strategies are expected to drive the development of high-efficiency, cost-effective solar cells, ultimately contributing to the advancement of renewable energy technologies [1].

Plasmonic materials primarily provide three exceptional properties to increase solar cell conversion efficiency. Initially, the subwavelength scattering processes of the plasmonic nanoparticles lengthen the optical channel and introduce more photons into the semiconductor. This process significantly enhances the absorption of light by ensuring that photons spend a longer time interacting with the active layers of the solar cell, thus increasing the probability of electron–hole pair generation. The enhanced light absorption directly translates to improved power conversion efficiency, making plasmonic materials a key area of research in photovoltaic technology.

Second, the metal nanoparticles on top of the semiconductor can be thought of as subwavelength antennas that enhance absorption into the semiconductor slab by coupling light into guided modes. These guided modes increase the path length of photons within the semiconductor, effectively boosting the likelihood of their absorption. By acting as antennas, these nanoparticles concentrate and redirect light into the semiconductor, leading to a more effective utilization of incident solar energy. Furthermore, the shape, size, and material composition of these nanoparticles play a crucial role in determining their efficiency in light trapping. Researchers are actively investigating the optimization of these parameters to maximize the absorption of a broad spectrum of sunlight.

Finally, the patterned metal back contact maintains the Surface Plasmon Polariton (SPP) modes supported by the back metal/semiconductor interface while coupling the red light from the plasmonic layer into the guided modes in the semiconductor. The presence of SPP modes at the metal–semiconductor interface leads to a redistribution of electromagnetic fields within the solar cell structure. This redistribution ensures that longer-wavelength photons, which typically have lower absorption coefficients in semiconductors, are more effectively trapped and converted into useful electrical energy. Additionally, the back contact's periodic structuring further aids in reducing reflectance losses and promotes efficient photon recycling within the cell.

As will be covered in the upcoming sections, metal and dielectric nanoparticles are known to exert four main forms of light-trapping phenomena. These include scattering, near-field enhancement, coupling to waveguide modes, and excitation of SPPs. Scattering increases the optical path length, near-field enhancement amplifies local electromagnetic fields, coupling to waveguide modes allows deeper penetration

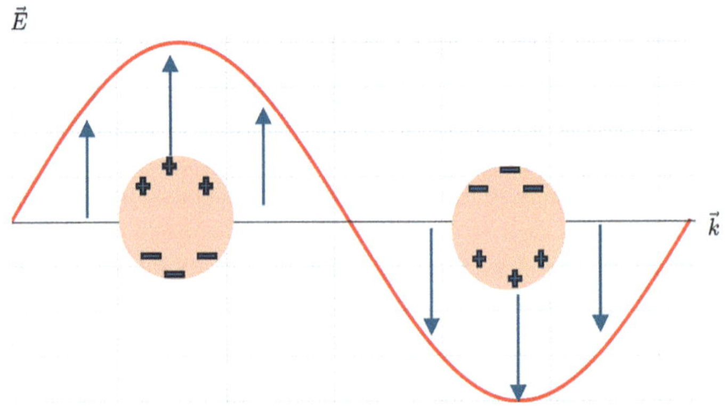

Fig. 1 Spherical nanoparticle illustration of a confined surface plasmon

of light into the semiconductor, and SPP excitation enables the trapping of specific wavelength ranges. Each of these mechanisms contributes to the overall efficiency enhancement in plasmonic solar cells.

Figure 1 illustrates a spherical nanoparticle confined surface plasmon, highlighting how localized surface plasmons interact with incident light to enhance optical absorption. The integration of such plasmonic nanostructures represents a promising avenue for next-generation solar cell designs, driving the evolution of highly efficient and cost-effective photovoltaic devices.

2 Light Scattering by Metal Nanoparticles

This idea is based on the use of high angle and multiple scattering from metal nanoparticles, which lengthens the optical path in the absorber layer and improves photon injection into the junction. By leveraging the scattering properties of metal nanoparticles, light can be redirected and effectively trapped within the active layer of a photovoltaic cell, leading to enhanced absorption and increased energy conversion efficiency.

It has been noted that the majority of the light that enters the cell is dispersed light with an angle greater than 60°, which is the crucial angle for reflection in the Si/air interface. When light enters at such steep angles, it encounters multiple internal reflections that prolong its presence in the absorber layer, maximizing the probability of photon absorption and electron–hole pair generation. This phenomenon is particularly important for thin-film solar cells, where the active layer is relatively thin and would otherwise suffer from inadequate light absorption. The scattering induced by nanoparticles ensures that light is not merely transmitted through the cell but is rather redirected within the material, increasing the overall absorption.

The injection of photons into the cell junction can be further enhanced by incorporating a back reflector, typically a thin metal coating, on the back side of the cell. This reflective layer serves a dual purpose: first, it redirects light that has not been absorbed back into the absorber layer for another opportunity at photon absorption, and second, it prevents light from escaping the cell, thereby improving overall efficiency. The presence of the back reflector ensures that dispersed light undergoes multiple bounces within the cell, effectively increasing the optical path length and allowing for better light trapping.

Additionally, the combination of high-angle scattering and multiple reflections from the back reflector results in more effective utilization of the solar spectrum. Certain wavelengths that would otherwise be lost due to reflection at the Si/air interface are instead retained within the cell, thereby enhancing absorption across a broader spectral range. This approach is particularly advantageous for improving performance under diffuse illumination conditions, where light is not incident at a single angle but rather arrives from multiple directions.

Recent advancements in nanophotonic engineering have further refined this concept, incorporating plasmonic structures that enhance localized electromagnetic fields around the nanoparticles. These localized fields boost absorption efficiency even in ultra-thin absorber layers, making the approach viable for next-generation solar technologies. The integration of these techniques with traditional photovoltaic designs represents a promising pathway toward more efficient and cost-effective solar energy harvesting.

In summary, by utilizing high-angle scattering, multiple light reflections, and a back reflector, the efficiency of photon injection into the cell junction is significantly improved. This method not only enhances light absorption but also contributes to the overall performance of solar cells, making them more effective in harnessing solar energy.

(i) **Light trapping using metal nano-antennas:**

To enhance photon injection into the junction and extend the optical path within the absorber layer, this concept leverages the use of metal nanoparticles to induce high-angle scattering and multiple scattering events. By strategically incorporating these nanoparticles into the design, incoming light can be redirected in a manner that maximizes absorption and minimizes reflection losses. The key mechanism at play involves the scattering of light at angles greater than 60°, which exceeds the critical angle for total internal reflection at the silicon (Si)/air interface. This means that the majority of the incoming light is converted into diffused light with high angular distribution, effectively increasing the probability of light trapping within the absorber layer.

A significant improvement in light absorption is achieved by incorporating a back reflector in the form of a thin metal coating on the rear side of the solar cell. This reflective layer ensures that photons that would otherwise escape are redirected back into the absorber layer, undergoing multiple bounces before ultimately being absorbed. The presence of metal nanoparticles enhances this effect further by promoting plasmonic interactions, which help guide the displacement current efficiently. Surface plasmon modes generated at the metal interfaces facilitate the propagation of electromagnetic waves along the surface, leading to a more effective light-trapping mechanism.

The interaction between the scattered light and plasmonic resonances plays a crucial role in enhancing the spontaneous emission from the metal nanoparticles. As light undergoes multiple scattering events, localized surface plasmons are excited, which can contribute to the amplification of electromagnetic fields near the absorber layer. This results in an increase in the overall optical absorption, leading to greater photon injection into the junction. By harnessing the properties of plasmonic modes, the displacement current is efficiently guided, ensuring optimal energy conversion.

Overall, this approach significantly improves the efficiency of light absorption in the solar cell by reducing reflection losses and extending the optical path within the absorber layer. The combination of high-angle scattering, plasmonic enhancement, and multiple reflections within the cell provides a promising strategy for boosting photon absorption. This ultimately leads to better carrier generation and an increase in the overall performance of the device, making it a viable technique for improving solar cell efficiency. The integration of metal nanoparticles and a back reflector presents a cost-effective and scalable method to enhance light trapping without requiring substantial modifications to conventional photovoltaic structures.

(ii) **Surface—Plasmon—Polariton (SPP) coupling mechanism by metal nanostructures:**

This behaviour is particularly beneficial when the absorption of surface plasmon polaritons (SPPs) in the semiconductor material significantly exceeds that of the metal. In such a scenario, the role of metal nanostructures becomes crucial as they serve as an intermediary to convert incident light into SPPs. These SPPs are then tightly confined at the metal/semiconductor interface due to the plasmon resonance frequency, which effectively traps and directs the electromagnetic waves through the contact region. The key factor enabling this process is the precise design of the nanostructures, which must have dimensions smaller than the wavelength of the incident light to ensure efficient coupling and confinement of the electromagnetic waves.

By leveraging advanced optical engineering techniques, it is possible to optimize the interaction between the trapped light and the absorber layer of the semiconductor. This optimization leads to enhanced photon absorption and, subsequently, improved electron generation, which is crucial for applications such as photovoltaic energy conversion. The ability to manipulate and confine light in this manner allows for the development of more efficient energy harvesting systems, particularly in the field of solar cell technology.

For improved solar cell efficiency, this plasmonic light-trapping approach can be integrated with traditional metallization techniques. This is especially advantageous in the context of thin-film silicon-based solar cells, where maximizing light absorption within a limited absorber layer thickness is a persistent challenge. Traditional solar cells often suffer from poor absorption in thin semiconductor layers due to minimal optical path lengths. However, by incorporating metal nanostructures to facilitate SPP excitation and confinement, the effective absorption path length can be significantly increased without requiring additional material thickness.

Furthermore, this plasmonic enhancement mechanism can be effectively combined with other well-established light-trapping strategies, such as anti-reflective coatings and textured surfaces, to further boost absorption efficiency. The synergy between these approaches can lead to substantial improvements in the overall power conversion efficiency of solar cells, making them more viable for widespread implementation in renewable energy systems.

In summary, the utilization of metal nanostructures to excite and guide SPPs offers a promising route for enhancing light absorption in semiconductors. When properly engineered, this technique enables efficient photon-electron harvesting and is particularly well-suited for thin-film solar cells. By integrating this approach with conventional metallization techniques, the potential for high-performance, cost-effective photovoltaic devices is greatly increased. The three primary light-trapping mechanisms discussed can be effectively visualized in Fig. 2, providing further insight into their implementation and impact on solar cell performance [2–7].

(iii) **Light trapping through dielectric nanoparticles:**

The earlier discussions highlighted the significance of metal nanoparticles and nanostructures in enhancing light harvesting in silicon solar cells, regardless of their thickness. However, the efficiency improvements achieved using metal nanoparticles come with inherent drawbacks, primarily due to Joule heating. This heating effect

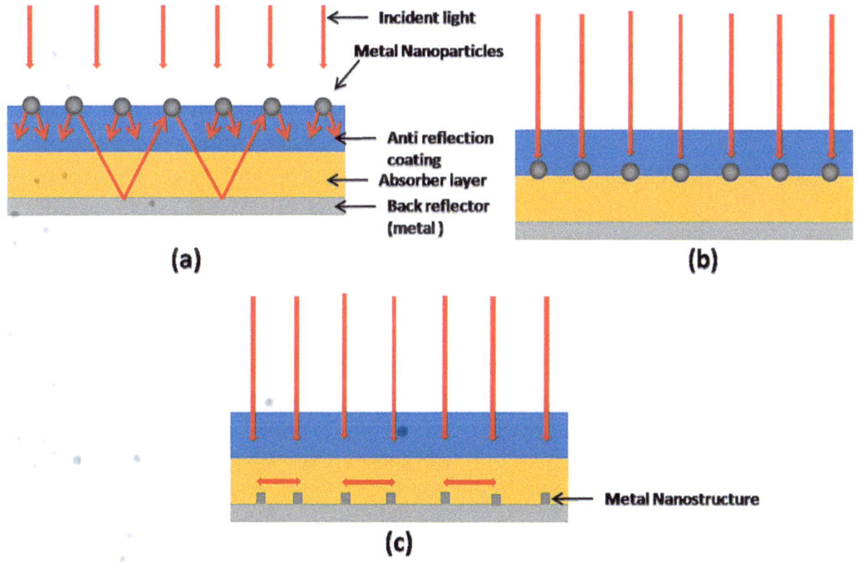

Fig. 2 Different types of light trapping schemes through metal nanostructures: **a** light scattering through plasmonic nanoparticles **b** light trapping using nano-antennas and **c** photon generation through Surface—Plasmon—Polariton mode [1]

arises when metal nanoparticles absorb light or electromagnetic waves, leading to localized temperature increases. Such heating can reduce overall device efficiency and introduce stability concerns in long-term applications. Additionally, the presence of metal nanoparticles on the semiconductor surface can influence passivation techniques, potentially degrading the cell's performance over time.

Given these challenges, contemporary research has shifted towards employing dielectric nanoparticles as an alternative plasmonic material. Dielectric nanoparticles offer significant advantages, primarily due to their inherently low absorption coefficients, which result from their large bandgap values. Unlike metal nanoparticles, which can contribute to self-absorption losses, dielectric nanoparticles excel at scattering incident light while minimizing internal absorption. This property ensures that more photons reach the junction of the device, thereby improving the overall efficiency of light harvesting.

Another crucial advantage of dielectric nanoparticles is their compatibility with passivation techniques. Since these materials are typically insulating, their incorporation into the solar cell structure does not interfere with established passivation methods. This characteristic helps maintain the surface recombination velocity at a minimal level, thereby enhancing carrier lifetime and ultimately boosting the open-circuit voltage of the device.

The lossless nature of dielectric nanoparticles also plays a pivotal role in managing the thermal profile of silicon solar cells. By reducing the amount of heat generated within the device, these nanoparticles contribute to maintaining a lower operational temperature. Lower device temperatures not only enhance efficiency but also prolong the lifespan of the solar cell by mitigating thermal degradation effects. Consequently, dielectric nanoparticles emerge as an ideal solution for optimizing both thin and thick crystalline silicon (c-Si) solar cells.

Recent studies have demonstrated that the strategic placement of dielectric nanoparticles can significantly enhance light trapping mechanisms. By leveraging Mie resonances and other optical phenomena, researchers have achieved substantial improvements in photon management. These advancements ensure that a higher proportion of incident sunlight is absorbed and utilized effectively within the photovoltaic device. As a result, solar cells embedded with dielectric nanoparticles exhibit remarkable efficiency gains compared to conventional designs that rely on metallic structures.

Furthermore, dielectric materials provide the added benefit of long-term stability. Unlike metal nanoparticles, which may undergo oxidation or morphological changes over time, dielectric nanoparticles maintain their structural integrity even under prolonged exposure to solar radiation. This durability translates to improved reliability and reduced maintenance costs for solar energy systems, making them an attractive option for large-scale deployment.

Overall, the integration of dielectric nanoparticles in silicon solar cells represents a transformative approach to improving photovoltaic performance. By mitigating the challenges associated with metal nanoparticles—such as Joule heating, passivation interference, and self-absorption losses—dielectric nanoparticles pave the way for more efficient, stable, and commercially viable solar energy solutions. As research in

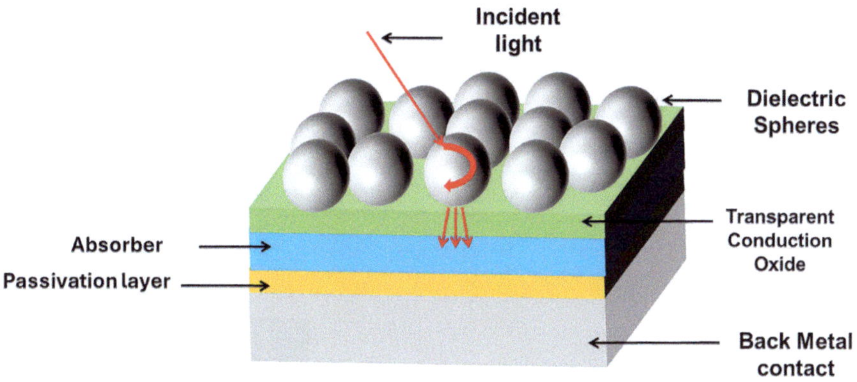

Fig. 3 Light trapping into the absorber layer through dielectric resonating spheres [5]

this field continues to evolve, the development of optimized nanoparticle designs and integration strategies will further enhance the potential of silicon-based photovoltaic technologies [8, 9].

Figure 3 may serve as an illustration of this phenomenon.

Whispering Gallery Modes (WGM) are another distinctive characteristic of these dielectric nanoparticles. Around 1878, Lord Rayleigh made the first observation of the WGM modes at St. Paul's Cathedral. Initially identified for acoustic waves in sufficiently big dome-like structures, this phenomenon was later discovered to be present in electromagnetic waves such as radio waves, microwaves, terahertz radiations, infrared radiations, and X-rays. Certain specific diffraction modes of the electromagnetic wave travel along the inner periphery of the dome wall and become imprisoned as a result of the complete internal reflection from the inner peripheral surface of the dielectric sphere-like structure. It is evident that the resonator's cavity quality (Q) factor plays a crucial role in determining the waves' losses and resonating frequency. The dielectric spheres, which resonate at various frequencies and modes and boost absorption at various spectrum wavelengths, can likewise produce this effect. Once more, the refraction grating at the interface causes the trapped WGMs to be channelled into the high refractive indexed absorber layer (such as silicon) (Fig. 4). Maximum photon injection into the silicon substrate may be guaranteed by the dielectric material's appropriate size and distribution. For this reason, the dielectric lossless nanospheres are also being studied by the modern research community as an anti-reflection covering for silicon solar cells. Mie resonances are also used by the dielectric nanospheres to efficiently store and link light into the silicon substrate, according to some recent research. Additionally, the particle's cavity quality factor (Q factor) and geometrical cross section play a significant role in this event. Additionally, it is becoming quite popular to use dielectric spheres as nano or micro lenses to inject the maximum amount of photons into the solar cell junction.

For better light management in the solar cell, the focal length of the lens can be adjusted by altering the size and shape of the particle. These novel findings, which

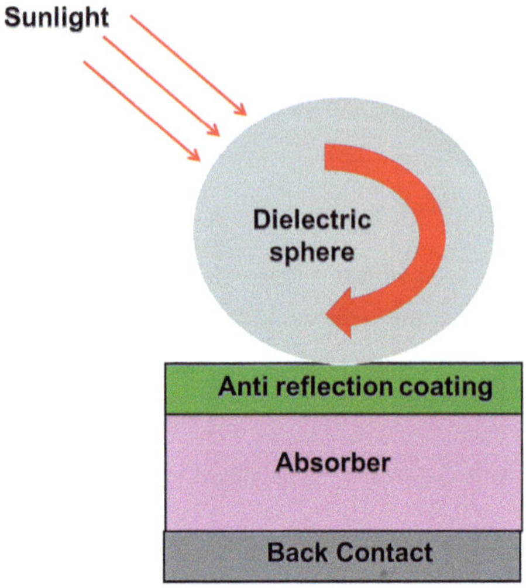

Fig. 4 Enhancement of light trapping into the thin Si solar cell using dielectric spheres through WGM modes [5]

have been carefully investigated and validated by multiple research groups [10–12], clearly demonstrate the value of dielectric particles for the construction of high efficiency thick or thin c-Si solar cells.

(iv) **Light trapping through semiconductor nanoparticles:**

Plasmonic nanoparticles operate based on the principle of localized surface plasmon resonance (LSPR), a phenomenon where conduction electrons within the nanoparticles oscillate collectively in response to incident light. When the frequency of the incoming photons matches the natural oscillation frequency of these electrons, a resonance condition is established, leading to a significant enhancement of the local electromagnetic field around the nanoparticle. This enhancement facilitates stronger light absorption and scattering, effectively increasing the efficiency of solar energy conversion.

Unlike traditional metallic plasmonic nanoparticles, which exhibit LSPR primarily in the visible spectrum, semiconductor plasmonic nanoparticles can be engineered to support plasmon resonances over a broader range of wavelengths, including the infrared (IR) region. This tunability is particularly beneficial for solar cell applications, as a significant portion of the solar spectrum falls within the near-infrared range. By carefully selecting the composition, doping level, and size of the nanoparticles, their plasmonic properties can be optimized to match the spectral response of the solar cell, ensuring efficient photon harvesting. The integration of semiconductor plasmonic nanoparticles into solar cells can significantly improve their performance through several key mechanisms:

1. **Light Scattering and Trapping**

One of the primary ways in which semiconductor plasmonic nanoparticles enhance solar cell performance is by increasing light scattering and trapping within the absorber layer. When these nanoparticles are embedded in the active region of a solar cell or positioned at the front surface, they cause incident sunlight to be scattered in multiple directions. This scattering increases the optical path length of light within the absorber layer, thereby enhancing photon absorption and reducing transmission losses.

In thin-film solar cells, where light absorption is often a limiting factor due to the reduced thickness of the absorber layer, the use of plasmonic nanoparticles can be particularly beneficial. By extending the effective optical path length, these nanoparticles compensate for the limited material thickness, enabling higher efficiency without increasing material costs.

2. **Near-Field Enhancement**

Another important mechanism by which semiconductor plasmonic nanoparticles improve solar cell efficiency is through near-field electromagnetic enhancement. When LSPR is excited in these nanoparticles, a strong local electromagnetic field is generated in their immediate vicinity. If the nanoparticles are positioned near the active junction of the solar cell, this enhanced field can significantly boost the generation of electron–hole pairs by increasing the absorption of incident photons.

This near-field enhancement effect is particularly useful in boosting absorption in the ultraviolet (UV) and visible spectral ranges, making it highly relevant for solar cells based on materials such as silicon (Si), gallium arsenide (GaAs), and perovskites, which have strong absorption in these regions.

3. **Plasmon-Exciton Coupling**

In addition to enhancing light absorption, semiconductor plasmonic nanoparticles can facilitate plasmon-exciton coupling, a phenomenon where the localized surface plasmon resonance interacts with the excitonic states of the absorber material. This interaction leads to enhanced charge carrier generation and improved carrier dynamics, resulting in higher photocurrent and overall solar cell efficiency.

By carefully tuning the plasmonic resonance of the nanoparticles to match the excitonic absorption bands of the absorber material, researchers can achieve strong plasmon-exciton coupling, leading to more efficient light harvesting and energy conversion.

4. **Hot Electron Injection**

Another promising mechanism enabled by semiconductor plasmonic nanoparticles is hot electron injection. When LSPR is excited, some of the absorbed photon energy is converted into high-energy electrons, known as hot electrons. These hot electrons can be injected into the conduction band of the semiconductor, contributing directly to photocurrent generation.

Unlike conventional charge carrier generation, which relies solely on bandgap absorption, hot electron injection provides an additional pathway for energy conversion, allowing solar cells to utilize sub-bandgap photons that would otherwise be lost. This process has the potential to significantly enhance the efficiency of photovoltaic devices, particularly in tandem and multi-junction solar cells.

3 Materials and Design Considerations

To fully harness the benefits of semiconductor plasmonic nanoparticles in solar cells, several key factors must be carefully considered:

A Choice of Semiconductor Material

The selection of an appropriate semiconductor material for plasmonic nanoparticles is crucial for optimizing their optical and electrical properties. Materials such as TiO_2 and ZnO are widely used due to their high refractive index, strong light-scattering capabilities, and stability. Other materials, such as doped semiconductors (e.g., aluminum-doped ZnO or indium tin oxide), can be engineered to exhibit tunable plasmonic resonances across different spectral regions.

B Particle Size and Shape

The size and shape of the nanoparticles play a significant role in determining their plasmonic properties. Smaller nanoparticles (typically below 50 nm) exhibit stronger near-field enhancement effects, while larger nanoparticles (above 100 nm) provide better light scattering. Additionally, non-spherical shapes such as nanorods, nanostars, and core–shell structures can offer enhanced plasmonic behaviour compared to spherical nanoparticles.

C Placement Within the Solar Cell

The positioning of plasmonic nanoparticles within the solar cell is another critical factor. Front-surface integration enhances light scattering and anti-reflection properties, while embedding nanoparticles within the absorber layer maximizes near-field enhancement and hot electron injection. Back-surface placement, often combined with a reflective layer, can further improve light trapping and internal reflections.

4 Future Perspectives and Challenges

Despite their promising potential, several challenges remain in the practical implementation of semiconductor plasmonic nanoparticles in commercial solar cells:

1. **Material Stability**: Many semiconductor plasmonic nanoparticles degrade under prolonged exposure to sunlight and environmental conditions, necessitating protective coatings or alternative stable materials.
2. **Manufacturing Scalability**: Large-scale fabrication and integration of nanoparticles into solar cells remain complex and costly. Scalable and cost-effective synthesis methods are needed for widespread adoption.
3. **Optimized Plasmonic Design**: Achieving the right balance between plasmonic enhancement and optical losses requires further optimization of nanoparticle size, shape, and composition. Recent studies have demonstrated that the incorporation of semiconductor plasmonic nanoparticles in devices such as dye-sensitized solar

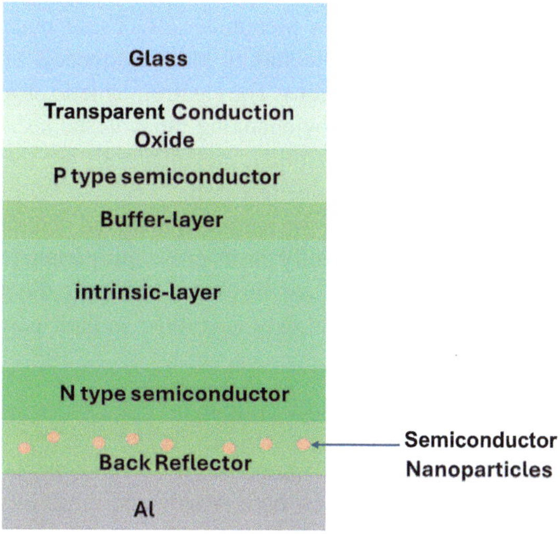

Fig. 5 Nanomirror-embedded Back Reflector Layer (BRL) for advanced light management in thin silicon solar cells [17]

cells (DSSCs), perovskite solar cells, and organic solar cells can lead to significant performance enhancements. For example, the addition of TiO_2 nanoparticles in DSSCs has been shown to increase light absorption, while ZnO nanoparticles can improve charge transport properties in perovskite solar cells.

Overall, semiconductor plasmonic nanoparticles hold great promise in the field of solar energy conversion, offering an environmentally friendly and cost-effective solution to improve solar cell performance [13–19].

Using a nontraditional back reflector layer (BRL) in amorphous silicon (a-Si:H) solar cells, Fig. 5 shows a technique for advanced light control. Because of its low parasitic absorption loss over a wide wavelength range (300–1100 nm), silver sulfide (Ag_2S) semiconductor nanoparticles with a diameter of around 50 nm have been selected as the nanomirror to achieve this. To improve light trapping within the active layers of an a-Si:H solar cell, the Ag_2S NPs were positioned as the back reflector layer between two indium tin oxide (ITO) layers. High reflectance (up to 93%) was demonstrated by the embedded structure in the red and near-infrared spectrum.

5 Metamaterials for Solar Cell Application

5.1 Background

Man-made materials with optical characteristics not typically seen in nature are known as metamaterials. A new phase of material research has emerged because of the development of these kinds of materials. Combinations of several distinct materials, such as metals, dielectrics, and semiconductors having multifaceted electromagnetic properties, are known as metamaterials. These materials work well with several electromagnetic bands, including radio frequency, microwave, terahertz, infrared, and visible spectrum regions. The electromagnetic properties of a given metamaterial can be changed simply by changing the material's size and shape. Because of their special qualities and flexibility, metamaterials are the most sought-after options in the field of material science research.

The characteristics of the materials involved determine the electromagnetic field. These characteristics specify the macroscopic parameters of a material's permeability (μ) and permittivity (ε). As illustrated in Fig. 6, the metamaterials are categorized into the following four classes according to their permittivity (ε) and permeability (μ):

A **Double Positive (DPS) Material:**

Double Positive (DPS) materials are a fundamental category of electromagnetic materials characterized by both positive permittivity (ε) and positive permeability (μ). These materials allow electromagnetic waves to propagate in a conventional manner, ensuring that the wave vector and energy flow, represented by the Poynting vector, move in the same direction. This property aligns with traditional electromagnetic wave behaviour, making DPS materials crucial in various applications in optics, telecommunications, and material science. In the field of electromagnetism, materials are classified based on their response to electric and magnetic fields. Permittivity (ε) is a measure of how a material stores electrical energy when exposed to an electric field, while permeability (μ) describes how a material responds to magnetic fields. DPS materials, where both of these parameters are positive, support normal wave propagation, meaning that light and other electromagnetic waves travel through them without exhibiting unusual behaviors such as negative refraction. This contrasts with Double Negative (DNG) materials, where both permittivity and permeability are negative, leading to unconventional phenomena like negative refraction and reverse Doppler effects. Most naturally occurring materials fall into the DPS category. Common

Fig. 6 Classification of metamaterials

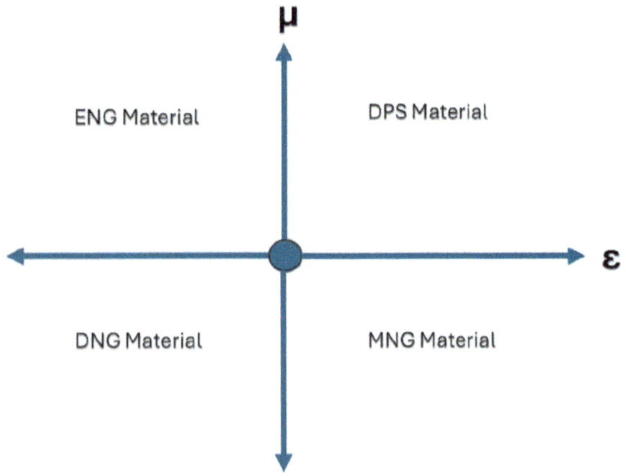

examples include dielectrics such as water, glass, and quartz, which have positive permittivity and permeability values and are widely used in optical and electronic applications. Additionally, common optical materials such as silica and sapphire are considered DPS materials because they allow light to pass through them and refract in a predictable manner. Even certain metals at low frequencies behave as DPS materials, although their response can change at higher frequencies where plasma-like effects become significant. Because of their predictable interaction with electromagnetic waves, DPS materials are essential in a variety of applications. In optics, they are used in lenses, prisms, and other optical components that rely on standard refraction principles to focus and direct light. In telecommunications, DPS materials are commonly used in the construction of antennas and transmission lines, ensuring efficient signal transmission with minimal distortion. These materials are also integral to waveguides, Fiber optic cables, and dielectric substrates in electronic circuits, where they help guide electromagnetic waves with precision and minimal loss. In biomedical imaging, DPS materials play a crucial role in technologies such as MRI and ultrasound, where biological tissues, which exhibit DPS properties, allow for the effective transmission of imaging signals. DPS materials follow the standard rules of wave propagation as described by Snell's Law, meaning that when an electromagnetic wave encounters a boundary between two DPS materials, it refracts at a positive angle. This property makes DPS materials ideal for conventional optical and electromagnetic applications, as they do not introduce anomalies such as negative refraction, which occurs in DNG materials. Due to their ability to support standard electromagnetic behaviour, DPS materials form the foundation for many scientific and engineering advancements in optics, communication, and material sciences. Their widespread availability and well-understood properties ensure that they will continue to be an essential component of technological innovation in various fields.

B **Epsilon Negative (ENG) Material:**

Epsilon Negative (ENG) materials are a class of metamaterials characterized by having negative permittivity ($\varepsilon < 0$) while maintaining a positive permeability ($\mu > 0$). This means that while they respond conventionally to magnetic fields, their interaction with electric fields is unconventional, leading to unique electromagnetic properties. These materials do not naturally exist in most common substances but can be engineered using specific structures and compositions, making them a key area of research in metamaterials and advanced optics.

In traditional materials, permittivity measures how a substance responds to an applied electric field, influencing how electric charges are stored and how electromagnetic waves propagate. Most dielectric materials have positive permittivity, meaning they allow electric field lines to penetrate them to some degree. However, in ENG materials, permittivity is negative, which means the material strongly opposes the formation of an electric field within it. As a result, electromagnetic waves behave differently compared to their propagation in conventional materials. The most notable effect is that transverse electromagnetic (TEM) waves cannot propagate freely through an ENG medium, leading to evanescent waves that decay exponentially rather than traveling through the material. This property makes ENG materials useful in applications that require wave manipulation, such as shielding and wave filtering.

A well-known example of an ENG material is plasmas at certain frequencies. In particular, metals such as silver and gold exhibit ENG behaviour at optical frequencies due to their free electron response, which leads to negative permittivity below the plasma frequency. At microwave and radio frequencies, engineered structures such as wire array metamaterials can exhibit ENG properties by controlling the effective permittivity using carefully designed periodic arrangements of conductive elements. These artificial structures allow researchers to tailor electromagnetic responses for specific applications.

One of the key implications of ENG materials is their role in metamaterials, especially when combined with other material classes. When ENG materials are paired with materials that have negative permeability ($\mu < 0$), forming Double Negative (DNG) materials, they can exhibit negative refraction, which allows for exotic optical effects such as perfect lensing and cloaking. Additionally, the interaction between ENG and Mu-Negative (MNG) materials can lead to unique resonances and guided wave phenomena, making them useful in antenna design, waveguides, and electromagnetic filters.

Applications of ENG materials are primarily found in advanced electromagnetic and optical systems. They are used in designing compact antennas with enhanced bandwidth and improved efficiency, as their properties allow for better impedance matching and reduced wave reflection. They also play a crucial role in waveguiding applications, where controlling the propagation of waves is necessary, such as in plasmonic circuits and subwavelength imaging. Another significant use is in electromagnetic shielding, where ENG materials help block specific frequency ranges by preventing wave transmission.

The study of ENG materials has opened new avenues in metamaterials research, offering innovative solutions in optics, wireless communication, and electromagnetic engineering. By enabling precise control over wave propagation, reflection, and transmission, these materials contribute to the development of next-generation technologies in imaging, sensing, and communication systems. Their unique properties continue to drive research in advanced material design and practical applications in various scientific and technological domains.

C **Mu Negative (MNG) Material:**

Mu-Negative (MNG) materials are a class of metamaterials characterized by having negative permeability ($\mu < 0$) while maintaining positive permittivity ($\varepsilon > 0$). This means they respond unconventionally to magnetic fields while interacting with electric fields in a standard manner. Unlike most naturally occurring materials, which have positive permeability, MNG materials must be engineered using specific structural designs and compositions to achieve their unique electromagnetic properties. These materials play a crucial role in the field of metamaterials and advanced wave manipulation applications.

Permeability (μ) is a fundamental property that describes how a material interacts with a magnetic field and determines the ease with which a magnetic flux can pass through it. Most conventional materials, such as air, glass, and common dielectrics, exhibit positive permeability, allowing electromagnetic waves to propagate normally. However, in MNG materials, negative permeability means that the material strongly opposes the formation of a magnetic field within it, leading to unusual wave behaviours. One of the most important effects of MNG materials is their inability to support transverse electromagnetic (TEM) wave propagation alone, resulting in evanescent waves that decay exponentially instead of propagating through the material. This characteristic makes them useful in specialized electromagnetic applications such as waveguides, filters, and resonance-based devices. MNG behaviour is typically observed in artificial structures rather than natural materials. One of the most common ways to achieve MNG properties is through periodic arrangements of resonant structures, such as split-ring resonators (SRRs). SRRs are small, engineered metallic loops that resonate at specific frequencies and induce a negative magnetic response. By designing these resonators at appropriate scales, researchers can tailor the permeability of a material to be negative over a desired frequency range. Ferrites and certain composite materials can also exhibit MNG behaviour under controlled conditions, particularly when subjected to external magnetic fields. A key aspect of MNG materials is their interaction with other types of metamaterials. When combined with Epsilon-Negative (ENG) materials, which have negative permittivity but positive permeability, the resulting composite can support unique guided wave propagation modes that do not exist in conventional materials. This combination is particularly useful in antenna technology, as it enables the design of compact and highly efficient antennas with enhanced radiation properties. Additionally, the pairing of MNG and ENG materials can lead to enhanced wave focusing and resonance effects, making them valuable in imaging systems and signal processing applications. Applications of MNG materials are found in various advanced electromagnetic and optical systems. They are used in designing novel antennas with improved performance, allowing for better impedance matching and increased bandwidth. In waveguiding applications, MNG materials enable precise control over electromagnetic wave propagation, leading to the development of highly efficient filters and transmission lines. They are also crucial in electromagnetic shielding and cloaking technologies, where the ability to manipulate magnetic responses helps in controlling the interaction of waves with different surfaces.

Research into MNG materials continues to expand, offering new opportunities in areas such as wireless communication, optical sensing, and stealth technology. Their ability to manipulate electromagnetic fields in unconventional ways makes them essential for next-generation metamaterials, leading to breakthroughs in imaging, communication, and energy-efficient device design. As scientists refine fabrication techniques and develop new composite structures, the potential applications of MNG materials are expected to grow, driving innovation across multiple scientific and engineering disciplines.

D **Double Negative (DNG) Material:**

DNG materials derive their properties from artificial structuring rather than chemical composition, meaning they are engineered at the micro or nanoscale to achieve their distinctive behaviour. Typically, they are constructed using periodic arrays of resonant elements such as split-ring resonators (SRRs) and conducting wires, which interact with electromagnetic waves in a way that produces negative permittivity and permeability over a specific frequency range. This artificial structuring allows precise control over wave propagation, leading to applications that challenge traditional limits in optics and electromagnetism.

One of the most remarkable implications of DNG materials is their ability to support negative refraction. In conventional materials, the refractive index is positive, meaning that light bends in the expected direction when passing from one medium to another. However, in a DNG material, the refractive index is negative, which causes light to bend in the opposite direction. This phenomenon has been experimentally demonstrated and has led to the development of super lenses, which can overcome the diffraction limit of traditional optical systems and enable imaging of objects at resolutions beyond what was previously thought possible.

Another key application of DNG materials is in electromagnetic invisibility cloaks. By carefully designing a metamaterial structure that manipulates the path of electromagnetic waves around an object, it is possible to create a cloaking effect where the object appears invisible to certain frequencies of light or other electromagnetic radiation. While perfect invisibility across all wavelengths remains a challenge, significant progress has been made in narrowband cloaking technologies, with potential applications in defence, security, and stealth technologies.

In addition to cloaking and superlensing, DNG materials have shown promise in enhancing antenna designs, waveguides, and filters in wireless communications. Their unique dispersion properties allow for the miniaturization of devices while maintaining high efficiency, making them valuable in the development of compact and high-performance communication systems. Moreover, DNG materials enable novel waveguiding mechanisms, such as backward wave propagation, which can be exploited to create more efficient signal transmission methods in microwave and optical systems.

Despite their potential, challenges remain in the practical implementation of DNG materials. Fabrication techniques are complex and often require precision engineering at the nanoscale, making large-scale production difficult. Additionally, many existing DNG materials operate within limited frequency ranges, requiring further research to extend their functionality to broader bandwidths and different electromagnetic spectra. Nevertheless, ongoing advancements in nanotechnology, fabrication techniques, and theoretical modeling continue to push the boundaries of what is possible with DNG materials, making them a critical area of research with far-reaching implications across multiple scientific and technological disciplines.

Veselago originally wrote about natural positive refractive indexed materials in 1968. Following a protracted hiatus, Smith et al. presented an artificial structure in 2000 that showed potential for use as a thermal photovoltaic system absorber. This amazing invention drew attention to metamaterials in the field of solar cell research. Following that, some research teams created amazing metamaterials for use in solar cell applications. Among these, the most well-known designs were polarization-insensitive metamaterial antennas (Cheng et al. 2011), metallic nanostructures with strong absorption properties (Diem et al. 2009), and metamaterial-based wide-angle absorbers (Avitzour et al. 2009). However, the use of metamaterials in solar cells is still rather limited. The author has attempted to present some excellent research on metamaterial-based solar cells in this review, which gives us some ideas on how to use the multifaceted properties of metamaterials in solar cell applications to increase conversion efficiency.

(i) **The Unique Properties of Metamaterials**

Metamaterials are known for their ability to manipulate electromagnetic waves in unconventional ways. This manipulation is primarily due to their engineered structures rather than their composition. One of the most remarkable properties of metamaterials is negative refraction, which enables novel applications such as superlenses, cloaking devices, and high-efficiency energy conversion. Additionally, metamaterials can exhibit properties like perfect absorption, which is crucial for solar energy harvesting applications.

(ii) **Metamaterials in Solar Energy Applications**

The field of solar energy has greatly benefited from metamaterial research, as these materials provide promising solutions to overcome limitations in conventional photovoltaic (PV) technology. The ability of metamaterials to enhance light absorption and tailor electromagnetic responses makes them ideal candidates for improving solar cell efficiency. Some of the most important applications of metamaterials in solar energy include:

1. **Enhanced Light Absorption** Metamaterials can be engineered to absorb nearly all incident light across a broad spectrum, reducing reflection losses and increasing the efficiency of solar panels. Metallic nanostructures, for example, have demonstrated exceptional absorption characteristics that enable better photon harvesting.
2. **Broadband Absorption** One of the key limitations of traditional solar cells is their inability to efficiently capture light across the entire solar spectrum. Metamaterials designed with gradient-index structures or multilayered configurations can significantly expand the absorption bandwidth, ensuring more photons are converted into usable energy.
3. **Antireflection Coatings** Conventional solar cells often suffer from reflection losses, which reduce their efficiency. Metamaterials can serve as highly effective antireflective coatings by manipulating refractive indices to match air and semiconductor interfaces, thereby minimizing reflection and maximizing absorption.
4. **Plasmonic Enhancement** Plasmonic metamaterials leverage surface plasmon resonance to amplify electromagnetic fields at the nanoscale. This amplification boosts the interaction of light with the photovoltaic material, leading to higher charge carrier generation and improved solar cell performance.

5. **Thin-Film Solar Cells** The integration of metamaterials with thin-film solar cells has shown promise in reducing material costs while maintaining high efficiency. Metamaterials enable ultra-thin absorbers to capture and utilize light more effectively, making them a viable alternative to traditional bulkier photovoltaic devices.

6 Absorption Enhancement in Solar Cell Using Metamaterials

The capacity to absorb the greatest number of incident photons close to the junction is the primary need for the realization of an efficient solar cell. Therefore, compared to traditional anti-reflection (AR) coatings on solar cells, placing metamaterial structures on top of the cell may be an option for greater photon collecting.

For improved light absorption, Wang et al. showed a metamaterial structure (Fig. 7) with an ultrathin semiconductor layer encircled by a dielectric film and capped with a solid nanopatterned metallic film. According to Wang et al., the baseline efficiency of an a-Si solar cell can be increased by 15% as a result of the structure's good electromagnetic absorption.

Additionally, metal structures that have been specially constructed and implanted with dielectric films may be considered metamaterials with a negative refractive index. According to modelling results by Chen et al. (2012), a Ni strip surrounded by SiN can provide the best absorption efficiency of about 77%. The establishment of metamaterials in solar cell applications was spurred by this discovery.

Metamaterial as back reflector in solar cells:

A portion of light that strikes a metallic coating is reflected back with a 180-degree phase shift. The reflected wave may become less intense as a result of the incident

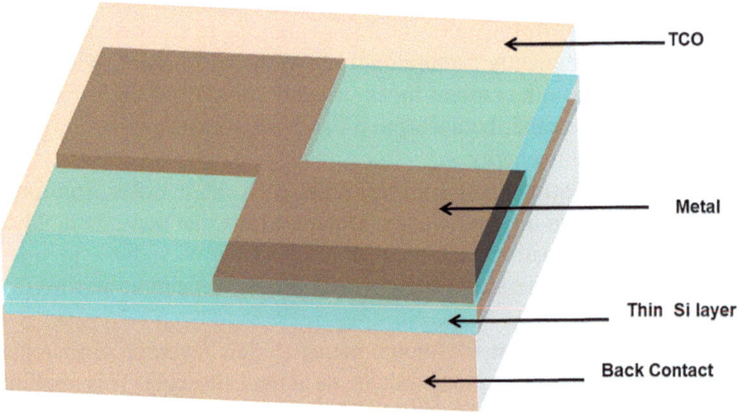

Fig. 7 Checker board like metamaterial structure as EM absorber [20]

and reflected waves' phase reversal. The effect is highly undesirable for solar cells. The primary reason for not employing metallic films as back reflectors for solar cells is this. Effective metamaterials, however, can prevent the reflected wave's phase reversal. Understanding the behaviour of the reflected wave is crucial to comprehending the dispersion of electromagnetic (EM) fields at the mirror surface. For light reflected from a surface, the complex reflection coefficient is as follows:

$$r = r_0 e^{i\phi} = \frac{Z_s - \eta_m}{Z_s + \eta_m} \tag{1}$$

where r_0 and φ are the reflection amplitude and phase, η_m is the characteristic impedance of the incident medium, and Z_S is the surface impedance of the reflector surface. As we can see from the equation; lower value of Z_S will lead to complete phase reversal i.e. $\varphi = \pi$. But by judicious patterning, the surface impedance and the reflection phase from the metal surface can be tailored. As in the groove array structure current has to follow convoluted path, the surface impedance will increase. The impedance Z_S can be accurately controlled by varying groove dimension resulting in negligible phase reversal.

A mirror-like structure capable of preventing phase reversal was first proposed by Fedotov et al. [22] (Fig. 8), marking a significant development in the field of metamaterials. This concept was further explored and successfully fabricated by M. L. Brongersma et al., who demonstrated its practical application as a back reflector in solar cells. Their research revealed that the incorporation of the metamaterial back reflector led to an increase of nearly 20% in the absorption efficiency of organic solar cells. This finding underscored the potential of metamaterials in enhancing solar cell performance, confirming their ability to manipulate light in ways that traditional materials could not. The improved light absorption was attributed to the unique optical properties of metamaterials, which can be engineered to control electromagnetic waves at subwavelength scales. By effectively increasing the optical path length within the solar cell, these structures enhanced photon absorption and improved overall energy conversion efficiency. However, despite this promising enhancement, the real-world application of metamaterial back reflectors in organic solar cells remains limited. One of the primary reasons for this limitation is the inherent instability of organic solar cells, which suffer from material degradation over time and lower power conversion efficiency compared to other photovoltaic technologies. Due to these challenges, the practical impact of metamaterial back reflectors in organic solar cells has been relatively insignificant, restricting their adoption in commercial applications.

Given the limitations of organic solar cells, the potential for metamaterial mirrors to make a meaningful impact lies in their application to crystalline silicon (c-Si) solar cells. Crystalline silicon remains the dominant material in commercial photovoltaic technology due to its superior stability, high efficiency, and well-established manufacturing processes. The integration of metamaterial-based back reflectors into c-Si solar cells could provide a substantial boost in performance by improving light management and reducing optical losses. However, an extensive review of existing

Fig. 8 Metamaterial mirror employed as black reflector of solar cell [21]

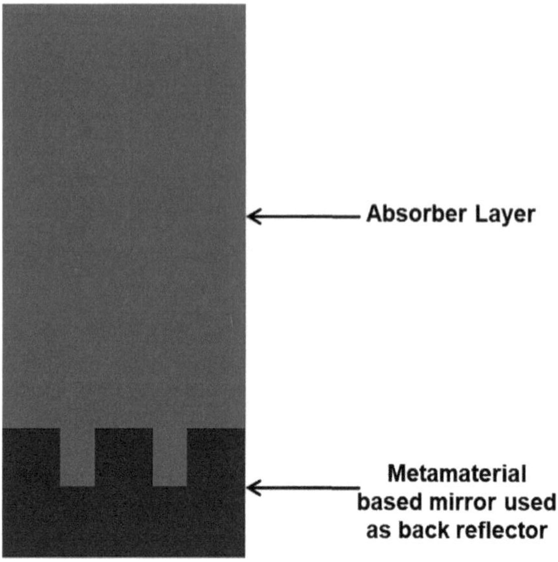

literature indicates that the majority of metamaterial designs still rely heavily on metals. While metallic metamaterials have been widely studied due to their strong plasmonic properties and ability to enhance light trapping, their use in solar cell applications presents several critical challenges. The most significant of these challenges is the increased recombination of minority carriers, which negatively affects charge carrier collection and reduces the overall efficiency of the solar cell. Additionally, metallic materials introduce Joule heating, where absorbed energy is dissipated as heat rather than being utilized for electricity generation. This heat generation not only reduces the efficiency of the solar cell but can also lead to thermal degradation over time, further compromising device performance.

The presence of optical losses due to non-radiative decay is another major drawback of metal-based metamaterials. When localized surface plasmon resonances are excited within these structures, a significant portion of the absorbed energy is lost through non-radiative decay processes, reducing the effective utilization of incoming photons. These combined factors make metal-based metamaterials less than ideal for solar energy applications despite their strong optical capabilities. As a result, researchers have turned their attention to dielectric metamaterials, which offer an alternative solution that overcomes many of these limitations. Dielectric metamaterials, composed of high-refractive-index materials such as silicon (Si), titanium dioxide (TiO_2), and gallium phosphide (GaP), exhibit negligible ohmic losses and do not suffer from the high recombination rates associated with metals. Their ability to support Mie resonances enables efficient light trapping and scattering across a broad spectral range, making them highly suitable for enhancing absorption in solar cells.

Furthermore, dielectric metamaterials can be designed to provide strong electromagnetic field confinement without introducing excessive energy dissipation, ensuring that more photons contribute to charge carrier generation.

In addition to their superior optical properties, dielectric metamaterials are also highly compatible with existing c-Si solar cell manufacturing processes. Unlike metal-based designs, which often require complex fabrication techniques and precise nanostructuring, dielectric materials can be more easily integrated into conventional photovoltaic systems. This compatibility makes them a practical choice for large-scale implementation, paving the way for next-generation solar cells with enhanced efficiency and stability. Recent advancements in computational design and nanofabrication techniques have further accelerated the development of dielectric metamaterials, allowing researchers to fine-tune their properties for optimal performance in photovoltaic applications. By leveraging these advancements, scientists aim to create metamaterial-based back reflectors that not only improve light absorption but also contribute to overall device reliability and longevity.

The transition from metal-based to dielectric metamaterials represents a crucial step toward the realization of high-efficiency solar cells. While metals have traditionally been favoured for their plasmonic capabilities, their inherent drawbacks limit their effectiveness in practical solar energy applications. Dielectric metamaterials, on the other hand, provide a more promising alternative by minimizing recombination losses, reducing heat generation, and enabling superior light trapping. As research continues to evolve, dielectric metamaterials are expected to play a key role in the next generation of photovoltaic technology, offering a scalable and efficient solution for improving solar energy conversion [20–35].

7 Summary

The use of plasmonic nanoparticles and metamaterials holds significant potential for improving the performance of silicon solar cells, whether they are thin or thick. By optimizing the size and distribution of these materials, it is possible to enhance the optical path length, which in turn increases photon absorption and improves carrier generation by overcoming the limitations posed by thin substrates and bulk recombination in thicker substrates. Plasmonic can significantly improve photon collection and carrier generation. This results in a more efficient conversion of light to photocurrent, thereby improving the overall performance of solar cells. These findings underscore the potential of plasmonic materials and metamaterials in advancing solar energy technologies and overcoming the challenges of using silicon in photovoltaic applications.

Acknowledgements The author would like to acknowledge the Department of Electronics and Communication Engineering and Nanotechnology Research Centre (SRM Institute of Science & Technology, Kattankulathur), India for the financial and infrastructural support.

References

1. H.A. Atwater, A. Polman, Plasmonics for improved photovoltaic devices. Nat. Mater. **9**, 205–213 (2010). https://doi.org/10.1038/nmat2630
2. P. Spinelli, V.E. Ferry, J.V. Groep, M.V. Lare, M.A. Verschuuren, R.E.I. Schropp, H.A. Atwater, A. Polman, Plasmonic light trapping in thin-film Si solar cells. J. Opt. **14**, 024002 (2012). https://doi.org/10.1088/2040-8978/14/2/024002
3. F. Monticone, A. Alù, Do cloaked objects really scatter less? Chin. Phys. B **23**, 047809 (2014). https://doi.org/10.1088/1674-1056/23/4/047809
4. S. Das, A. Kundu, H. Saha, S.K. Dutta, Investigation of plasmonic effects in thin-film solar cells. J. Renew. Sustain. Energy **5**, 031603 (2013). https://doi.org/10.1063/1.4812995
5. J. Grandidier, D.M. Callahan, J.N. Munday, H.A. Atwater, Light absorption enhancement in thin-film solar cells using whispering gallery modes. IEEE J. Photovoltaics **1**, 99–104 (2011). https://doi.org/10.1109/JPHOTOV.2011.2180512
6. M.R. Khan, X. Wang, P. Bermel, M.A. Alam, Photonic band engineering for high-efficiency light trapping in solar cells. Opt. Express **22**, A973–A985 (2014). https://doi.org/10.1364/OE.22.00A973
7. J.B. Khurgin, How to deal with the loss in plasmonics and metamaterials. Nat. Nanotechnol. **10**, 2–6 (2015). https://doi.org/10.1038/nnano.2014.310
8. S. Molesky, C.J. Dewalt, Z. Jacob, High-temperature epsilon-near-zero and epsilon-near-pole metamaterial emitters for thermophotovoltaics. Opt. Express **21**, A96–A110 (2012). https://doi.org/10.1364/OE.21.000A96
9. M.A. Ordal, L.L. Long, R.J. Bell, S.E. Bell, R.R. Bell, R.W. Alexander, C.A. Ward, Optical properties of the metals Al Co, Cu, Fe, Pb, Ni, Pd, Pt, Ag, Ti, and W in the infrared and far infrared. Appl. Opt. **22**, 1100–1120 (1983). https://doi.org/10.1364/AO.22.001100
10. F. Dincer, O. Akgol, M. Karaaslan, E. Unal, C. Sabah, Investigation of novel plasmonic structures. Prog. Electromagn. Res. **144**, 93–101 (2014). https://doi.org/10.2528/PIER13100809
11. G.V. Naik, B. Saha, J. Liu, S.M. Saber, E.A. Stach, J.M.K. Irudayaraj, T.D. Sands, V.M. Shalaev, A. Boltasseva, Titanium nitride as a plasmonic material for visible and near-infrared wavelengths. Proc. Natl. Acad. Sci. USA **111**, 7546–7551 (2014). https://doi.org/10.1073/pnas.1321618111
12. J.B. Khurgin, G. Sun, In search of the elusive lossless metal. Appl. Phys. Lett. **98**, 211106 (2011). https://doi.org/10.1063/1.3590265
13. R. Shashanka, H. Esgin, V.M. Yilmaz, Y. Ceylan, A study on the optical properties of hybrid materials. J. Sci. Adv. Mater. Devices **5**, 185–191 (2020). https://doi.org/10.1016/j.jsamd.2020.03.007
14. A.E. Suliman, Y. Tang, L. Xu, Optical characterization of thin films for solar applications. Sol. Energy Mater. Sol. Cells **91**, 1658–1662 (2007). https://doi.org/10.1016/j.solmat.2007.05.007
15. A.B. Roy, K.V. Kumar, M. Saha, Plasmonic properties of metal nanoparticles. Appl. Phys. A **127**, 119 (2021). https://doi.org/10.1007/s00339-021-04436-6
16. A. Baral, G. Das, A.B. Roy, A. Kole, N. Mukherjee, S. Bose, Design and optimization of plasmonic structures for photovoltaics. IEEE J. Photovoltaics **10**, 99–104 (2020). https://doi.org/10.1109/JPHOTOV.2020.2991832
17. S. Banerjee, S. Mandal, S. Dhar, A.B. Roy, N. Mukherjee, Engineering plasmonic nanostructures for chemical applications. Ind. Eng. Chem. Res. **58**, 28–35 (2019). https://doi.org/10.1021/acs.iecr.9b04678
18. S. Nakade, M. Matsuda, S. Kambe, Y. Saito, T. Kitamura, T. Sakata, Y. Wada, H. Mori, S. Yanagida, Electrolyte effects on dye-sensitized solar cells. J. Phys. Chem. B **106**, 10004–10010 (2002). https://doi.org/10.1021/jp020078v
19. Q.A. Alsulami, Z. Arshad, M. Ali, G. Wageh, Novel gels for photovoltaic applications. Gels **9**, 101 (2023). https://doi.org/10.3390/gels9020101
20. Y. Wang, T. Sun, T. Paudel, Y. Zhang, Z. Ren, K. Kempa, Nano Lett. **12**, 440–445 (2012). https://doi.org/10.1021/nl203763k

21. M. Esfandyarpour, E.C. Garnett, Y. Cui, M.D. McGehee, M.L. Brongersma, Nat. Nanotech. **10**, 1038/NNANO.2014.117 (2014). https://doi.org/10.1038/nnano.2014.117
22. V.A. Fedotov, A.V. Rogacheva, N.I. Zheludev, P.L. Mladyonov, S.L. Prosvirnin, Appl. Phys. Lett. **89**, 091119 (2006). https://doi.org/10.1063/1.4921969
23. Y. Liu, X. Zhang, Metamaterials: a new frontier of science and technology. Chem. Soc. Rev. **40**, 2494–2507 (2011). https://doi.org/10.1039/C0CS00184H
24. C. Wu, B. Neuner, J. John, A. Milder, B. Zollars, S. Savoy, G. Shvets, J. Opt. **14**, 024005 (2012). https://doi.org/10.1088/2040-8978/14/2/024005
25. F. Monticone, A. Alù, Chin. Phys. B **23**, 047809 (2014). https://doi.org/10.1088/1674-1056/23/4/047809
26. B.X. Wang, L.L. Wang, G.Z. Wang, W.Q. Huang, X.F. Li, X. Zhai, IEEE Photonics Tech. Lett. **26**, 111–114 (2014). https://doi.org/10.1109/JPHOT.2014.2381633
27. Y. Liu, Y. Chen, J. Lib, T.C. Hung, J. Li, Sol. Energy **86**, 1586–1599 (2012). https://doi.org/10.1016/j.solener.2012.02.021
28. R. Niepelt, J. Hensen, A. Knorr, V. Steckenreiter, S. Kajari-Schoder, R. Brendel, Plasmonic effects on solar cell efficiency. Energy Procedia **55**, 570–577 (2014). https://doi.org/10.1016/j.egypro.2014.08.028
29. R.L. Wallace, J.I. Hanoka, S. Narasimha, S. Kamra, A. Rohatgi, in *Proceedings of the 26th IEEE Photovoltaic Specialists Conference (PVSC)* (Sept. 30–Oct. 3, 1997), pp. 99–102. https://www.nrel.gov/docs/legosti/fy98/25232.pdf
30. J. Kim, J. Hwang, K. Song, N. Kim, J.C. Shin, J. Lee, Enhanced light absorption in silicon solar cells. Appl. Phys. Lett. **108**, 253101 (2016). https://doi.org/10.1063/1.4954039
31. R.L. Wallace, J.I. Hanoka, A. Rohatgi, G. Crotty, High-efficiency thin-film solar cells. Sol. Energy Mater. Sol. Cells **48**, 179–186 (1997). https://doi.org/10.1016/S0038-092X(03)00187-7
32. G.P. Willeke, Advances in photovoltaic materials. Sol. Energy Mater. Sol. Cells **72**, 191–200 (2002). https://doi.org/10.1016/S0927-0248(01)00164-7
33. S. Wang, B.D. Weil, Y. Li, K.X. Wang, E. Garnett, S. Fan, Y. Cui, Nanostructured solar cells with enhanced light trapping. Nano Lett. **13**, 4393–4398 (2013). https://doi.org/10.1021/nl402230v
34. A.B. Roy, Metamaterial embedded light trapping structure for p-Si/n-ZnO based thin c-Si solar cell. Mater. Res. Innov. **27**, 2 (2022). https://doi.org/10.1080/14328917.2022.2087026
35. F.C. Krebs, M. Jorgensen, K. Norrman, O. Hagemann, J. Alstrup, T.D. Nielsen, J. Fyenbo, K. Larsen, J. Kristensen, Stability and performance of organic photovoltaics. Sol. Energy Mater. Sol. Cells **93**, 422–441 (2009). https://doi.org/10.1016/j.solmat.2008.12.001

Integrating ML Approaches for Photodetectors

Sayantani Bhattacharya, Sukanya Ghosh, Debajit Deb, Praveenkumar Hiremath, and Debarati Nath

Abstract Optical sensor technologies are central to the advancement of Internet of Things (IoT) ecosystems and smart systems. Among these, photodetectors—a key subset of optical sensors—find extensive applications in areas such as smart sensing, home automation, autonomous vehicles, wearable electronics, and AI-driven platforms. The demand for highly sensitive, cost-effective, flexible, and lightweight photodetectors continues to grow, especially for seamless integration with CMOS processors and wearable systems. However, developing high-performance photodetectors presents several challenges, including optimizing device lifetime, operational speed, and selecting efficient material systems. Addressing these issues requires a combination of comprehensive literature review, fabrication process optimization, and detailed material characterization. In recent years, machine learning (ML) techniques have emerged as powerful tools to overcome these limitations. ML not only enhances fabrication efficiency by reducing processing time, cost, and human intervention, but also plays a crucial role in predicting material behavior, optimizing device parameters, and accelerating design cycles. By leveraging ML, researchers can gain deeper insights into material–device interactions and significantly improve device performance. This chapter presents a comprehensive overview of the latest developments in integrating machine learning into photodetector design. It covers the foundational concepts of ML, classifications and operating principles of various photodetector types, and illustrates how ML contributes to performance enhancement, smart design, and real-time optimization in modern photodetector systems.

S. Bhattacharya · S. Ghosh · D. Nath (✉)
Department of ECE, College of Engineering and Technology, Faculty of Engineering and Technology, SRM Institute of Science and Technology, Kattankulathur, Chennai, Tamil Nadu, India
e-mail: debaratiphd@gmail.com

D. Deb
Department of ECE, Koneru Lakshmaiah Education Foundation, Vaddeswaram, AP, India

P. Hiremath
Mechanics, Materials and Component Design, Lund University, Lund, Sweden

B. Ramachandran et al. (eds.), *Advanced Optical Sensors*, Progress in Optical Science and Photonics 38, https://doi.org/10.1007/978-981-95-1687-2_9

Keywords Photodetector · Machine learning · Optoelectronic devices · Smart sensor · Material design and optimization

1 Introduction

Photodetectors are integral components in numerous applications, including but not limited to optical communication, quantum computing, augmented reality (AR), and artificial intelligence (AI) as well [1]. Photodetectors have a wide range of applications in the Internet of Things (IoT) enabled medical devices, industrial systems, environmental monitoring, autonomous vehicles, smart homes, smartphones, and flexible electronics [1, 2]. The fundamental operation of a photodetector is to identify optical signals and transmute the optical signal into electrical signals. This functionality enables IoT systems to detect, monitor, and respond to their surroundings, making photodetectors requisite in Information and Communications Technology (ICT). Various semiconductor-based optoelectronics materials are used for fabricating the photodetector [2]. Halide perovskite, graphene, quantum dots, organic semiconductors, and various flexible semiconductors show great interest in the photodetector field due to their flexibility, low cost, lightweight, optical tunability, and broader spectral response [3]. Various research groups reported that a photodetector consisting of such materials offers extraordinary detectivity, superior responsivity, large dynamic range, and high efficiency [3, 4].

A flexible ultrathin perovskite-based wearable photodetector was designed, offering remarkable responsivity of 5600 A/W, and fast response times of 3.2 μs [5]. A photodetector was designed on a flexible PET substrate, revealing extraordinary flexibility after one thousand cycles of bending at an angle of 120° [6]. A flexible organic photodetector was developed for real-time heart rate measurement systems, providing an ultrafast response of 220 ns and ultrahigh selectivity of 3.1 $\times 10^{13}$ Jones [7]. Organic pulse oximetry (OPO) consisting of the organic photodetector (OPD) and organic light-emitting diodes (OLED) was developed in which OLEDs and OPDs integrated on the identical surface of the anatomical structures for identifying human vital signs [8]. Lee et al. designed a prototype of flexible OPD integrated with a rigid printed circuit board (PCB) which can increase the mechanical flexibility of the integrated circuit (IC) system [8]. Arias et al. documented a reflectance oximeter system consisting of four red and NIR-printed OLEDs alongside eight OPDs which is analogous to an image sensor system [9]. Such highly sensitive, cost-effective, flexible, and lightweight photodetectors are required for integrating the detector with the CMOS processor and wearable devices to meet the increasing demands of modern technologies. A photodetector with fast response time and good sensitivity is necessary for better camera performance. Highly developed wearable photodetector is required for health and environmental monitoring systems. Lifetime, stability, dark current, speed, and choice of efficient materials are some common challenges in developing the photodetector. As, they hold great promise for applications in wearable devices, healthcare, and IoT, addressing these challenges is critical for their widespread adoption. To overcome such challenges, many literature surveys, proper choice of material, improved fabrications, and characterization techniques must be followed. In addition to this, for achieving a high-performance flexible photodetector, lots of trial and error as well as improvements are essential. There are various computational approaches including density functional theory (DFT), the Monte Carlo method, the drift–diffusion method, and molecular dynamics to analyze material properties, including structural, electronic, and optical characteristics, among others [2, 10]. These methodologies are too costly, complex, and time-consuming for analyzing material behavior.

To address the above-mentioned limitations, machine learning (ML), a subset of AI was incorporated into the field of optoelectronic research due to the growing complexity of modern applications, the need for high-performance systems, and the demand for intelligent, adaptive, and efficient solutions [2, 10, 11]. ML is one of the powerful tools to enhance the accuracy of analyses, simulations, and predictions of optoelectronic properties in organic materials. ML enhances photodetector-based IoT systems by improving accuracy, prediction, and performance. The integration of ML with photodetectors allows photodetectors to not only sense light but also process, analyze, and make data-driven decisions in real-time, which is critical in advanced systems such as IoT, healthcare, autonomous vehicles, and more. ML allows computers to improve the performance of the device with the help of various tasks from the data [2, 10, 11]. ML predicts optimal designs and configurations, minimizing the need for extensive physical experiments. Hence, ML can reduce the computational time and effort of general trial and error methodologies for photodetector performance measurement [2, 10, 11]. ML integrated with a photodetector can analyze the huge amount of data and identify the hidden correlations which can further predict the photodetector's material properties and behavior [2]. Photodetectors equipped with ML can self-tune their parameters, reducing design and manufacturing time. Therefore, ML provides various approaches for predicting and designing of photodetector. In addition to this, ML accelerates the discovery of new materials for photodetectors with enhanced optical and electronic properties. Researchers are already growing their attention to ML integrated with photodetector-based sensor technologies [12–14]. Researchers have incorporated ML concepts in various domains such as material properties prediction, device optimization, experimental process optimization, and so on [13, 14]. In order to build an ML model, an appropriate dataset needs to be built by collecting data from experimental results or some previous literature. However, the generation of experimental data is again time-consuming and will slow down the processing power of the ML model in the optoelectronics area. Therefore, the ML model integrating with some computational or simulated data provides new solutions to optoelectronic problems in organic and inorganic hybrid devices.

1.1 ML with Material Science

ML is a branch of AI that has gained immense popularity due to its capability to process large datasets that traditional methods find challenging to handle [15]. ML enables the efficient automation of complex tasks such as self-driving cars, fraud detection, medical diagnosis, image recognition, and speech processing without any human effort [16, 17]. It is highly used in biological and chemical sciences but nowadays its application reaches also in the field of solid-state materials science. Finding and characterizing crystal structures based entirely on the composition of new materials are the main challenges in this domain. Experimental research is highly time-consuming and costly and it is possible with a limited number of materials [18]. However, ML algorithms are successfully applicable for the prediction of new materials and also their properties such as compositions [19], and structures that enable the researchers to accelerate the discovery and optimization of materials [20]. The field of ML is generally divided into three primary categories: supervised learning, unsupervised learning, and reinforcement learning [21]. Figure 1 shows the basic workflow of the highly used supervised ML model.

In materials science, features must be capable of encapsulating all the essential information required to differentiate between various atomic or crystal structures [22]. Feature extraction is an essential part where in materials science, the most common features such as nuclear charges and atomic positions are used to describe the general properties of descriptors [23]. To solve the nature of the problem in materials science, the best algorithms have been proposed such as regression models, and classification models. The linear regression model is presented to identify the linear correlation between features and targets. Decision trees (DTs), random forests (RF), and gradient boosting (GB) are also useful for nonlinear models that can capture complex patterns [23]. Support Vector Machines (SVMs) find the hyperplane that best separates the dataset into classes [23]. Neural Networks are used for complex problems like image or speech recognition [23]. The model is trained according to

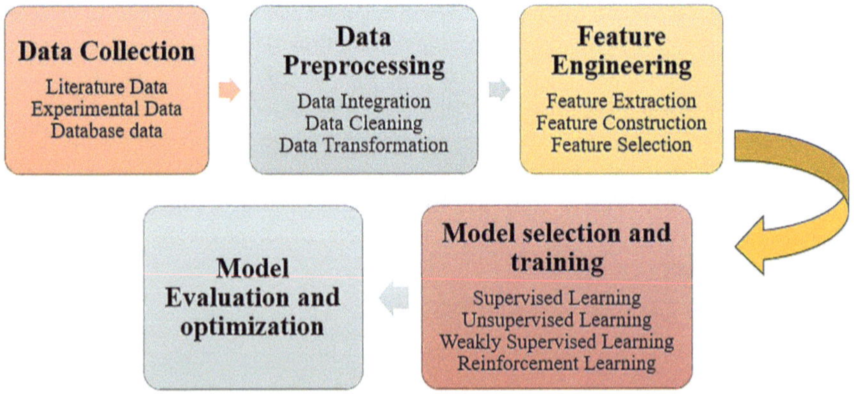

Fig. 1 Basic workflow of machine learning model

the training dataset to predict the material [24, 25]. Objectively assessing model performance is essential for enhancing its generalization and adaptability to unseen data. ML methods successfully predict a large number of material properties such as band gap [26, 27], bulk and shear moduli [26–28], topological states [29], superconductivity [30], the density of states [31], fermi energy, and Poisson ratio [26], thermal expansion coefficient [27], Curie temperature [32] are few of them. Table 1 presents details of material properties predicted using ML methods along with their corresponding references.

Table 1 Details of material properties predicted with ML algorithm and corresponding references

Material property	References
Magnetic transition temperature	[32]
Energy and entropy contributions from atomic vibrations	[33]
Forbidden energy gap	[26, 27]
Dielectric failure threshold	[34]
Lattice dimension	[35]
Characteristic temperature and heat capacity	[26, 27]
Softening temperature	[36]
Coefficient of thermal expansion (CTE)	[27]
Interface heat resistance	[37]
Material heat conductivity	[38]
Intrinsic magnetic moments	[39]
Phase transition temperature	[23]
Magnetically induced thermal effects	[23]
Intergranular boundaries	[40]
Intergranular energy	[41]
Intergranular mobility	[42]
Boundary energy	[35]
Thermoelectric sensitivity	[43]
Thermoelectric performance index	[23]
Elastic bulk modulus and shear modulus	[26–28]
Material resistivity	[43, 44]
Density of quantum states	[45]
Fermi level energy and elastic strain ratio	[26]
Impurity solution energy	[46]
Electronic phase classification	[47]
Invariant topological quantities	[29]
Threshold temperature for superconductivity	[23]
Lithium-ion mobility in electrolytes	[47]

1.2 ML with Optoelectronics

Optoelectronic materials are essential to numerous advanced technologies, offering diverse applications such as fiber optic communication, ultraviolet (UV) detectors, light-emitting diodes (LEDs), power devices, solar cells, optical sensors, and lasers [48–51]. The research and advancement of conventional optoelectronic materials demand rigorous experimentation, iterative refinements, and continuous advancements to optimize their material properties and structures to enhance performance and functionality. Various computing technologies including DFT, molecular dynamics simulations, and Monte Carlo simulations [2] have been extensively utilized to predict the electronic structure of material and physical movements of atoms and molecules [52–54]. However, each step of these computational methods is time-consuming, repetitive data acquisition, and highly complex for data analysis significantly extending the research timeline. ML focuses on creating algorithms and models to learn from and make predictions or decisions based on data. Unlike traditional programming where rules are explicitly specified, ML algorithms extract patterns and relationships from large datasets, enabling them to make decisions or predictions about patterns, correlations, and structures within the data [55, 56]. By minimizing conventional trial-and-error approaches in optoelectronic material research, ML algorithms can help researchers focus on the most promising candidates, saving time and reducing costs in the research and development phase. The performance of the ML models is highly dependent on the quality and quantity of the data are trained on. The basic workflow of the ML process for optoelectronics material is the same as the basic ML process. First, identify the object that needs to be addressed as an optoelectronic material. Then gather high-quality data from reliable sources like experimental results, computational simulations, and research papers. The experimental data such as properties of optoelectronic material are sourced from COD and ICSD databases. Computational material databases include AFLOW, MP, NOMAD, and CMR [2]. The microscopic images of optoelectronic materials obtained from scanning electron microscopy (SEM) and atomic force microscopy act as another dataset. The noise reduction is an essential part of data preprocessing. The feature engineering objective is to extract and construct meaningful features to represent a material. Various feature construction methods are used for optoelectronic materials such as compositional features, structural descriptors, graph-based representations, and image features. The compositional features mean components, crystal structure parameters, band gap, electronic density of states, and other related properties. The structural information of the material plays a vital role in predicting and understanding atomic symmetry functions, and crystal fingerprints. Using graph neural networks (GNNs) to extract connectivity network features is a powerful approach to describing the atomic arrangement and crystal symmetry of materials. To characterize the size, shape, distribution, and other information of materials SEM and transmission electron microscope (TEM) images are used for image feature extraction. ML algorithms are typically classified into the following categories: supervised, unsupervised, weakly supervised, and reinforcement learning [2, 57]. Various sophisticated ML algorithms have progressively been utilized to numerous applications such as material designing, optimization of processes, property prediction, structure analysis, and more. Such as SVM, and RF have been significant tools in predicting material properties like band gaps, formation energies, magnetic moments based on atomic or molecular descriptors, lattice parameters, cell defect properties, and optical properties [58, 59]. DL has a significant role in designing and optimizing next-generation optoelectronic devices. The key contributions of DL in optoelectronic materials are predicting the bandgap, refractive index, and carrier mobility of material. Convolutional neural networks (CNN), recurrent neural networks (RNN), generative adversarial networks (GAN), and GNN are utilized for predicting properties such as electronic structure, optical characteristics, and mechanical properties [60]. These models are equally used for predicting the arrangement at the atomic or molecular level [61]. CNNs and GANs are particularly suited for image processing tasks collected from electron microscopy or spectroscopy to identify defects, classify materials, or enhance the quality of the images. The optimization of the material compositions, device architectures, and synthesis processes to improve performance in optoelectronic devices is another property like GANs and RNNs [2]. Transfer learning (TL) is indeed a highly valuable technique that enables deep learning models to perform effectively even with limited data availability [60]. Active learning has proven useful applications in predicting properties [38], identifying novel materials, detecting defects from microscopic images, and capturing atomic force fields [2]. Reinforcement Learning is an effective approach for improving the efficiency of material optimization by enabling the discovery of optimal materials, configurations, or synthesis processes in a systematic and data-driven manner [41, 42]. ML Interatomic Potentials (ML-IAPs) and force fields (MLFFs) represent advanced computational ML models that are trained to predict the potential energy and interatomic forces acting on atoms within a material using ML techniques [62, 63]. Support Vector Regression (SVR) has emerged as a valuable tool in the field of optoelectronic materials, which is used to predict the bandgap of a material, charge carrier mobility, and defect states in materials [2].

1.3 ML in Sensing

The rapid advancements in AI have generated immense curiosity about its role in the digital transformation of human life. The generation of flexible sensors is highly demanding as it creates a crucial link between the physical world and digital signals. These flexible sensors can efficiently acquire, analyze, and interpret extensive and diverse data. Advanced ML algorithms [64, 65] are capable of designing various intelligent flexible sensing systems that present significant opportunities such as medical diagnosis, monitoring of health, user interaction with machines, and smart homes to develop a smarter and more efficient world for humanity [65]. ML involves creating algorithms that enable computers to learn from large raw datasets, further enabling predictions or decisions without explicit programming while ensuring high computational efficiency and accuracy [2]. Before applying ML to flexible sensing systems, the raw data needs to be transformed into suitable input features involving pre-processing tasks like outlier removal, denoising, and normalization [64, 65]. Different ML algorithms are used such as SVM, ANN, CNN, and RNN [64, 66] for flexible sensor systems. SVM is highly effective for high-dimensional datasets and proves effective when there is a clear decision boundary between different classes [67, 68]. SVM can handle small datasets but is sensitive to noise, outliers, and the selection of parameters. An ANN is another machine-learning model inspired by the human brain to recognize structures, process information, and predict results [69]. Instead of the requirement of large amounts of data, its application area is wide, useful for high adaptivity, and strong fault detection. At present CNN and RNN represent the two predominant structures under the deep learning algorithm. A CNN is a specifically designed model for image processing [70]. It performs tasks like classification of images, object detection, and facial recognition manual feature extraction without any manual feature extraction. RNN is designed to process sequential data by maintaining the memory of previous inputs [64, 66]. By incorporating feedback loops, RNNs can preserve past information across time steps, and make them ideal for tasks involving NLP, speech processing, and predicting time-dependent data. Data pre-processing techniques, including principal component analysis (PCA), linear discriminant analysis (LDA), and t-distributed stochastic neighbor embedding (t-SNE), are also convenient for flexible sensing [71]. Nowadays machine learning-based smart flexible sensing is highly demanding for sensing, large-scale data analysis, high efficiency, multimodal information processing and decoupling, and intelligent environment perception. SML (Statistical machine learning) based ML algorithm is used to design a textile triboelectric sensor for the application of ambulatory cardiovascular condition assessment and highly reliable pulse monitoring [72]. The innovative air-permeable and biodegradable smart mask has successfully been developed [73]. It uses ML algorithms to analyze respiratory signals, enabling accurate diagnosis and monitoring of chronic respiratory diseases. The combined effect of flexible piezoelectric acoustic sensors and ML techniques (GMM, DNN, CNN, t-SNE) [74, 75] are possible for speaker recognition and voice communication. The low-cost triboelectric smart sock, utilizing 1D CNN-based deep learning analysis, was successfully developed for gait patterns to understand human movements and postures through real-time data processing and analysis [76]. Kim et al. [77] developed skin-like stretchable strain sensors for gesture recognition where sectional CNN and sparse neural networks were used for sensor data integration and pattern recognition at the feature level. Using CNN, ML algorithms [78], piezoelectric tactile sensors are developed for acquiring human tactile information such as the sensation of pressure and temperature, rough surfaces or objects with unique textures, and sliding speed. Sahatiya et al. [78] presented a versatile sensor combined with ML algorithms (KNN, DT, RF, ET) for health monitoring without medical interventions. A flexible, wireless electronic system has been developed to continuously analyze ECG data and monitor the user's heart condition in real-time. CNN technique is used here to classify different user activities directly from the raw ECG data. SVM, KNN, and DT ML algorithms were utilized to develop an epidermal electronic system by extracting the key features from signals including ECG, GSR signals, and the respiration rate of humans [79]. Lee et al. [80] developed an AI-powered system for sign language recognition and communication, featuring smart sensing gloves equipped with triboelectric sensors, a deep learning module, and a virtual reality interface for interaction. An intelligent chair-based sitting recognition system was introduced by Hu et al. [81], using six flexible piezoresistive sensors, and a two-layer ANN for ML to distinguish different health-related sitting postures. Machine learning-based CNN algorithm is also applicable for making smart toilets to identify user information while seated on the toilet with over 90% accuracy [82]. Combining AI technology with advanced flexible sensing systems will be essential in soft robots, humanoid robots, and human–machine interfaces in the future [83, 84].

1.4 ML in Photodetector

The domain of ML has emerged as a revolutionary influence across multiple disciplines, with photodetector technology being a notable instance. By using advanced algorithms and data-focused methods, researchers can improve how well photodetectors work, resulting in major gains in their sensitivity, speed, and overall performance. These advancements not only facilitate more precise identification of light signals but also unveil novel opportunities for implementation in domains such as telecommunications, medical imaging, and environmental surveillance [85]. Photodetectors are electronic components that transform incoming light into corresponding electrical signals, enabling optical sensing and detection across various applications. Conventional photodetection techniques are based on core physical phenomena such as the photoelectric effect, photoconductivity, and the photovoltaic effect [86, 87]. Advancements in ML are making photodetectors more intelligent, allowing for real-time data processing, improved noise reduction, and adaptive optimization to enhance overall performance. Oh et al. [85] explores the use of ML to enhance the performance of silicon photodiodes (SiPDs), with a particular focus on optimizing their efficiency for near-infrared wavelengths. It compares the performance of XGBoost and light gradient-boosting machine (LGBM) algorithms by utilizing the simulation data to streamline the design process. The research emphasizes the high accuracy of the LGBM algorithm in forecasting photodiode responsivity, showcasing the potential of ML to improve the efficiency and performance of photodetector design. The decision tree-based ML models are effectively proposed in predicting the performance of two-dimensional metal halide perovskite photodetectors using various material properties and device performance metrics like responsivity and detectivity [88]. The applications of ML for predicting band gaps in polymers used for photodetectors are proposed in Ref. [89]. Around fifteen regression models, the LGB and hist gradient boosting models emerged as the most effective for targeted predictions. The design of near-IR organic semiconductors for photodetectors utilizes Extra Trees Regressor and Random Forest Regressor ML models, which showcase their predictive abilities by producing over 1000 novel organic semiconductors

[90]. Ocaya et al. [91] introduced an artificial neural network (ANN) approach to analyze and determine internal parameters, including barrier height, ideality factor, and series resistance, for Schottky photodiodes under different doping levels and illumination intensities, eliminating the need for the thermionic emission equation. The study emphasizes the significant difficulties in synthesizing complex materials like borophene, which require precise performance predictions. By incorporating advanced ML methods, such as artificial neural networks and generative adversarial networks for data enhancement, researchers [92] aim to overcome these obstacles. Multiple linear regression and polynomial regression, both supervised learning algorithms, are employed to predict the responsivity of graphene-based photodetectors, especially in the challenging deep UV wavelength range [93]. The results revealed that polynomial regression offered a considerably superior fit compared to multiple linear regression. The maximum likelihood ML detection estimation approach experiences significant nonlinear phase noise in Ref. [94] for understanding the reliability of the communication system. The researchers assess the efficiency of their suggested detector through symbol error rate (SER), which reflects the frequency with which the identified symbols diverge from the original transmitted symbols. The advancement of ML in photodetectors is set to revolutionize material discovery, enhance performance optimization, enable real-time adaptability, and refine intelligent signal processing. As ML technologies progress, photodetectors will become increasingly efficient, adaptable, and well-suited to meet the evolving requirements of next-generation applications in imaging, communication, and sensing.

2 Understanding of ML

The resurgence of AI has had a positive impact on science and technology as is becoming apparent. In the field of materials science, ML which is a specific area of AI that deals with computer algorithms, also referred to as models, to make predictions by learning patterns in large datasets, is proving to be an indispensable asset. In the case of optoelectronics devices, ML can be used to achieve high device performance and discover new materials thereby leading to reduced device development costs [2]. This is possible because the use of ML enables the identification and optimization of factors that affect device performance [2, 10].

Building an ML model requires data collection, data pre-processing, feature engineering (if needed), training, testing, and performance evaluation. The first step in implementing the ML model is to collect data. Collected data is to be pre-processed to get a suitable format. Collected data will be split into two sections: training and testing data. The training data is utilized to train the model whereas testing data is for measuring the model's performance. Various algorithms are trained by the training data however, the best model can be identified by evaluating some performance parameters. After that, the trained model can perfectly identify the patterns and that model is applied to new test data for prediction purposes. The steps involved in developing ML are as follows.

2.1 Data Collection

It is one of the important steps in the entire ML process. The general procedure for collecting a dataset is as follows: (1) Identify the purpose of the ML model to be built and types of the data to be collected. For example, for the prediction of the optoelectrical properties of the device, the model needs to be built using the related data. This data can be obtained either from experiments or from numerical computations. With regards to numerical computations, techniques based on quantum mechanical approximations—DFT (for electronic ground state properties), time-dependent DFT, GW approximation combined with Bethe–Salpeter Equation [95] (for electronic excited state properties) can be adopted. Another such example can be predicting the surface morphology of the device for which morphological pictures are required. Data can also be collected from various public databases for doing research in ML [96]. Widely used computational material science databases are the Cambridge Structural Database (CSD) [13], Crystallography Open Database (COD) [13], Inorganic Crystal Structure Database (ICSD) [13], and so on. It is also feasible to collect optoelectronics-related data from various research articles through literature surveys [2 and therein]. (2) Collection of sufficient amounts of data for training and testing. Large enough data is a requirement for the optimization of an ML model. Although a dataset can be prepared from the experiments, this approach will be very challenging if a large dataset is required to build an ML model [96]. Such experimental datasets may contain image data or CSV files. But, computational or simulation results can produce large and efficient datasets without as much effort as from experiments which further helps to provide an optimal solution [2, 13, 96].

2.2 Data Preprocessing

The data preprocessing step helps convert unstructured data to a suitable format. This process involves data cleaning, standardization, and data segmentation. Duplicate data and outliers can be removed during the data-cleaning process [97]. Standardization converts all parameters into the same format. Data segmentation divides the dataset into training and testing datasets [97].

2.3 Feature Engineering

A dataset contains two types of variables—dependent (target) and independent variables. Independent variables include attributes such as material structure properties, fabrication details, and some experimental parameters. Among all the attributes, there may be a strong correlation between a few of them and the target variable [23]. Therefore, the role of feature engineering is to select suitable attributes to enable the ML model to predict the target variable. This is particularly important because the presence of some irrelevant attributes can reduce the ML model performance. Thus, feature engineering is performed, if needed, after the data pre-processing step. Filtering and embedding algorithms are normally used in feature engineering [97]. Harikrishna Sahu et al. used feature engineering concepts to select a proper descriptor for predicting the efficiency of organic solar cells [14]. Srimanta Mitra et al. used feature engineering techniques to select suitable structural properties for predicting the target variable [97]. Baian Chen et al. also used the feature engineering concept for choosing suitable atomic parameters and fabrication conditions of perovskite devices [13].

2.4 Model Selection

The common ML methodologies are categorized into supervised learning, unsupervised learning, and reinforcement learning [96]. Depending on the type of data, supervised and unsupervised algorithms are usually employed in ML-based materials research [2, 96].

2.4.1 Supervised Learning

Supervised learning is used for classification and regression tasks. The frequently utilized supervised learning algorithms are linear regression, logistic regression, SVM, RF, and DT. Algorithms must be chosen based on specific applications and behavior to get accurate predictions. In the material science concept view, supervised learning is used to predict the material properties of new material if those properties are already known for some previous attributes [2, 13].

2.4.2 Unsupervised Learning

Tasks such as dimensionality reduction and clustering are performed with the help of unsupervised learning [96]. Unlike the data in supervised learning, the data in unsupervised learning is unlabelled. The dimensionality reduction is particularly important when feature elicitation i.e. the features of the data that are strongly correlated with the dependent variable [96]. As an example, in terms of optoelectronics, this would facilitate the identification of factors that strongly influence device performance. Similarly, clustering can be used to group materials based on their impact on device performance.

2.5 Learning Algorithms (For Supervised and Unsupervised)

2.5.1 Linear Regression

It is an algorithm that develops a linear relationship between independent and dependent variables, as shown in Eq. 1. The slope parameter (m) can be calculated with the help of known independent (x) and dependent parameters (y). After getting the slope, a new dependent value can be calculated from a known independent variable with the help of Eq. 1. It is used for regression analysis.

$$y = mx + C \tag{1}$$

This algorithm is used in optoelectronics if any properties of the material change linearly with some parameters. Therefore, it is useful for determining the linear relationship between device parameters and device performance. For example, defect properties of optoelectronics material can be analyzed with a linear regression algorithm [2]. Performance metric parameters of a photodetector such as avg. Photocurrent, the impulse response is supposed to be predicted using the algorithm [12].

2.5.2 K-Nearest Neighbour (KNN)

KNN algorithm is used for regression and classification tasks also. The role of the KNN regression is to average the values of the nearest neighbour to a given point. KNN classification algorithm classifies the target label based on the nearest pattern [96]. Various distance metrics, such as Euclidean, Manhattan, or Minkowski, are used for finding the similarity between data points. k-NN mainly prefers the Euclidean distance which is the Euclidian distance between a training point and a given test point [96]. The formula for Euclidian distance is expressed in Eq. (2).

$$dist(x_1, x_2) = \sqrt{\sum_{i=1}^{n} (x_{1i} - x_{2i})^2} \tag{2}$$

where x_1 and x_2 are the two points, $x_1 = (x_{11}, x_{12}, \ldots, x_{1n})$ and $x_2 = (x_{21}, x_{22}, \ldots, x_{2n})$

KNN model is used for fitting the behavior photodetector as it requires very little timing for testing and does not need an explicit format for training data. Hyperparameter tuning is easily possible by varying distance metric and k values. The algorithm can be used to study the current–voltage behavior of the photodetector and predict the average value of photocurrent [24, 27]. It can identify potential materials for optoelectronic devices by comparing new materials with known high-performance ones.

2.5.3 Support Vector Machine (SVM)

It uses an optimal hyperplane to divide the data points into various groups. The hyperplane is drawn as a line in such a way that the margin between closed data points of various groups should be maximum. Input data is mapped into a higher-dimensional space with the help of kernel function ϕ. Suppose the training set is denoted as $D = \{(x^m, y^m)\}m = 1^M$, with xm $\in$ Rn, ym $\in \{-1,1\}$, an SVM satisfies the discriminations represented in Eq. (3).

$$\begin{cases} w^T\phi(x^m) + b \geq 1 \quad if \ y^m = +1 \\ w^T\phi(x^m) + b \leq -1 \ if \ y^m = +1 \end{cases} \tag{3}$$

It is applicable for small and high-dimensional datasets and properly fits to handle nonlinear dependencies in organic optoelectronic devices. In the optoelectronics field, the algorithm is suitable for optimizing small-molecule material to achieve highly efficient photodetector as well as minimizing process parameters [13]. This also helps to predict electronic structural properties such as atomic number and radius, molecular percentage etc. [2]. Moreover, it is widely used in the field of flexible sensor systems for monitoring various real-time activities [11].

2.5.4 Decision Tree (DT)

This model is built to predict the target value based on previous data. The algorithm is a tree-like structure in which root nodes test on attributes and branches corresponding to attribute values. In the structure, leaf nodes are used for final prediction based on the decision of previous nodes.

It can handle both linear and nonlinear data and is broadly used in the optoelectronics field. By using the DT algorithm, the dark current, photocurrent, and quantum efficiency of the photodetector can be modeled and predicted [12]. It performs well for selecting suitable materials based on band gaps and other structural parameters.

2.5.5 Random Forest (RF)

It is also one of the ensemble techniques used for regression and classification analysis. In RF, many DTs are used to form a strong classifier. Each DT runs parallelly to predict separate results and lastly final predicted result will be estimated by averaging all predicted results, as shown in Fig. 2.

RF model works well when material properties and device processing parameters need to be analyzed at the same time. With the help of the RF model, prediction of chemical descriptors, electronic structural properties and defect properties can be possible [2]. RF is one of the important tools for the use of selecting the proper

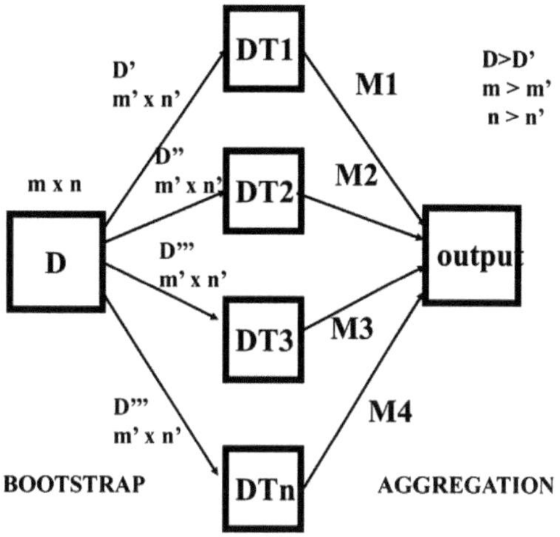

Fig. 2 Random forest regression algorithm

organic material for making high-performance photodetector [13]. It also helps to predict the photocurrent and efficiency of the photodetector [12]

2.5.6 Gradient Boosting Machine (GBM)

Gradient Boosting Machine (GBM) represents a formidable ensemble methodology that integrates various weak learners DTs to form a strong predicted model. The predicted model can be formed by using the iteration process of a non-parametric process. Here, a new tree has to be added to the previous model based on an error generated by the previous iteration and then slowly get a model with high predictive power. Therefore, a weak learner DT combines into a strong leaner DT iteratively. The final predicted result will be estimated by the sum of all predicted results.

GBM shows extraordinary performance for parse data with high dimensionality. GBM is strong for material discovery and bandgap prediction. It is applicable to predict the HOMO, and LUMO energy levels of organic semiconductors, photocurrent, and efficiency of organic optoelectronic devices [2, 12]. It has the capability to analyze the electronic structure of the material and some parameters that impact the bandgap of the material [13]. Feature important analysis means most influential parameters affecting the device performance, can be possible with the help of GBM.

2.5.7 Gaussian Processes Regression (GPR)

Gaussian processes regression (GPR) is a non-parametric, probabilistic regression model that helps us to build complex relationships between input features and output [98, 99]. By varying covariance (kernel) functions, the kernel function identifies the similarity between various input features and further estimates the behavior of the output in the system [98, 99].

GPR is one of the most powerful tools for dealing with small datasets. To optimize the process parameters, the proper combination of process parameters can be predicted with the help of the GPR model which can reduce the repetition of experimental processes [99]. The GPR model is widely used to predict the bandgap, doping concentration, material properties, device performance, and stability based on some other processing and material properties [2]. It can predict the device performance of unexplored data with associated uncertainties.

2.5.8 K-Mean

K-mean is one of the types of clustering algorithms used for making groups [96]. Initially, a set of centroids (k) are randomly selected. Depending on the distance (L2-norm) from the centroid, the algorithm randomly assigns all the data points to one of the k-centroids. After assigning each point to a particular centroid, the k-centroids are updated by taking the average of data points that were assigned to the individual centroids in the previous step. Again, the centroid (which can also be termed as cluster) to which a data point belongs is updated. This is repeated until the change in the location of the new centroids with respect to their previous location is smaller than some tolerance value.

It is applicable for grouping materials based on similar optical properties, structural properties, and electrical properties. It can also categorize the device into groups based on high and low device performance.

2.5.9 Principal Component Analysis (PCA)

PCA, is an approach used for dimensionally reduction problems, which helps to reduce large numbers of features into smaller ones. A dataset including small feature sets will be helpful for problem-solving. A covariance matrix is designed after converting all the data into standardization form. Then, the principal component of the PCA is designed from eigenvalues and vectors of the covariance matrix [96].

PCA typically helps identify key experimental parameters and establishes the correlation between various parameters. It is incorporated to determine the influence of various structural factors and compositional parameters on the device's performance [100].

Among all the supervised algorithms, KNN, SVM, GPR, and other ensemble models are used for material structural and electrical properties predictions as well as device output response predictions [23, 97–99]. All these algorithms are used in the optoelectronics field for the prediction of material properties, the discovery of new material, etc. [2, 10, 12–14]. ML concept in the optoelectronics field can improve device performance as well as research outcomes. Therefore, the proper choice of ML algorithm is a challenging factor if the behavior of material properties suddenly

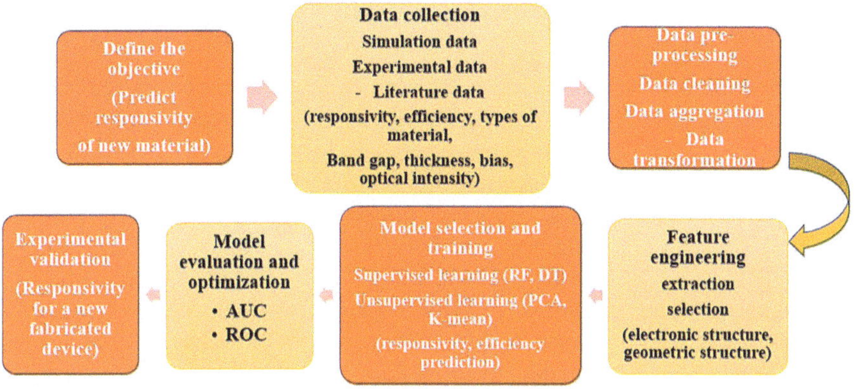

Fig. 3 The ML process for predicting photodetector performance

changes based on process parameters. Sometimes neural network plays a vital role in analyzing large amounts of data; however, it requires lots of computational resources. In addition, unsupervised learning is suitable for identifying of hidden nature of unlabelled data, still, it needs sometimes human intervention to validate output variables [96]. Therefore, supervised learning algorithms provide more accurate data mining problems without human intervention. Figure 3 represents the general ML process for predicting photodetector performance.

2.6 Model Training and Validation

The dataset is divided into two parts: the training dataset for model training and the test dataset for validation. A higher amount of training data provides a more accurate ML model and avoids over-fitting problems [13]. Usually, 10% of the original dataset is kept for testing purposes. During the training phase, the performance of the model can be determined by accuracy, F1 score from classification analysis, area under curve (AUC)-region of convergence (ROC) curve, root mean square error (RMSE), R^2 score, etc. from regression analysis. By achieving an R^2 value of 0.999 and RMSE of 0, a good predictive model can be determined [97]. By tuning the hyperparameter, model performance can be achieved through which good prediction should be possible. Under the model validation, new test data can be correctly predicted with the help of model performance evaluation.

3 Different Types of Organic and Hybrid Photodetectors

Optical sensors play a crucial role in modern life, found in various applications like monitoring food quality, measuring blood oxygen levels with wristbands, and facilitating automated industrial processes. Organic and hybrid photodetectors (OHPDs) offer remarkable advantages, including flexibility, transparency, biocompatibility, and cost-effectiveness. Recent advancements have enhanced both our understanding of their device physics and their performance, making OHPDs competitive with, or even superior to, traditional inorganic photodetectors like silicon. For example, OHPDs can be tailored for specific wavelengths, eliminating the need for bulky optical filters and enabling smaller devices. Furthermore, the development of stretchable and biocompatible detectors opens up new possibilities for medical applications.

3.1 Perovskite-Based Photodetectors

Lead-halide perovskites (ABX_3) have gained significant focus owing to their remarkable photonic properties, such as elevated absorption rates, extended carrier diffusion distances, and adjustable bandgaps [100, 101]. Their production techniques—including vapor deposition, solution processing, and spin coating—facilitate cost-effective and large-scale manufacturing, rendering them suitable for large-scale photodetector systems [102–104]. Recent efforts have concentrated on enhancing the stability of these materials, as they are prone to degradation from moisture, heat, and light [105]. Researchers have improved the resilience of perovskite photodetectors through the use of enhancers, passivation treatments, and better encapsulation methods. Additionally, the tunable bandgap of perovskites enables detection across a wide range of light wavelengths, from UV to near-infrared, which positions them as flexible options for multi-wavelength photodetectors. Hong et al. [103] have also developed a novel crystallization technique focused on surface tension, which produces single perovskite crystals for use in vertically oriented photodetectors, depicted in Fig. 4. Halide perovskites, specifically Ruddlesden-Popper (RP) type, have unique optoelectronic properties that make them ideal for high-performance photodetectors. By carefully tuning the quantum well index, researchers found that perovskites with $n = 4$ exhibited exceptional performance. These photodetectors showed improved sensitivity, lower dark current, and a remarkable on/off ratio of over 3.5 orders of magnitude. Most notably, they were able to maintain elevated sensitivity and minimal power consumption in self-powered mode, qualifying them for emerging optoelectronic applications. Feng et al. [1] have presented a differential spray-coated perovskite hemispherical photodetector designed for smart image capture and position determination. Its low noise level ($\sim 10^{-13}$A $Hz^{-0.5}$), high external quantum efficiency (EQE) of around 1000%, and hemispherical design facilitate the capture of significant differential signals, allowing for more extensive retrieval of relevant information. To enhance location tracking capabilities, a differential photodetector has been developed that utilizes two types of differential electrodes, as shown in Fig. 5a. This photodetector with a bottom electrode is divided into eight segments (spaced at 45° intervals) to gather light signals from various angles, functioning as differential pixels. This design includes two types of electrodes aimed at amplifying signal differences from distinct positions, thereby enhancing the accuracy of the location tracking algorithm. Additionally, the hemispherical photodetector is mounted on a PCB to serve as a transportable model for location tracking. Figure 5b illustrates the ML and data recovery processes utilizing a differential perovskite hemispherical photodetector. While training, signals (x_1, x_2,..., x_8) from different positions of the light source were accumulated by each differential pixel to fit a Bayesian regularization mathematical model. To simulate location tracking, signals ($x_1\prime, x_2\prime, \ldots x_8\prime$) from the light source at various unidentified locations were gathered to reconstruct the object's position and trajectory. An LED was mounted on an accurate X–Y electric multi-axis positioning table for accurate light source positioning. The differential hemispherical photodetector was then positioned beneath the light source to capture the LED's trajectory, covering a $6 \times 6\ cm^2$ area in a grid of 20×20 steps, each with a 3 mm movement increment, to develop a comprehensive signal matrix for the eight differential pixels.

3.2 *Flexible and Transparent Photodetectors*

Recent progress in adaptable and transparent photodetectors has enabled their integration into future wearable electronics, transparent displays, and medical devices. These cutting-edge sensors integrate mechanical pliability with optical transparency, allowing for innovative applications e.g. continuous health monitoring through skin-like sensors, safeguarding technologies, and engaging visual interfaces integrated into transparent surfaces [106]. Emerging photodetector materials like organic polymers, perovskites, and two-dimensional materials (graphene and MoS_2) have played a crucial role in these advancements [107]. To create flexible photodetectors, researchers have used substrates like Polyethylene Terephthalate (PET), Polyethylene naphthalate (PEN), and Polydimethylsiloxane (PDMS) that are durable enough to absorb repeated mechanical stresses without affecting device functionality. These substrates, integrated with thin layers of innovative materials, enable photodetectors that can be curved, elongated, or contorted while withstanding prolonged stress but maintaining peak efficiency. This adaptability creates new applications in wearable technology, such as health-monitoring devices that monitor essential indicators such as heart rate and oxygen saturation continuously without limiting the user movement [108, 109].

Huang et al. [110] developed an adaptable and transparent UV photodetector using Copper Iodide (CuI) thin films, depicted in Fig. 6a. This device retains its transparency and functionality even after repeated flexing on a flexible polyimide

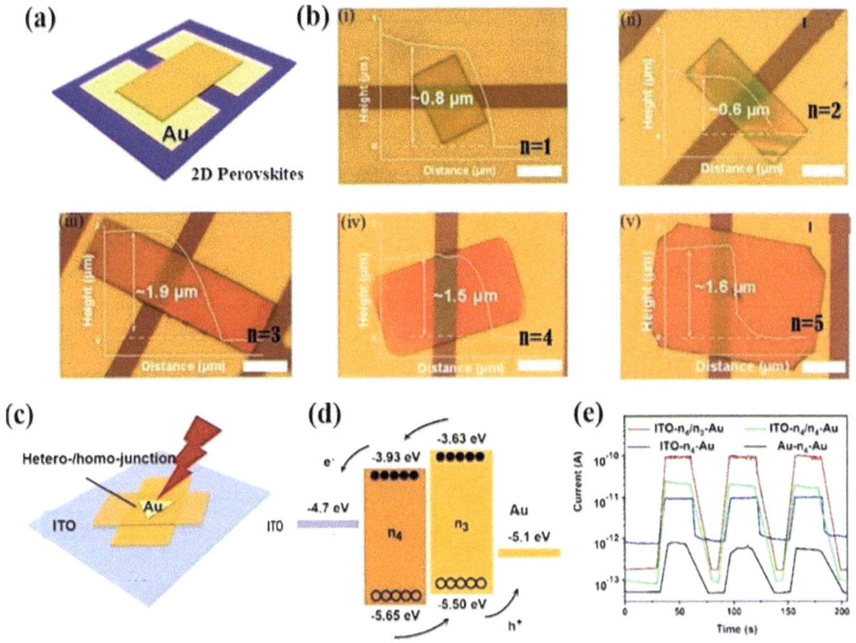

Fig. 4 Device microscopic images including the 2D, single-crystalline microplate photodetectors performance, reproduced from [4]. **a** Device illustration, **b** $(BA)_2(MA)_{n-1}Pb_nI3_{3n+1}$ stacked microplate photodetectors on Au electrodes, **c** hetero and homo-structure photodetectors schematic, **d** $(BA)_2(MA)_3Pb_4I_{13}/(BA)_2(MA)_2Pb_3I_{10}$ hetero-structure band diagram, **e** self-powered properties of different photodetectors: ITO-n_4/n_3-Au, ITO-n_4/n_4-Au, ITO-n_4-Au, and Au-n_4-Au under 600 nm light at 0 V

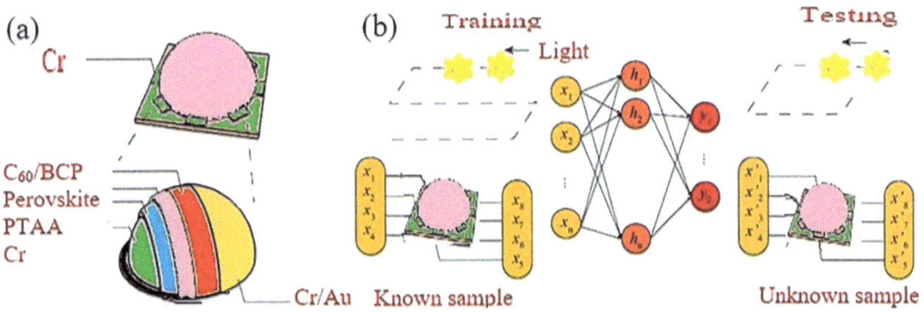

Fig. 5 **a** The hemispherical photodetector device schematic, **b** methodology for constructing the model using non-negative factorization (NNF) and trace reconstruction, reproduced from [1]

substrate, demonstrating its durability. It exhibits excellent performance, including a high responsivity of 123.3 mA/W, a high detectivity of 5.7×10^{12} Jones, and fast response times. The photodetector's performance remains resilient under mechanical pressure, rendering it appropriate for use in wearable technology, transparent

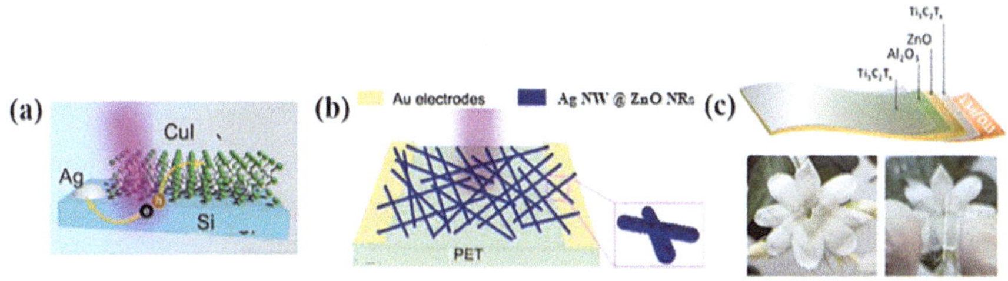

Fig. 6 **a** CuI/Si photodiode schematic, reproduced from [110], **b** Ag nanowire @S-ZnO nanorod-based UV photodetector, reproduced from [6], **c** $Ti_3C_2T_x$ (Titanium Carbide)/Al_2O_3/ZnO/$Ti_3C_2T_x$/ITO/PET-based ultrafast photodetector, reproduced from [111]

displays, and ecological observation systems. Zhang et al. [6] have developed a transparent and flexible UV photodetector using a hybrid structure of silver nanowires and Zinc Oxide nanorods, shown in Fig. 6b. This innovative design enhances light absorption and mechanical flexibility, allowing the device to detect UV light with high sensitivity. Fabricated a flexible PET substrate, the device preserves high transparency and endures repeated bending, making it ideal for application in wearable electronics and flexible displays. MXene (transition metal carbides, nitrides, and carbonitrides)-based flexible and ultrafast photodetector [111] is reported to offer both flexibility and transparency, crucial for wearable electronics and optical communication, shown in Fig. 6c. Utilizing an integrated pyro-phototronic effect, it operates in a self-powered mode, eliminating the requirement for an external power supply. The device features an impressive photo-response time of 8 μs and a responsivity of 0.34 A/W, making it ideal for ultra-fast optical communication applications, including the handling of encrypted optical signals, such as Morse code.

3.3 Organic Photodiodes and Recent Advancements

Organic photodiodes are one of the prevalent types of photodetectors in recent times where the vertical two-terminal organic photodiodes consist of an anode, a hole-transporting layer (HTL), an organic photoactive layer, an electron-transporting layer (ETL), and a cathode. The organic photoactive layer is typically composed of a mixture of electron-donating and electron-accepting materials in a bulk heterojunction (BHJ). Photocurrent generation in the BHJ involves four main processes: exciton generation, exciton dissociation, charge transport, and charge collection. The HTL and ETL are essential in facilitating charge extraction at the interfaces between the BHJ and the electrodes, leading to photocurrent generation. A photodiode's response efficiency relies on its capability to collect photogenerated charge carriers, enabling operation in either self-powered mode or under reverse bias [112]. Jinno et al. [113] developed the first washable and flexible organic photodiodes by enclosing a free-standing ultrathin organic photodiode within two pre-stretched acrylic elastomers.

This design attained a solar cell efficiency of 7.9% and a stretchability of 52%, even when submerged in water. Additionally, after undergoing 20 cycles of compression and exposure to water for 100 min, the photodiode retained 80% of its original photovoltaic efficiency. Increasing pixel densities is crucial for sensor arrays, but reducing pixel sizes has been difficult. The pixel fill factor, expressed as the ratio of a sensor's active area to its total pixel area, significantly affects sensitivity and costs; fill factors below one necessitate using micro-lenses to focus light on the sensor. Monolithic integration is desirable as it can achieve a fill factor of one, maximizing pixel density. However, the main challenge lies in ensuring compatibility during the fabrication of all layers. In 2018, Zalar et al. [114] introduced a passively addressed photodiode matrix that exemplifies this integration, achieving high pixel densities of up to 262 pixels/inch (ppi) with pixel areas of only 2.5×10^{-3} mm^2 and 50 μm pitches. This structure maintains a fill factor of one, enabling real-time imaging for various applications, including spatial PPG measurements and heart rate monitoring in conjunction with ECG data. Table 2 summarizes the operational qualities of modern adaptable organic photodiodes, highlighting variations in active materials, flexible substrates, and device performance.

3.4 Organic Phototransistors and Recent Advancements

A phototransistor is a device that is sensitive to light, allowing incident light to modulate the density of charge carriers in its channel. This enables light to function as an additional control terminal alongside the standard three terminals, allowing for both electrical and optical modulation without interference. Photonic memory outperforms electric programming memory in establishing hierarchical data storage while avoiding destructive reading issues, as the processes for reading and programming are orthogonal. High-efficiency phototransistor memory is characterized by a large memory window (the threshold voltage difference between ON and OFF states) and a significant memory ratio (the current difference between states), which enhances data reliability [121]. Incorporating light for charge storage expands the functionality of phototransistors, enabling tunable performance across different wavelengths and intensities. This hysteresis feature supports non-volatile information storage, allowing phototransistors to transform optical signals into electrical ones, thus facilitating optical signal processing and photonic neuromorphic applications. Combining responses to electrical fields and light enhances capabilities for light-assisted encryption, multi-bit data storage, dual-mode operation, and photo-recovery. As a result, recent investigations into organic phototransistors have shifted towards improving their memory characteristics and mimicking synaptic behavior [122].

Single crystals with small molecules exhibit higher mobility, making them ideal for organic phototransistors (OPTs). However, OPT development faces challenges, such as the rigidity and fragility of commonly used silicon and silicon dioxide as gate electrodes and dielectric layers. Additionally, channel layers must find a balance between effective charge transport and light absorption. To address these

Table 2 An overview of the latest advancements in the development of organic photodiodes

Device structure	Flexibility	Responsivity (A/W)	Detectivity (Jones)	Response time	Reference
PCDTBT: PCBM/PI	–	0.21	3×10^{12}	–	[115]
TQ1: PNDI-T10/PET	–	0.17	10^{11}	–	[116]
D18:Y6 and D18/Y6/PET	Bending with a radius of 11 mm	0.35	9.39×10^{11}	0.013 ms	[117]
PBDTTT OFT: IEICO-4F/PI	1000 bending cycles with a radius of 25 μm	11.9	1.7×10^{15}	0.037 s	[118]
P3HT: PCBM/ 3DG/PET	100 bending cycles with a radius of 5 mm	5.8×10^5	3×10^{15}	24 ms	[119]
PTB7-Th: COTIC-4F/PEN	100 bending with R = 7.5 mm	≈ 0.42	1.25×10^{13}	0.021 ms	[120]

PCDTBT: poly[N-9′-heptadecanyl-2,7-carbazole-alt-5,5-(4′,7′di-2-thienyl-2′,1′,3′-benzothia diazole)],
PCBM: phenyl-C61-butyric acid methyl ester,
PI: polyimideTQ1: Poly[2,3-bis(3-octyloxyphenyl) quinoxaline-5,8-diyl-alt-thiophene-2,5-diyl],
PNDI-T10: Polynaphthalene diimide-based polymer acceptor,
D18: Poly[(2,6-(4,8-bis(5-(2-ethylhexyl-3-fluoro)thiophen-2-yl)-benzo[1,2-b:4,5-b′] dithiophene))-alt-5,5′-(5,8-bis(4-(2-butyloctyl)thiophen-2-yl)dithieno[3′,2′:3,4;2″,3″:5,6] benzo[1,2c]thiadiazole)],
Y6: (2,2′-((2Z,2′Z)-((12,13-bis(2-ethylhexyl)-3,9-diundecyl-12,13-dihydro-[1,2,5] thiadiazolo [3,4-e]thieno[2″,3″:4′,5′]thieno[2′,3′:4,5]pyrrolo[3,2-g] thieno [2′,3′:4,5] thieno[3,2-b]indole-2,10-diyl)bis(methanylylidene))bis(5,6-difluoro-3-oxo-2,3-dihydro-1H-indene-2,1-diylidene))dimalononitrile),
PBDTTT: Poly(bis (alkylthienyl)thiophene), OFT: Organic Field Effect Transistor,
IEICO-4F: Indacenodithiophene-Quaterfluorodihaloanthracene,
P3HT: Poly(3-hexylthiophene),
3DG: 3-deoxyglucosone,
PTB7-Th: Polymer Thermal Benzothiadiazole-7-Thiophene,
COTIC-4F: Cyclic Olefin Thermoplastic Intermediate Copolymer 4-Functional,
PEN: Polyethylene Naphthalate

concerns, Huang et al. [123] successfully created a phototransistor using a high-mobility, unipolar polymer called D18, which serves as both the light absorption layer and the charge transport channel, thereby simplifying the device architecture and enhancing stability. The researchers also substituted traditional SiO_2 dielectric layers with a hybrid of Tantalum Pentoxide (Ta_2O_5) and polyacrylonitrile (PAN), known for their flexibility and high capacitance. Recent years have seen extensive research into the progression of conjugated polymers, including bulk heterojunctions

(BHJs) for photodetectors and flexible phototransistors produced through several fabrication techniques.

Bilayer heterostructure (BHS) phototransistors are emerging as effective device designs by integrating BHS with a BHJ film for exciton generation and separation, along with a high-mobility semiconductor layer facilitating carrier transport. Nevertheless, creating BHS OPTs using solution processes poses challenges due to the fragility of the underlying layers during deposition [123–125]. In 2021, Li et al. [126] developed near-infrared (NIR) polymeric phototransistors that utilized a Solution-processable BHS featuring cross-linked polymeric semiconductors as the lower conductive channel, demonstrated in Fig. 7a. These devices featured BHJs made from low-bandgap polymer donors and fullerene acceptors, with a specific focus on diketopyrrolopyrrole-based polymer (PDPP-DTT) for enhanced sensitivity to 808 nm light. Additionally, an indanone-condensed thiadiazoloquinoxaline polymer was used for effective NIR light absorption (including wavelengths of 808, 1064, and 1550 nm). The resulting phototransistors demonstrated significantly improved performance compared to control devices, addressing the restrictions of low mobility and high off-currents typical of low-bandgap polymer devices. A flexible array of NIR phototransistors was successfully created on a polyimide substrate, maintaining stability at bending radii as small as 1 mm, with a dark current remaining below 0.2 nA even when folded and achieving a highest output power exceeding 10,000 at a 1 mm bending radius. OPTs typically rely on p-channel materials owing to the limited availability of stable and high-performing n-channel organic materials. Developing conformable OPTs using n-channel materials is crucial for creating complementary electronics and optoelectronic circuits, offering benefits like exceptional operational reliability, manageable photo-switching voltages, and increased photosensitivity [127]. In 2017, Nam et al. [128] introduced an ultrathin conformable OPT array supported by polyvinyl alcohol (PVA), shown in Fig. 7b. This array features an air-stable n-type active layer made from N, N′-Ditridecyl-3,4,9,10-perylenedicarboximide (PTCDI-C13H27), with the entire device measuring approximately 830 nm thick. PTCDI-C13H27, a parylene derivative, enhances the stability of n-type organic transistors. The wide-area OPT array demonstrated impressive electrical performance in dark conditions, achieving a mobility of 0.58 cm^2/Vs, an on/off ratio exceeding 10^9, and noteworthy air stability. While the device's performance varied when bent in different orientations, the key parameters such as on-state current, off-state current, ON/OFF ratio, and mobility remained consistent. Notably, as the bending radius decreased, both dark and light currents exhibited only slight reductions, yet the photosensitivity remained stable at over 10^4, indicating strong performance consistency under strain on various curved surfaces.

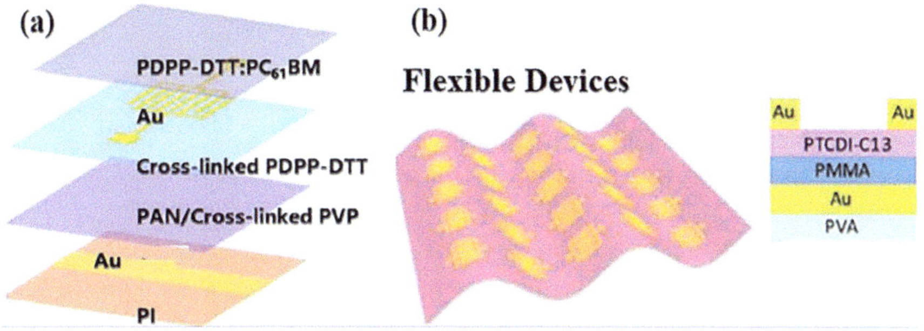

Fig. 7 **a** Flexible OPT schematic diagram, reproduced from [126], and the **b** conformal device, reproduced from [128]

3.5 Organic Photodetectors for Flexible and Wearable Applications

The growth of the IoT has driven the need for smaller, more flexible sensing components. As a result, flexible photodetectors have been developed, enabling non-invasive and accurate health monitoring through wearable devices or direct attachment to the body. These detectors have been used in applications e.g. monitoring heart rate, blood pressure, and blood oxygen levels. A highly sensitive and low-power epidermal and flexible photoplethysmography (PPG) sensors were developed to operate in the NIR range in 2017 by Xu et al. [120, 129], shown in Fig. 8a. The sensor design combines an OPT with an inorganic LED, which boasts a power conversion efficiency of 14.3%, significantly higher than that of NIR organic LEDs. This LED can easily be attached to flexible substrates via transfer printing. Noninvasive 2D oxygenation mapping could revolutionize real-time and post-surgery care, as well as the tracking wounds, organs, and tissues. A real-time in vivo device for mapping spatial oxygenation is valuable for evaluating tissue damage and vulnerability to injury, as depicted in Fig. 8b. For instance, a flexible optoelectronic sensor array can be utilized to visualize oxygenation distribution levels in skin grafts. In 2018, Lee et al. [8] introduced a reflectance oximeter array (ROA), featuring a flexible printed electronic system that integrates organic optoelectronics with standard silicon circuits for measuring blood and tissue oxygenation. The ROA includes four red and four near-infrared OLEDs and eight OPDs, and its design on flexible plastic substrates enhances comfort and improves signal-to-noise ratio by providing an effective sensor-skin interface.

The integration of thin-film transistors (TFTs) into processing circuitry has emerged as an effective strategy for developing advanced image sensors. A notable example is the cutting-edge image sensor created by Yokota et al. [130] which utilizes polycrystalline silicon processed at low temperatures—TFTs to detect small photocurrents below 10 Pa, depicted in Fig. 9. This design has significantly minimized device size while achieving a spatial resolution of 508 pixels per inch (PPI), all while maintaining mechanical flexibility. As TFT technology continues to advance,

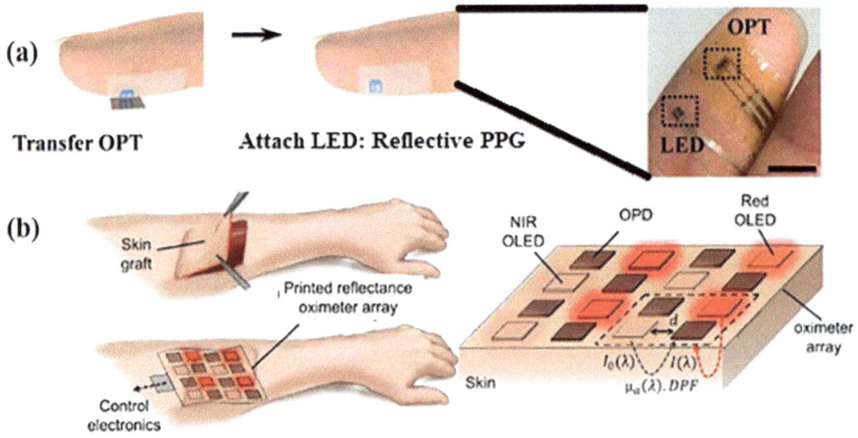

Fig. 8 **a** Finger image with an epidermal hemodynamic photoplethysmography (hPPG) sensor attached, with a scale reference of 5 mm, reproduced from [129], and **b** an illustrative example of ROA involving a 2D oxygenation mapping scenario: monitoring the oxygen levels in a skin graft placed on the forearm, reproduced from [8]

it is anticipated that stiffer components in processing circuits will be substituted with highly compliant, high-performance TFTs. For instance, Li et al. [131] developed large-scale flexible ICs based on MoS_2, featuring a high density of 1518 transistors per square cm (approximately 905 PPI for imaging). While non-silicon-based ICs still require enhancement to satisfy practical demands, the faster evolution of flexible and high-mobility materials, such as organic and two-dimensional semiconductors, promises significant applications in wearable technology. In addition to connecting frontend sensing components to processing circuitry directly, wireless data transmission via antennas enhances advanced wearable systems. Chung et al. [132] introduced a wireless, battery-free PPG sensor featuring a near-field communication (NFC) antenna for neonatal intensive care applications. This design enables simultaneous wireless data transmission and power supply to the host system. They also increased the effective communication range from 25 cm with NFC to 20 m using Bluetooth technology.

Wearable systems are electronic devices worn directly on the skin or integrated into textiles, typically characterized by their compact size and low power consumption. These devices often deal with low-intensity physiological signals that can be easily disrupted by motion artifacts. Since these weak signals require additional processing, such as amplification and filtering, they also increase power demands. Furthermore, the continuous real-time tracking of physiological signals generates a substantial amount of data, necessitating significant storage and transmission capabilities.

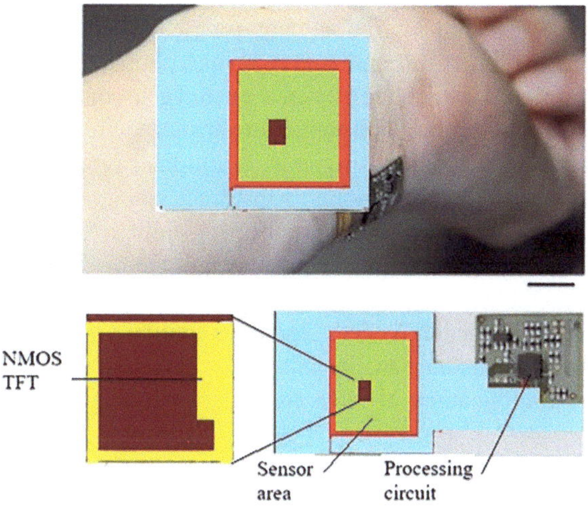

Fig. 9 Flexible imager with TFT and its processing circuitry, reproduced from [130]

3.6 Emerging Application Fields of Photodetectors

Photodetectors are becoming vital elements in optical communication networks, as they transform optical signals into electrical signals, enhancing data transmission speed and efficiency. With the increasing need for high-capacity networks driven by cloud computing, 5G, and the IoT, the need for advanced photodetectors has escalated. Recent innovations have led to the creation of high-speed photodetectors designed to cope with rising data speeds in contemporary communication systems. As conventional systems face surging data traffic, these advanced photodetectors offer improved responsivity, increased bandwidth, and reduced noise levels, which are essential for supporting 5G networks, data centers, and edge computing solutions. These progresses facilitate faster data transfer, reduce latency, and promote energy efficiency, all crucial for rapid processing and data transmission [133]. Photodetectors play a crucial role in LIDAR (Light Detection and Ranging) systems that assess air quality. They measure the light reflected back from airborne contaminants like dust, smoke, and pollutants, delivering real-time information on particulate matter concentrations in the atmosphere. In climate research, photodetectors are crucial for monitoring the radiation balance between the Earth and the atmosphere. These sensors track the Earth's albedo effect, which refers to the surface's reflectivity, and its influence on climate change.

Photodetectors are becoming crucial in healthcare and biometric technologies by facilitating non-invasive, real-time monitoring with high accuracy. Such sensors are fundamental to medical devices that track vital signs, improving diagnosis and treatment methods in contemporary medicine. Their use in wearable technologies has revolutionized patient care, facilitating continuous monitoring away from hospitals, which reduces the need for frequent visits and allows for prompt identification of health concerns [134]. Photodetectors are widely used in pulse oximetry, measuring blood oxygen saturation. This process involves light emitting from LEDs at two wavelengths—infrared and red—and photodetectors gauging the light absorbed by the blood, which shifts based on oxygen concentrations. This innovation delivers crucial real-time information in various environments, such as hospitals, home care settings, and during surgeries. Additionally, photodetectors are integral to heart rate and blood pressure monitors that track changes in blood flow. Photodetectors gather important health data by collecting light reflected off the skin and converting it into analog electrical signals in health monitoring devices. Recent breakthroughs in adaptable and translucent photodetection technologies have broadened their use in wearable health devices, enabling continuous vital sign monitoring and enhancing personal health management while alleviating pressure on healthcare systems. Prominent applications of OPDs are also introduced in health monitoring as well as in image sensors.

4 Role of ML in Designing the Photodetector

4.1 Material Optimization

Material optimization is an essential step in the photodetector-integrated smart technology field, improving the photodetector's performance in IOT sensor applications. Earlier, lots of trial-and-error methods were used for identifying suitable material in simulation and experimental ways, which slowed down the breakthrough process of new materials. With the help of the ML concept, suitable materials with good photodetecting properties can be optimized from large numbers of organic and inorganic optoelectronics materials which further can reduce unnecessary experimental costs. Various ML algorithms such as RF, KNN, and GPR are used to select proper optoelectronic material and predict the material's properties [14, 97–99]. These studies help optimize material properties including bandgap, optical absorption coefficient, carrier mobilities, and so on [14, 97–99]. Multiple linear regression (MLR), polynomial regression (PR), and ANN have been incorporated to observe the impact of various structural factors on the performance of nanocomposite devices [135]. An RF model from ML was used to forecast the structural and magnetic characteristics of Heusler alloy (HA) families [97]. The structural and magnetic properties have been predicted by considering various input features such as the total number of outer-shell electrons, ionic radii, and electronegativity of i atom. The correlation matrix is employed to determine the relationship between the predictor input features and the output variable.

Therefore, the ML model can allow us to determine new HA material by predicting the structural and magnetic properties. Bayesian optimization algorithm was incorporated to determine the optimal structure of the magnetic tunnel junction consisting of $MgAl_2O_4$ (MAO) [136]. Structure modification plays an important role in achieving high tunneling magnetoresistance (TMR). More than 100 structures of disordered

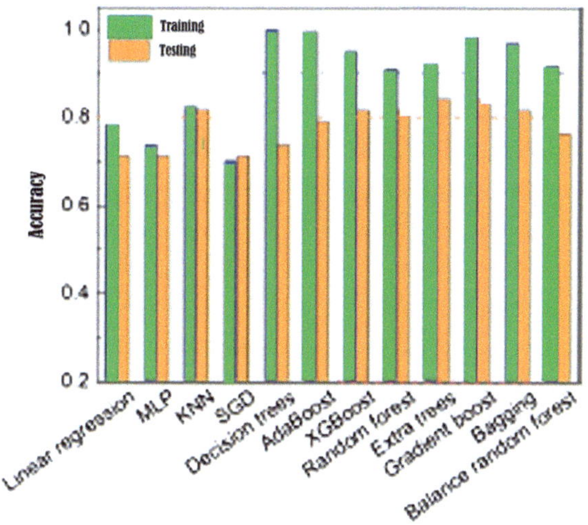

Fig. 10 Accuracies of 12 classification models on training and test sets, with the red dotted box marking five models exceeding 90% accuracy on the training set and accuracies of > 80% on the test set, reproduced from [137]

MAO are considered for predicting high-performance MTJ-based spintronic devices. ML-based approaches were used for determining experimental parameters to develop high-performance ferroelectric tunnel junctions (FTJs) [137]. Various ferroelectric materials are chosen for predicting a high on/off ratio. The GB algorithm from ML is most suitable for predicting the accurate on/off ratio of a new FTJ structure Fig. 10b. The association rule mining approach is used to choose suitable ferroelectric materials resulting higher on/off ratio from FTJ. Hence, the ML concept is supposed to predict the properties of novel organic semiconductors or acceptor materials. Proper donor–acceptor combinations for enhanced charge separation can be chosen by analysis of organic semiconductor properties using ML, reducing the time needed to identify materials and optimize device designs.

4.2 Device and Its Fabrication Optimization

Optimization of the device design, including composition ratios, layer thickness, choice of electrodes, interfacial area, etc. is important for getting good performance from the photodetectors. ML approaches help to optimize the device design by investigating the various design parameters. In this case, usually, parameter optimization algorithms such as Bayesian algorithms, GPR, etc. are used to obtain optimal device structure for achieving higher efficiency from the photodetector [98, 99]. Apart from that, fabrication parameters such as substrate temperature, deposition methods, and ambient pressure are crucial as these parameters directly influence the device's structure and performance. The proper tuning of fabrication parameters is mandatory for improving the device's quality and performance. Bayesian optimization algorithm helps to optimize fabrication parameters (e.g., spin-coating speed, annealing

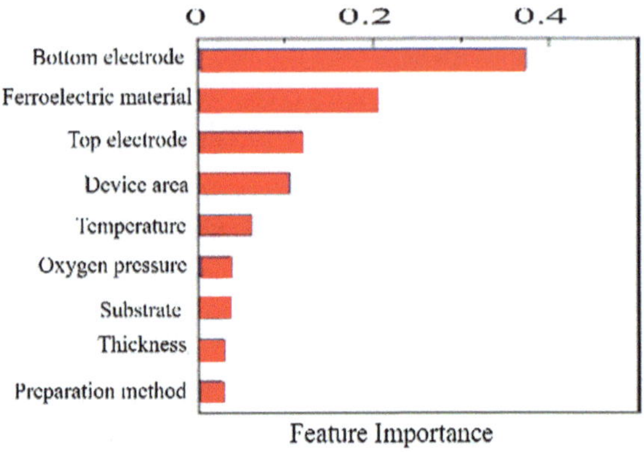

Fig. 11 Importance ranking obtained from the GB classification model, reproduced from [137]

temperature) to maximize optoelectronics device performance. The impact of process parameters on the device structure can be investigated using ML-related models such as SVM, XGBoost (XGB), etc. [13]. MLR, PR, and ANN have been incorporated to observe the impact of various compositional factors concerning the performance of nanocomposite devices [135]. The author suggested that the proper choice of doping concentration impacts the performance of composite devices and this can be investigated with the help of ML approaches. An ML model is developed for predicting the efficiency of organic photovoltaic (OPV) by considering 13 important features [14]. The impact of various structural parameters on photo physics mechanism in solar cells has been reported which will be helpful in designing new materials Fig. 11.

The GPR model from ML was used for predicting the structural properties of organic molecular systems [98, 99]. Structural properties such as HOMO–LUMO level, chemical potential, softness, etc. can be predicted with the help of GPR and DFT which further provide ideas to identify the suitable device structure for optoelectronics material. Raghvendra Kumar Mishra et al. used Bayesian analysis for optimizing material thickness and composition ratio to achieve good electrical and thermal properties from the devices [138]. It was observed that the thickness of the active layer and the composition ratio of the material highly impact the performance of the composite devices. XGB and LightGBM models were used to optimize design and process parameters such as substrate thickness, substrate length, and active layer thickness of Si photodetector [12]. LightGBM offers more accurate performance prediction than XGB with lower RMSE values. Therefore, ML algorithms allow to identify optimal layer thickness, electrode materials, and interfacial layers of OPD architecture. Fine-tuning of fabrication parameters and device architecture is also possible by the using ML concept for getting reproducible results and superior performance.

4.3 Feature Importance and Performance Optimization

Feature importance from ML techniques allows us to find the crucial parameters that mainly affect the behavior of the photodetector such as responsivity, efficiency, and response speed. Features such as band gap, absorption coefficient, heterostructure, thickness, bias voltage, and ambient noise influence the photodetector. Various ML concepts are used to identify which features are more important among all features. RF, GB Models, and regularization techniques are used for the feature-important task [13]. Baian Chen et al. reported feature important concepts for developing perovskite-based optoelectronics materials [13]. Out of 55 features, only 25 are considered more important for getting the best results. After feature selection, the researcher can easily decide to optimize the photodetector. The optimized device then provides better performance and quality output. An ML model is developed for predicting the efficiency of OPV by considering 13 important features [14]. The importance of 13 features for the GB and RF models has been investigated, which will help to identify a new material. A RF model from ML was used to estimate the structural and magnetic characteristics of HA families [97]. Forecasting structural and magnetic behavior has been done by considering various input features such as the total no of outer-shell electrons, ionic radii, and electronegativity of i atom. Feature importance analysis for all the input features has been carried out and it is to be concluded that no of valence electron's role is more important than other features. XGB and LightGBM models were used for extracting feature importance parameters such as substrate thickness, substrate length, and active layer thickness for inorganic photodetector [12]. The author highlighted the integration of ML in the photodetector, offering good performance of the photodetector with the least resources. In one study, KNN, ANN, and Quantum Neural Network were incorporated to predict the I-V characteristics of the photodetector which will help to determine the trend of the I-V curve of the photodetector composing new material [139]. ML algorithms can predict the conductivity behavior and some spin response functions for a device implemented by varying magnetic impurities and configuration [140]. The Quantum ML model was used to get more accurate results for calculating the behavior of conductivity and spin response function. These approaches were suitable for designing new solid-state devices by predicting the performance of the molecular systems. Various ferroelectric materials along with types of electrodes, substrate temperature, device area, thickness, ambient pressure, etc. are chosen for predicting a high on/off ratio from FTJs [137]. The GB algorithm from ML is the optimal algorithm for forecasting the accurate on/off ratio of a new FTJ structure. After feature extraction, it was investigated whether the bottom electrode had a greater influence on the efficiency of the FTJ structure. The association rule mining approach is used to choose suitable fabrication parameters, resulting in a high on/off ratio from FTJ. Various algorithms, such as RF and GB models, were included in a model for performance optimization of the HgCdTe homojunction photodetector [141]. Various algorithms, including ANN, were used for predicting the performance prediction of the photodetector, in which ANN shows more accurate results [142]. Henceforth, forecasting the efficiency of

OPD under different light intensities or reverse bias can be possible. The stability and lifetime of OPD under various environmental conditions can be identified using ML algorithms. With the help of important feature analysis, a data-driven design of OPD is possible, further reducing experimental trials and costs. Therefore, integrating ML with photodetector plays a vital role in developing photodetector technology in the smart IoT field. Table 3 shows some progress in ML-integrated photodetector and solar cells.

5 Summary

Photodetectors are gaining significant traction in next-generation electronics, including quantum computing, augmented reality (AR), the Internet of Things (IoT), artificial intelligence (AI), and wearable technologies. These devices convert optical signals into electrical responses via the photoelectric effect, enabling IoT systems to perceive and interact with their environments intelligently. Emerging materials such as perovskites, graphene, organic semiconductors, and flexible hybrid materials are being actively explored for photodetector applications due to their lightweight nature, low cost, mechanical flexibility, optical tunability, and broad spectral response. However, flexible and organic photodetectors still face several technical, environmental, and commercial challenges that hinder large-scale deployment. To overcome these limitations, the integration of artificial intelligence (AI) and machine learning (ML) into photodetector development is proving transformative. ML algorithms can optimize material selection, improve device architecture, and predict performance metrics under varying operational conditions. For example, ML can assist in selecting suitable donor–acceptor pairs in organic semiconductors to enhance charge separation efficiency, ultimately improving device responsiveness. Furthermore, ML enables rapid performance forecasting, reducing the time and computational resources required for simulations. By embedding ML-driven intelligence, photodetectors evolve from simple sensors into smart components capable of real-time sensing, processing, and adaptive decision-making—vital for applications in healthcare, autonomous systems, smart environments, and beyond. In summary, the convergence of flexible photodetector technology with machine learning is shaping a new class of intelligent optical sensors, offering greater functionality and adaptability for a wide range of advanced IoT-driven ecosystems.

Table 3 Some progress in ML-integrated photodetector and solar cell

Topic	Types of devices	ML	Feature/ Contribution	Reference
Regression analysis for photocurrent	NTCDA/p-Si OPD	KNN QNN	Predict the behavior of I-V characteristics under different intensities of light	[139]
Predicting efficiency of organic solar cells	(SM-OPV), benzodithiophene-(BDT), diketopyrrolopyrrole-(DPP), quinoxaline-, Zn–porphyrin-based	LR KNN GB RF ANN	Analyze the importance of various descriptors on efficiency and build a model for predicting the efficiency of OPV (Best model-GB and RF)	[14]
Photodetector performance prediction	Uni-traveling carrier (UTC) photodetectors	LR KNN RF ANN	Computed the phase noise, mean output current, and impulse response for 1750 designs (by considering thickness, doping concentration, etc.)	[142]
Assessment and prediction of optoelectronic capabilities of photodetector	HgCdTe based homojunction photodetector	DT RF Extra tree (ET) GBM	Determine the interdependencies between input factors (voltage and incident light wavelength) and resulting optoelectronic parameters and predict the output characteristic parameters (efficiency, dark, and photocurrent)	[141]
Development and evaluation of a near-infrared photodetector	Si- photodetector	XGB and LGBM	predict responsivity based on variations in input parameters (Horizontal and vertical electrode configuration, substrate electrode etc.)	[12]

(continued)

Table 3 (continued)

Topic	Types of devices	ML	Feature/ Contribution	Reference
Analysis of J–V curve for solar cell	Mn-doped CMZTSSe solar cells	GPR	Development of 5% Mn-doped CMZTSSe solar cells with a recorded efficiency of 8.9%	[2]
The proactive support of ML techniques for discovering new materials	Sn-based perovskite solar cell	Deep neural networks (DNN	Suggestions provided by the ML algorithm, optimize the design of perovskite-based solar cells. As a result, the PCE of the fabricated device, guided by ML recommendations, achieves 5.57%	[10]
Impact of small molecule mechanisms on perovskite solar cell performance	Various small molecule	RF SVR XGBoost	Developed a predictive model aimed at facilitating the screening of optimal small molecules for high-efficiency solar cells	[13]
Evaluation of the impact of input features on PCE	Halide bis perovskite solar cell	RF SVR XGBoost GB LR DT	Optimization of device architecture of solar cell by analysis of influence of material compositions and fabrication parameters on solar cell	[13]

References

1. H. Guo, S. Saifi, K. Fukuda, H.M. Cheng, Z. Lou, X. Xu, Flexible organic photodetectors and their use in wearable systems. Digit. Signal Process. **125**, 103145 (2022)
2. C.Z. Zhang, X.Q. Fu, Applications and potentials of machine learning in optoelectronic materials research: an overview and perspectives. Chinese Phys. B **32**(12), 126103 (2023)
3. J. Zou, S. Zhang, X. Tang, Recent advances in organic photodetectors. Photonics **11**(11), 1014 (2024)
4. T. Ma, N. Xue, A. Muhammad, G. Fang, J. Yan, R. Chen, J. Sun, X. Sun, Recent progress in photodetectors: from materials to structures and applications. Micromachines **15**(10), 1249 (2024)

5. H. Jing, R. Peng, R.M. Ma, J. He, Y. Zhou, Z. Yang, C.Y. Li, Y. Liu, X. Guo, Y. Zhu, et al. Flexible ultrathin single-crystalline perovskite photodetector. Nano Lett. **20**, 7144–7151 (2020)
6. L. Zhang, N. Li, Q. Ma, J. Ding, C. Chen, Z. Hu, W. Zhao, Y. Li, H. Feng, M. Li et al., Large-area flexible and transparent uv photodetector based on cross-linked ag nw@zno nrs with high performance. J. Mater. Sci. Technol. **110**, 65–72 (2022)
7. Y.Y. Yu, C.I. Liu, Y.C. Peng, B.H. Jiang, Y.W. Su, S.J. Liu, C.P. Chen, Ambient-stable near-infrared organic photodetectors with ultrahigh detectivity and ultrafast response for biometric monitoring. Adv. Electron. Mater. **8**, 2200585 (2022)
8. H. Lee, E. Kim, Y. Lee, H. Kim, J. Lee, M. Kim, H.J. Yoo, S. Yoo, Toward all-day wearable health monitoring: an ultralow-power, reflective organic pulse oximetry sensing patch. Sci. Adv. **4**(11), eaas9530 (2018)
9. Y. Khan, D. Han, A. Pierre, J. Ting, X. Wang, C.M. Lochner, G. Bovo, N. Yaacobi-Gross, C. Newsome, R. Wilson, A.C. Arias, A flexible organic reflectance oximeter array, Proc. Natl. Acad. Sci. USA **115**(47), E11015–E11024 (2018)
10. B. Chen, R. Chen, B. Huang, Machine learning-accelerated development of perovskite optoelectronics toward efficient energy harvesting and conversion. Adv. Energy Sustain. Res. **4**(10), 2300157 (2023)
11. T. Sun, B. Feng, J. Huo, Y. Xiao, W. Wang, J. Peng, Z. Li, C. Du, W. Wang, G. Zou, L. Liu, Artificial intelligence meets flexible sensors: emerging smart flexible sensing systems driven by machine learning and artificial synapses. Nano-Micro Lett. **16**, 14 (2024)
12. N. Alfryyan, M. Saqib, S. Ali, T. Mubashir, M. Tahir, Z. Alrowaili, M. Buriah, Design and analysis of near-IR photodetector using machine learning approach. IEEE Sens. J. **24**(16) (2024)
13. X. Li, Y. Mai, C. Lan, F. Yang, P. Zhang, S. Li, Machine learning-assisted design of high-performance perovskite photodetectors: a review. Adv. Compos. Hybrid Mater. **8**, 27 (2025)
14. H. Sahu, W. Rao, A. Troisi, H. Ma, Toward predicting efficiency of organic solar cells via machine learning and improved descriptors. Adv. Energy Mater. **8**(24), 1801032 (2018)
15. S. Marsland, *Machine Learning: An Algorithmic Perspective*, 2nd edn. (Chapman and Hall/CRC, 2014)
16. D. Silver et al., Mastering the game of go with deep neural networks and tree search. Nature **529**, 484–489 (2016)
17. M. Bojarski, End to end learning for self-driving cars (2016). arXiv preprint arXiv:1604.07316
18. K. Rajan, Materials informatics. Mater. Today **8**(10), 38–45 (2005)
19. W. Li, R. Jacobs, D. Morgan, Predicting the thermodynamic stability of perovskite oxides using machine learning models. Comput. Mater. Sci. **150**, 454–463 (2018)
20. J. Graser, S.K. Kauwe, T.D. Sparks, Machine learning and energy minimization approaches for crystal structure predictions: a review and new horizons. Chem. Mater. **30**, 3601–3612 (2018)
21. R.S. Sutton, A.G. Barto, *Reinforcement Learning* (The MIT Press, Cambridge, MA, 2018)
22. L.M. Ghiringhelli, J.S. Vybiral, V. Levchenko, C. Draxl, M. Scheffler, Big data of materials science: critical role of the descriptor. Phys. Rev. Lett. **114**, 105503 (2015)
23. J. Schmidt, M.R.G. Marques, S. Botti, et al., Recent advances and applications of machine learning in solid-state materials science. npj Comput. Mater. **5**, 83 (2019)
24. A.Y.T. Wang, R.J. Murdock, S.K. Kauwe, A.O. Oliynyk, A. Gurlo, J. Brgoch, K.A. Persson, T.D. Sparks, Machine learning for materials scientists: an introductory guide toward best practices. Chem. Mater. **32**(12), 4954–4965 (2020)
25. J. Fang, M. Xie, X. He, J. Zhang, J. Hu, Y. Chen, Y. Yang, Q. Jin, N interface-reinforced rhombohedral Prussian blue analogue in semi-solid state electrolyte for sodium-ion battery. Mater. Today Commun. **33**, 104900 (2022)
26. T. Xie, J.C. Grossman, Crystal graph convolutional neural networks for an accurate and interpretable prediction of material properties. Phys. Rev. Lett. **120**, 145301 (2018)

27. O. Isayev et al., Universal fragment descriptors for predicting properties of inorganic crystals. Nat. Commun. **8**, 15679 (2017)
28. A. Furmanchuk, A. Agrawal, A. Choudhary, Predictive analytics for crystalline materials: bulk modulus. RSC Adv. **6**, 95246–95251 (2016)
29. Y. Zhang, E.-A. Kim, Quantum loop topography for machine learning. Phys. Rev. Lett. **118**, 216401 (2017)
30. V. Stanev et al., Machine learning modeling of superconducting critical temperature. Npj. Comput. Mater. **4**, 29 (2018)
31. K.T. Schütt et al., How to represent crystal structures for machine learning: towards fast prediction of electronic properties. Phys. Rev. B **89**, 205118 (2014)
32. P.V. Balachandran, B. Kowalski, A. Sehirlioglu, T. Lookman, Experimental search for high-temperature ferroelectric perovskites guided by two-step machine learning. Nat. Commun. **9**, 1668 (2018)
33. F. Legrain, J. Carrete, A. Roekeghem, S. Curtarolo, N. Mingo, How chemical composition alone can predict vibrational free energies and entropies of solids. Chem. Mater. **29**, 6220–6227 (2017)
34. C. Kim, G. Pilania, R. Ramprasad, From organized high-throughput data to phenomenological theory using machine learning: the example of dielectric breakdown. Chem. Mater. **28**, 1304–1311 (2016)
35. G. Pilania, X.-Y. Liu, Machine learning properties of binary wurtzite superlattices. J. Mater. Sci. **53**, 6652–6664 (2018)
36. D.R. Cassar, A.C. Carvalho, E.D. Zanotto, Predicting glass transition temperatures using neural networks. Acta Mater. **159**, 249–256 (2018)
37. T. Zhan, L. Fang, Y. Fang, Prediction of thermal boundary resistance by the machine learning method. Sci. Rep. **7**, 7109 (2017)
38. A. Seko et al., Prediction of low-thermal-conductivity compounds with first principles anharmonic lattice-dynamics calculations and Bayesian optimization. Phys. Rev. Lett. **115**, 205901 (2015)
39. T.-L. Pham et al., Learning structure-property relationship in crystalline materials: a study of lanthanide–transition metal alloys. J. Chem. Phys. **148**, 204106 (2018)
40. S. Kikuchi, H. Oda, S. Kiyohara, T. Mizoguchi, Bayesian optimization for efficient determination of metal oxide grain boundary structures. Phys. B **532**, 24–28 (2018)
41. S. Kiyohara, H. Oda, K. Tsuda, T. Mizoguchi, Acceleration of stable interface structure searching using a kriging approach. Jpn. J. Appl. Phys. **55**, 045502 (2016)
42. C.W. Rosenbrock, E.R. Homer, G. Csányi, G.L.W. Hart, Discovering the building blocks of atomic systems using machine learning: application to grain boundaries. npj Comput. Mater. **3**, 29 (2017)
43. M.W. Gaultois et al., Perspective: web-based machine learning models for real-time screening of thermoelectric materials properties. APL Mater. **4**, 053213 (2016)
44. A. Glasmann, A. Kyrtsos, E. Bellotti, Machine learning for analyzing and characterizing InAsSb-based nBn photodetectors. Mach. Learn.: Sci. Technol. **2**, 025006 (2021)
45. B.C. Yeo, D. Kim, C. Kim, S.S. Han, Pattern learning electronic density of states. Sci. Rep. **9**, 5879 (2019)
46. B. Meredig, C. Wolverton, Dissolving the periodic table in cubic zirconia: data mining to discover chemical trends. Chem. Mater. **26**, 1985–1991 (2014)
47. S. Shi et al., Multi-scale computation methods: their applications in lithium-ion battery research and development. Chin. Phys. B **25**, 018212 (2016)
48. Y. Xing, Y. Xu, Q. Wu, G. Wang, M. Zhu, Optoelectronic functional fibers: materials, fabrication, and application for smart textiles. J. Mater. Chem. **9**(2), 439–455 (2021)
49. X. Yan, W. Zhen, H. Hu, L. Pi, C. Zhang, W. Zhu, High-performance visible light photodetector based on BiSeI single crystal. Chin. Phys. Lett. **38**(6), 068103 (2021)
50. M.A. Schmidt, A. Argyros, F. Sorin, Hybrid optical fibers—an innovative platform for in-fiber photonic devices. Adv. Optical Mater. **4**(1), 13–36 (2016)

51. D. Wang, X. Liu, S. Fang, C. Huang, Y. Kang, H. Yu, Z. Liu, H. Zhang, R. Long, Y. Xiong, Y. Lin, Y. Yue, B. Ge, T.K. Ng, B.S. Ooi, Z. Mi, J.-H. He, H. Sun, Pt/AlGaN nanoarchitecture: toward high responsivity, self-powered ultraviolet-sensitive photodetection. Nano Lett. **21**(1), 120–129 (2021)
52. L.-S. Zhu, S.-J. Zhao, Influence of Ni on Cu precipitation in Fe–Cu–Ni ternary alloy by an atomic study. Chin. Phys. B **23**(6), 063601 (2014)
53. M. Zafara, S. Ahmed, M. Shakil, M.A. Choudhary, K. Mahmood, Theoretical investigation of sulfur defects on structural, electronic, and elastic properties of ZnSe semiconductor. Chin. Phys. B **24**(7), 076106 (2015)
54. A. Champagne, J.B. Haber, S. Pokawanvit, D.Y. Qiu, S. Biswas, H.A. Atwater, F.H. Jornada, J.B. Neaton, Quasiparticle and optical properties of carrier-doped monolayer $MoTe_2$ from first principles. Nano Lett. **23**(10) (2023)
55. X.-Y. Ma, J.P. Lewis, Q.-B. Yan, G. Su, Accelerated discovery of two-dimensional optoelectronic octahedral oxyhalides via high-throughput *Ab Initio* calculations and machine learning. J. Phys. Chem. Lett. **10**(21), 6734–6740 (2019)
56. F. Mayr, M. Harth, I. Kouroudis, M. Rinderle, A. Gagliardi, Machine learning and optoelectronic materials discovery: a growing synergy. J. Phys. Chem. Lett. **13**, 1940 (2022)
57. J. Fang, M. Xie, X. He, J. Zhang, J. Hu, Y. Chen, Y. Yang, Q. Jin, Machine learning accelerates the materials discovery. Mater. Today Commun. **33**, 104900 (2022)
58. S.P. Ong, Accelerating materials science with high-throughput computations and machine learning. Comput. Mater. Sci. **161**, 143–150 (2019)
59. M. Sun, T. Wu, A.W. Dougherty, M. Lam, B. Huang, Y. Li, C.-H. Yan, Self-validated machine learning study of graphdiyne-based dual atomic catalyst. Adv. Energy Mater. **11**, 2003796
60. M. Gastegger, J. Behler, P. Marquetand, Machine learning molecular dynamics for the simulation of infrared spectra. Chem. Sci. **8**, 6924 (2017)
61. B. Charbuty, A. Abdulazeez, Classification based on decision tree algorithm for machine learning. J. Appl. Sci. Technol. Trends **2**(1), 20 (2021)
62. H. Liu, J. Cheng, H. Dong, J. Feng, B. Pang, Z. Tian, S. Ma, F. Xia, C. Zhang, L. Dong, Screening stable and metastable ABO_3 perovskites using machine learning and the materials project. Comput. Mater. Sci. **177**, 109614 (2020)
63. A.E. Siemenn, Z. Ren, Q. Li, T. Buonassisi, Fast Bayesian optimization of Needle-in-a-Haystack problems using zooming memory-based initialization (ZoMBI). npj Comput. Mater. **9**, 79 (2023)
64. Y. Wang, M.L. Adam, Y. Zhao, W. Zheng, L. Gao, Z. Yin, H. Zhao, Machine learning-enhanced flexible mechanical sensing. Nano-Micro Lett. **15**, 55 (2023)
65. B. Shih, D. Shah, J. Li, T.G. Thuruthel, Y.-L. Park, F. Iida, Z. Bao, R.K. Bottiglio, M.T. Tolley, Electronic skins and machine learning for intelligent soft robots. Sci. Robot. **5**(41) (2020)
66. M. Wang, T. Wang, Y. Luo, K. He, L. Pan, Z. Li, Z. Cui, Z. Liu, J. Tu, X. Chen, Fusing stretchable sensing technology with machine learning for human-machine interfaces. Adv. Func. Mater. **31**, 2008807 (2021)
67. C. Du, G. Zou, B. Feng, J. Huo, A. Zhanwen, Y. Xiao, W. Wang, L. Liu, Predicting effective thermal conductivity of sintered silver by microstructural-simulation-based machine learning. J. Electron. Mater. **52**, 2347–2358 (2023)
68. M. Mahmood, D. Mzurikwao, Y.-S. Kim, Y. Lee, S. Mishra, R. Herbert, A. Duarte, C. Siang Ang, W.-H. Yeo, Fully portable and wireless universal brain-machine interfaces enabled by flexible scalp electronics and deep learning algorithms. Nat. Mach. Intell. **1**, 412–422 (2019)
69. C.C. Vu, J. Kim, Highly elastic capacitive pressure sensor based on smart textiles for full-range human motion monitoring. Sens. Actuators A: Phys. **314**, 112029 (2020)
70. C. Du, G. Zou, A. Zhanwen, B. Lu, B. Feng, J. Huo, Y. Xiao, Y. Jiang, L. Liu, Highly accurate and efficient prediction of effective thermal conductivity of sintered silver based on deep learning method. Int. J. Heat Mass Transf. **201**, 123654 (2023)
71. N. Dai, I.M. Lei, Z. Li, Y. Li, P. Fang, J. Zhong, Recent advances in wearable electromechanical sensors—moving towards machine learning-assisted wearable sensing systems. Nano Energy **105**, 108041 (2023)

72. Y. Fang, Y. Zou, J. Xu, G. Chen, Y. Zhou, M. Roustaei, T.K. Hsiai, J. Chen, Ambulatory cardiovascular monitoring via a machine-learning-assisted textile triboelectric sensor. Adv. Mater. **33**(41), 2104178 (2021)
73. K. Zhang, Z. Li, J. Zhang, D. Zhao, Y. Pi, Y. Shi, R. Wang, P. Chen, C. Li, G. Chen, I.M. Lei, J. Zhong, Biodegradable smart face masks for machine learning-assisted chronic respiratory disease diagnosis. ACS Sens. **7**(10), 3135–3143 (2022)
74. Y.H. Jung, S.K. Hong, H.S. Wang, J.H. Han, T.X. Pham, H. Park, J. Kim, S. Kang, C.D. Yoo, K.J. Lee, Flexible piezoelectric acoustic sensors and machine learning for speech processing. Adv. Mater. **32**(35), 1904020 (2020)
75. Z. Zhang, T. He, M. Zhu, Z. Sun, Q. Shi, J. Zhu, B. Dong, M.R. Yuce, C. Lee, Deep learning-enabled triboelectric smart socks for IoT-based gait analysis and VR applications. npj Flex. Electron. **4**, 29 (2020)
76. M. Wang, Z. Yan, T. Wang, P. Cai, S. Gao, Y. Zeng, C. Wan, H. Wang, L. Pan, J. Yu, S. Pan, K. He, J. Lu, X. Chen, Gesture recognition using a bioinspired learning architecture that integrates visual data with somatosensory data from stretchable sensors. Nat. Electron. **3**, 563–570 (2020)
77. K. Kim, M. Sim, S. Lim, D. Kim, D. Lee, K. Shin, C. Moon, J.-W. Choi, J.E. Jang, Tactile avatar: tactile sensing system mimicking human tactile cognition. Adv. Sci. **8**(7), 2002362 (2021)
78. N. Bokka, V. Selamneni, P. Sahatiya, A water destructible SnS_2 QD/PVA film based transient multifunctional sensor and machine learning assisted stimulus identification for noninvasive personal care diagnostics. Mater. Adv. **1**(8), 2818–2830 (2020)
79. Z. Zeng, Z. Huang, K. Leng, W. Han, H. Niu, Y. Yu, Q. Ling, J. Liu, Z. Wu, J. Zang, Nonintrusive monitoring of mental fatigue status using epidermal electronic systems and machine-learning algorithms. ACS Sens. **5**(5), 1305–1313 (2020)
80. F. Wen, Z. Zhang, T. He, C. Lee, AI enabled sign language recognition and VR space bidirectional communication using triboelectric smart glove. Nat. Commun. **12**, 5378 (2021)
81. Q. Hu, X. Tang, W. Tang, A smart chair sitting posture recognition system using flex sensors and FPGA implemented artificial neural network. IEEE Sens. J. **20**(14), 8007–8016 (2020)
82. Z. Zhang, Q. Shi, T. He, X. Guo, B. Dong, J. Lee, C. Lee, Artificial intelligence of toilet (AI-Toilet) for an integrated health monitoring system (IHMS) using smart triboelectric pressure sensors and image sensor. Nano Energy **90**, 106517 (2021)
83. Y. Pei, Z. Li, B. Li, Y. Zhao, H. He, L. Yan, X. Li, J. Wang, Z. Zhao, Y. Sun, Z. Zhou, J. Zhao, R. Guo, J. Chen, X. Yan, A multifunctional and efficient artificial visual perception nervous system with Sb_2Se_3/CdS-Core/Shell (SC) nanorod arrays optoelectronic memristor. Adv. Func. Mater. **32**(29), 2203454 (2022)
84. R. Bao, J. Tao, J. Zhao, M. Dong, J. Li, C. Pan, Integrated intelligent tactile system for a humanoid robot. Sci. Bull. **68**(10), 1027–1037 (2023)
85. S. Oh, H. Kim, M. Meyyappan, K. Kim, Design and analysis of near-IR photodetector using machine learning approach. IEEE Sens. J. **24**(16), 25565–25572 (2024)
86. C. Li, J. Lu, Y. Zhao, L. Sun, G. Wang, Y. Ma, S. Zhang, J. Zhou, L. Shen, W. Huang, Highly sensitive, fast response perovskite photodetectors demonstrated in weak light detection circuits and visible light communication system. NANO MICRO Small **15**(44), 1903599 (2019)
87. C. Bao, J. Yang, S. Bai, W. Xu, Z. Yan, Q. Xu, J. Liu, W. Zhang, F. Gao, High performance and stable all-inorganic metal halide perovskite-based photodetectors for optical communication applications. Adv. Mater. **30**(38), 1803422 (2018)
88. S.V. Pandey, N. Parikh, A. Kalam, D. Prochowicz, S. Satapathi, S. Akin, M.M. Tavakoli, P. Yadav, A machine learning framework for predicting device performance in 2D metal halide perovskite photodetector. Sol. Energy **270**, 112399 (2024)
89. F.M.A. Alzahrani, M. Sagir, M. Saqib, S. Bashir, T. Sarwar, S. Hussain, S. Murtaza, A. Mushtaq, R. Razzaq, Z.A. Alrowaili, M.S. Al-Buriahi, Machine learning assisted prediction of band gaps and designing of new polymers for photodetectors: a complete pipeline. Comput. Mater. Sci. **293**, 112961 (2024)

90. N. Alfryyan, M. Saqib, T. Mubashir, M.H. Tahir, Z.A. Alrowaili, M.S. Al-Buriahi, Designing of near-IR organic semiconductors for photodetectors: machine learning and data mining assisted efficient pipeline. Mater. Today Commun. **36**, 106556 (2023)
91. R.O. Ocaya, A.A. Akinyelu, A.G. Al-Sehemi, A. Dere, A.A. Al-Ghamdi, F. Yakuphanoglu, Machine learning models for efficient characterization of Schottky barrier photodiode internal parameters. Sci. Rep. **13**, 13990 (2023)
92. A. Arjmandi-Tash, A. Mansourian, F. Rahsepar, Y. Abdi, Predicting photodetector responsivity through machine learning. Adv. Theory Simul. **7**(6), 2301219 (2024)
93. V. Sorathiya, M. Kumar, B. Garai, J. Saha, Prediction of photodetector response using supervised learning, in *7th International Conference on Computation System and Information Technology for Sustainable Solutions (CSITSS)* (IEEE, 2023), (pp. 1–6)
94. A.S. Tan, H. Wymeersch, P. Johannisson, E. Agrell, P.A. Andrekson, M. Karlsson, An ML-based detector for optical communication in the presence of nonlinear phase noise, in *IEEE International Conference on Communications (ICC)* (IEEE, 2011), pp. 1–5
95. M. Rohlfing, P. Kruger, J. Pollmann, Quasiparticle band-structure calculations for C, Si, Ge, GaAs, and SiC using Gaussian-orbital basis sets. Phys. Rev. B **48**, 17791–17805 (1993)
96. K. Suganthi, R. Karthik, G. Rajesh, P.H. Ching, *Machine Learning and Deep Learning Techniques in Wireless and Mobile Networking Systems*, 1st edn. (CRC Press, 2022)
97. S. Mitra, A. Ahmad, S. Biswas, A.K. Das, A machine learning approach to predict the structural and magnetic properties of Heusler alloy families. Comput. Mater. Sci. **216**, 111836 (2023)
98. V.L. Deringer, A.P. Bartók, N. Bernstein, D.M. Wilkins, M. Ceriotti, G. Csányi, Gaussian process regression for materials and molecules. Chem. Rev. **121**, 10073−10141 (2021)
99. M.P. Bahlke, N. Mogos, J. Proppe, C. Herrmann, Exchange spin coupling from Gaussian process regression. J. Phys. Chem. A **124**(42), 8708–8723 (2020)
100. Z. Yin, Y. Shan, M. Yu, L. Yang, J. Song, P. Hu, F. Teng, Enhanced performance of UV photodetector based on ZnO nanorod arrays via TiO_2 as electrons trap layer. Mater. Sci. Semicond. Process. **148**, 106813 (2022)
101. Z. Liu, Z. Zhang, X. Zhang, X. Li, Z. Liu, G. Liao et al., Achieving high responsivity and detectivity in a quantum-dot-in-perovskite photodetector. Nano Lett. **23**(4), 1181–1188 (2023)
102. R.K. Upadhyay, S. Jit, Solution-processed ZnO nanoparticles (NPs)/$CH_3NH_3PbI_3$/PTB_7/MoO_3/Ag inverted structure-based UV–visible-Near Infrared (NIR) broadband photodetector. Opt. Mater. **135**, 113290 (2023)
103. E. Hong, Z. Li, T. Yan, X. Fang, Surface-tension-dominant crystallization of 2D perovskite single crystals for vertically oriented hetero-/homo-structure photodetectors. Nano Lett. **22**(21), 8662–8669 (2022)
104. S. Liu, S. Jiao, H. Lu, Y. Nie, S. Gao, D. Wang et al., Polycrystalline perovskite CH_3 NH_3PbCl_3/amorphous Ga_2O_3 hybrid structure for high-speed, low-dark current and self-powered UVA photodetector. J. Alloy. Compd. **890**, 161827 (2022)
105. C. Li, J. Li, Z. Li, H. Zhang, Y. Dang, F. Kong, High-performance photodetectors based on nanostructured perovskites. Nanomaterials **11**(4), 1038 (2021)
106. L. Chen, Y. Wang, J. Zhang, L. Chen, J. Zhai, J. Song, Inorganic semiconductor-based flexible UV photodetector arrays achieved by specific flip-chip bonding. ACS Appl. Mater. Interfaces. **16**(38), 51089–51096 (2024)
107. G. Yang, J. Li, M. Wu, X. Yu, J. Yu, Recent advances in materials, structures, and applications of flexible photodetectors. Adv. Electron. Mater. **9**(10), 2300340 (2023)
108. T. Yan, Z. Li, F. Cao, J. Chen, L. Wu, X. Fang, An all-organic self-powered photodetector with ultra-flexible dual-polarity output for bio signal detection. Adv. Mater. **34**(30), 2201303 (2022)
109. Y. Li, X. Wei, Q. Yang, J. Zhang, W. Wang, L. Dong et al., Fast-response self-powered flexible transparent ultraviolet photodetectors based on a CuO/ZnO nanowire array heterojunction. New J. Chem. **47**(13), 6397–6404 (2023)
110. Y. Huang, J. Tan, G. Gao, J. Xu, L. Zhao, W. Zhou et al., Transparent p-type CuI film based self-powered ultraviolet photodetectors with ultrahigh speed, responsivity and detectivity. J. Mater. Chem. C **10**(36), 13040–13046 (2022)

111. T.T. Nguyen, G. Murali, A.S. Nissimagoudar, P. Bhatnagar, S. Lee, M. Patel et al., Flexible and transparent MXene-platformed ultrafast photodetector for encrypted signal communication in self-powered operation. Nano Energy **109**, 108331 (2023)
112. Z. Lan, M.H. Lee, F. Zhu, Recent advances in solution-processable organic photodetectors and applications in flexible electronics. Adv. Intell. Syst. **4**(3), 2100167 (2022)
113. H. Jinno, K. Fukuda, X. Xu, S. Park, Y. Suzuki, M. Koizumi et al., Stretchable and waterproof elastomer-coated organic photovoltaics for washable electronic textile applications. Nat. Energy **2**(10), 780–785 (2017)
114. P. Zalar, N. Matsuhisa, T. Suzuki, S. Enomoto, M. Koizumi, T. Yokota et al., A monolithically processed rectifying pixel for high-resolution organic imagers. Adv. Elect. Mater. **4**(6), 1700601 (2018)
115. A.J. van Breemen, M. Simon, O. Tousignant, S. Shanmugam, J.L. van der Steen, H.B. Akkerman, et al., Curved digital X-ray detectors. npj Flex. Electron. **4**(1), 22 (2020)
116. Y. Xia, L.E. Aguirre, X. Xu, O. Inganäs, All-polymer high-performance photodetector through lamination. Adv. Elect. Mater. **6**(3), 1901017 (2020)
117. Y. Wei, H. Chen, T. Liu, S. Wang, Y. Jiang, Y. Song et al., Self-powered organic photodetectors with high detectivity for near infrared light detection enabled by dark current reduction. Adv. Func. Mater. **31**(52), 2106326 (2021)
118. Z. Jiang, K. Yu, H. Wang, S. Rich, T. Yokota, K. Fukuda, T. Someya, Ultra-flexible integrated organic electronics for ultrasensitive photodetection. Adv. Mater. Technol. **6**(1), 2000956 (2021)
119. Z. Ge, N. Xu, Y. Zhu, K. Zhao, Y. Ma, G. Li, Y. Chen, Visible to mid-infrared photodetection based on flexible 3D graphene/organic hybrid photodetector with ultrahigh responsivity at ambient conditions. ACS Photonics **9**(1), 59–67 (2022)
120. J. Simões, T. Dong, Z. Yang, Non-fullerene acceptor organic photodetector for skin-conformable photoplethysmography applications. Adv. Mater. Interfaces **9**(10), 2101897 (2022)
121. W. Li, F. Guo, H. Ling, H. Liu, M. Yi, P. Zhang et al., Solution-processed wide-bandgap organic semiconductor nanostructures arrays for non-volatile organic field-effect transistor memory. Small **14**(2), 1701437 (2018)
122. C. Liu, H. Chen, X. Hou, H. Zhang, J. Han, Y.G. Jiang et al., Small footprint transistor architecture for photoswitching logic and in situ memory. Nat. Nanotechnol. **14**(7), 662–667 (2019)
123. H. Huang, L. Jiang, J. Peng, Y. Qi, S. Bai, Q. Lin, High-performance organic phototransistors based on D18, a high-mobility and unipolar polymer. Chem. Mater. **33**(20), 8089–8096 (2021)
124. C. Fuentes-Hernandez, W.F. Chou, T.M. Khan, L. Diniz, J. Lukens, F.A. Larrain et al., Large-area low-noise flexible organic photodiodes for detecting faint visible light. Science **370**(6517), 698–701 (2020)
125. Y. Lei, L. Mao, H. Zhu, J. Yao, Development of catechol-functionalized chitosan/poly (vinyl alcohol) nanocomposite films incorporated with dual network coated layered clay for active packaging applications. J. Appl. Polym. Sci. **138**(42), 51251 (2021)
126. Q. Li, Y. Ran, W. Shi, M. Qin, Y. Sun, J. Kuang et al., High-performance near-infrared polymeric phototransistors realized by combining cross-linked polymeric semiconductors and bulk heterojunction bilayer structures. Appl. Mater. Today **22**, 100899 (2021)
127. M.Y. Lee, J. Park, J.H. Oh, High-performance Ambipolar organic phototransistors based on core-shell p–n junction organic single crystals. ACS Appl. Electron. Mater. **2**(1), 9–18 (2019)
128. S. Nam, J. Seo, H. Han, H. Kim, D.D. Bradley, Y. Kim, Efficient deep red light-sensing all-polymer phototransistors with p-type/n-type conjugated polymer bulk heterojunction layers. ACS Appl. Mater. Interfaces. **9**(17), 14983–14989 (2017)
129. H. Xu, J. Liu, J. Zhang, G. Zhou, N. Luo, N. Zhao, Flexible organic/inorganic hybrid near-infrared photoplethysmogram sensor for cardiovascular monitoring. Adv. Mater. **29**(31), 1700975 (2017)
130. T. Yokota, T. Nakamura, H. Kato, M. Mochizuki, M. Tada, M. Uchida et al., A conformable imager for biometric authentication and vital sign measurement. Nat. Electron. **3**(2), 113–121 (2020)

131. N. Li, Q. Wang, C. Shen, Z. Wei, H. Yu, J. Zhao et al., Large-scale flexible and transparent electronics based on monolayer molybdenum disulfide field-effect transistors. Nat. Electron. **3**(11), 711–717 (2020)
132. H.U. Chung, B.H. Kim, J.Y. Lee, J. Lee, Z. Xie, E.M. Ibler, et al., Binodal, wireless epidermal electronic systems with in-sensor analytics for neonatal intensive care. Science **363**(6430), eaau0780 (2019)
133. Z. Wang, Y. Gao, Y. Li, H. Yan, F. Kang, Y. Shen et al., High speed dual-band photodetector for dual-channel optical communications in wavelength division multiplexing and security enhancement. Adv. Func. Mater. **34**(17), 2310911 (2024)
134. W. Du, C. Li, Y. Huang, J. Zou, L. Luo, C. Teng et al., An optoelectronic reservoir computing for temporal information processing. IEEE Electron Device Lett. **43**(3), 406–409 (2022)
135. D. Ebrahimibagha, S. Armida, S. Datta, M. Ray, Machine learning based models to investigate the thermoelectric performance of carbon nanotube-polyaniline nanocomposites. Comp. Mater. Sci. **232**, 112601 (2024)
136. S. Ju, Y. Miura, K. Yamamoto, K. Masuda, K. Uchida, J. Shiomi, Machine learning analysis of tunnel magnetoresistance of magnetic tunnel junctions with disordered $MgAl_2O_4$. Phys. Rev. Res. **2**, 023187 (2022)
137. J. Rao, Z. Fan et al., Experimental search for high-performance ferroelectric tunnel junctions guided by machine learning. J. Adv. Die. **12**(3), 2250005 (2022)
138. R. Mishra, S. Goel, H. Nezhad, Computational prediction of electrical and thermal properties of graphene and $BaTiO_3$ reinforced epoxy nanocomposites. Biomater. Polym. Horiz. **1**(1), 1–14 (2022)
139. A. Mahalawy, K. Safty, Classical and quantum regression analysis for the optoelectronic performance of NTCDA/p-Si UV photodiode. Optik **246**, 167793 (2021)
140. K. Ghosh, S. Ghosh, Classical and quantum machine learning applications in spintronics. Digit. Disco. **2**, 512 (2023)
141. S. Bansal, A. Jain, S. Kumar, A. Kumar, P. Kumar, K. Prakash, M. Soliman, M. Islam, M. Islam, Optoelectronic performance prediction of HgCdTe homojunction photodetector in long wave infrared spectral region using traditional simulations and machine learning models. Sci. Rep. **14**, 28230 (2024)
142. E. Simsek, S. Mahabadi, T. Carruthers, C. Menyuk, Photodetector performance prediction with machine learning, in *Proceedings Frontiers in Optics + Laser Science* (2021). ISBN: 978-1-55752-308-2

Next Generation Optical Sensor Devices

Ritesh Kumar, Priyatam Kumar, and Amit Ranjan

Abstract Micro and nano structured optical sensors have improved performance over traditional optical sensors. They provide advance optical sensing techniques and integration over photonic integrated circuits. Various optical cavity structures have been designed in the past few decades to improve sensing performance of the sensors based on the materials choice and their physical properties. Choice of materials, sensing parameters, size, loss, robustness, selectivity and sensitivity define the application of sensors in the area of healthcare diagnosis, defense, remote sensing, and industrial applications. In this chapter, development and challenges in the next generation optical sensor devices are discussed along with their applications.

Keywords Optical sensors · Fiber patterning · Interferometry · Biosensors · Remote sensing

1 Introduction

Long-length optical fiber sensors were initially developed in the 1970s with the advent of low-loss fibers. While many fiber sensors use the amplitude of transmitted or back scattered signal, many powerful sensing technologies use the phase of the optical signal that interrogates the fiber. Early examples include the optical gyroscope, optical frequency domain reflectometry and coherent optical time domain reflectometry. With increased interest in the potential for distributed sensing for numerous applications [1–10]. In recent years, optical fiber based wearable sensors are commercially available in real time monitoring of heart bit, oxygen levels in blood, fire and explosion detection in energy applications like Lithium battery, detection of pathogen disease for example COVID-19, electrical signal measurements, constellations in global positioning system, real time position, gestures, and shape tracking

R. Kumar (✉) · P. Kumar · A. Ranjan
Department of Electronics and Communication Engineering, Shri Phanishwar Nath Renu Engineering College, Araria, India
e-mail: ritesh.iitdhanbad@gmail.com

B. Ramachandran et al. (eds.), *Advanced Optical Sensors*, Progress in Optical Science and Photonics 38, https://doi.org/10.1007/978-981-95-1687-2_10

in the area of robotics and automation. Apart from the above, optical sensors are advantageous over traditional sensors in terms of multiplexing of signals.

Optical sensor system comprises of four units namely, light source to have optical carrier signal, transducer unit that converts physical parameters superimpose in optical domain; an optical transmitting media for the transportation of optical signal, and a detection unit for the conversion of optical signal into electrical domain. Optical fibers with change in geometrical structure and doping of material can work as intrinsic sensors. Also, it works as an optical waveguide with the choice of different refractive indices and comes under the category of extrinsic type optical sensors. Intrinsic optical sensors can be distributed or quasi distributed in nature and point type [11]. Development of optical sensors with technology requirements have been shown in Fig. 1.

In case of extrinsic type optical sensors, they can be used for endoscopic purposes, mid infrared cameras with high pixel and as long wavelength optical source [12]. Continuous and unforming sensing for environmental monitoring can be done with the help of distributed sensing at any location through fiber. A light pulse is emitted into the extended sensing fiber and is thereafter constantly back reflected to the fiber input end via natural scattering processes, like in a radar system. Analyzing this back-reflected light (spectrum and amplitude) informs about the quantity to be measured and observing its time response translates into a position-resolved information, considering the finite speed of light and the specific time required by light for returning from a given point along the fiber [11]. It is logical to anticipate more effects in sensing given the growing interest in multimode and multicore fibers for telecom applications [13]. By quickly postprocessing the encoded incident signal and utilizing artificial intelligence more, more processing capacity will also make it easier to interpret backscattered signals [14]. The great majority of the quasi-distributed OFS market is made up of Fiber Bragg gratings (FBG). Because of their long lifespan, high accuracy, small size, quick reaction time, and most importantly, multiplexing, FBGs have continuously drawn a lot of attention. They have found important uses in the power sector, the oil and gas market, and structural health monitoring for civil and aerospace engineering [15]. A wide range of sensors are included in point sensors, such as temperature, vibration, chemical, biological, acceleration, and rotation, to name a few.

To date, fiber optic sensors have been implemented using specialty fibers, micro or nano fibers, waveguide structures, and resonant structures. Photonic crystal fiber, chalcogenide silica fiber, distributed fibers have been extensively used in development of optical sensors for various applications. In the past few decades successful design and implementation of optical sensors have been reported, however, more challenges may come from fabrication, practical applications, and sensing mechanism innovations. Cost of fabrication, precision of the sensor, and scalability to integrate varies with the lower subwavelength regions limit the scientific research and practical applications. For instance, when the detection limit of microforce is reduced to a specified level or below, the sensor can be employed to measure critical Casimir forces, optical scattering forces, and optical momentum. With the rapid development of health care, energy, robotics and artificial intelligence, there is an

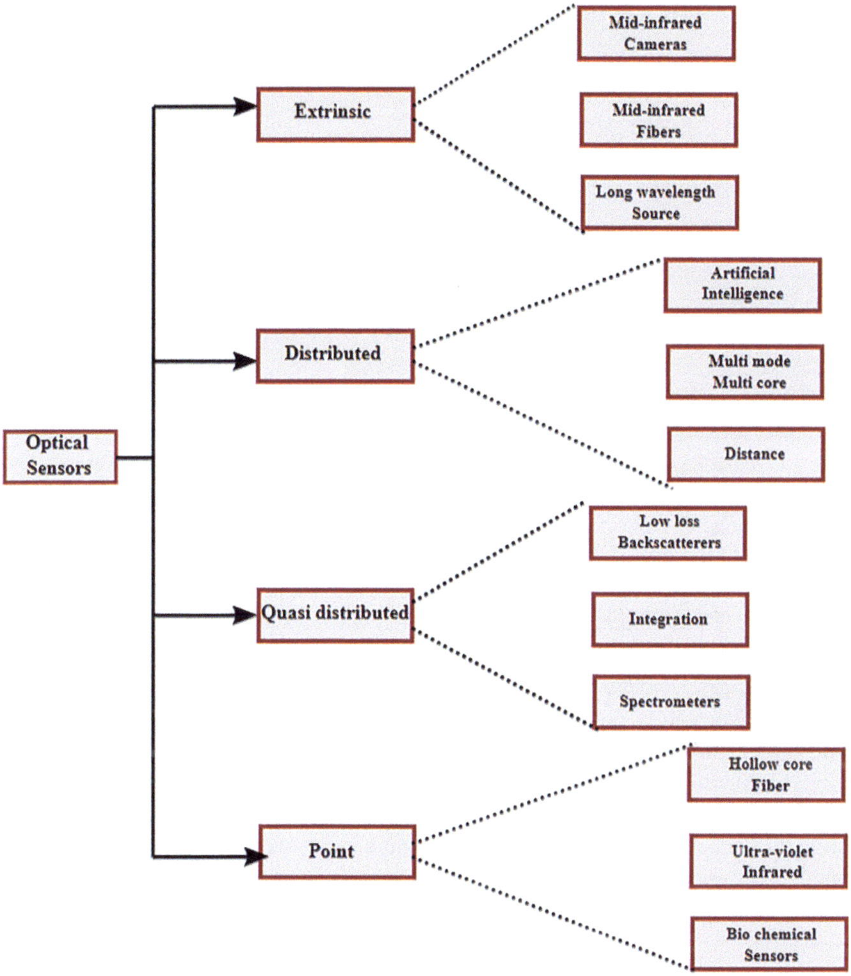

Fig. 1 Optical sensors development with technology requirements

increasing demand for novel sensors to meet the needs from these areas. To address the issues in the aforementioned advanced applications, innovative sensing structures and mechanisms are essential. For example, current leakage due to insufficient insulation, and high sensitivity to electromagnetic disturbances are still challenges for E-skin sensors. An alternative to E-skin is the detection of pressure, strain, bending, and temperature by optical sensors due to their inherent electrical safety, immunity to electromagnetic interference, and small size. Note that multiparameter (e.g. pressure, strain, and temperature) signals often mix together, how to realize an efficient decoupling of the output of a fibre-optic sensor should be considered for real applications. Fiber Bragg grating based optical sensors comes under the category of quasi-distributed type and are usually used for the detection of temperature and vibration.

2 Choice of Fabrication Platform

Optical sensors adjust one of its optical characteristics in response to a molecule's interaction with the analyte being measured. Research efforts aimed at resolving the difficulties in accurately on-site biosensing have increased dramatically in the last few decades. Modern fiber optic technology makes it possible to interact light with functionalized fiber surfaces at far-off places, creating a new, compact, and reasonably priced fiber lab for biosensing applications. There are countless functionalization techniques available to create selective bio-recognition components for label-free biosensors thanks to the recent astounding advancements in nanotechnology.

One can first distinguish optical fibers or waveguides among various types of sensors, which can be combined with a sensible coating to measure analytes, disease, or biological phenomena [16, 17]. In recent years, Lab-On-Chips (LOCs) have been demonstrated to be a crucial instrument for a comprehensive knowledge of sensing applications [18–20]. Integration of optical fiber waveguides on thin sensitive coating materials explores potential applications of sensors in diverse areas of chemical sensing [21, 22]. Depending on different transduction mechanisms, remarkable techniques can be readily coupled with fiber surfaces to produce extremely selective light-matter interaction [23, 24]. Three main stages make up the lab on fiber technology: the synthesis procedures for integrating functional nanomaterials on optical fibers, the physical principles guiding the different light-matter interaction processes in optical fiber sensors, and the selective capture components involved in the particular interactions in optical fiber sensors. These stages are discussed in following subsections.

2.1 Nanopatterning Techniques

The history of optical fiber patterning began in the 1990s when surface enhanced Raman scattering applications from an optical fiber were studied using an abrasive or self-assembly technique. The microstructure Fabry Perot interferometric cavity on the tip of the optical fiber like multi core fiber for temperature and CO_2 sensing [25], single mode fiber for refractive index and pressure sensing [26] are printed using 3-D micro-printing. Schematic diagram of 3-D micro-printing and printing process is depicted in Fig. 2 [25]. As shown in Fig. 2, the optical fiber was initially positioned beneath the ultrasonic nozzle for the spray coating of a thin layer of SU-8 in order to create the 3D microstructure seen in Fig. 2b. After that, the film was soft baked in place to eliminate the solvent. The film thickness and SU-8 solution concentration were used to optimize the soft-bake duration. With the aid of the digital camera-based

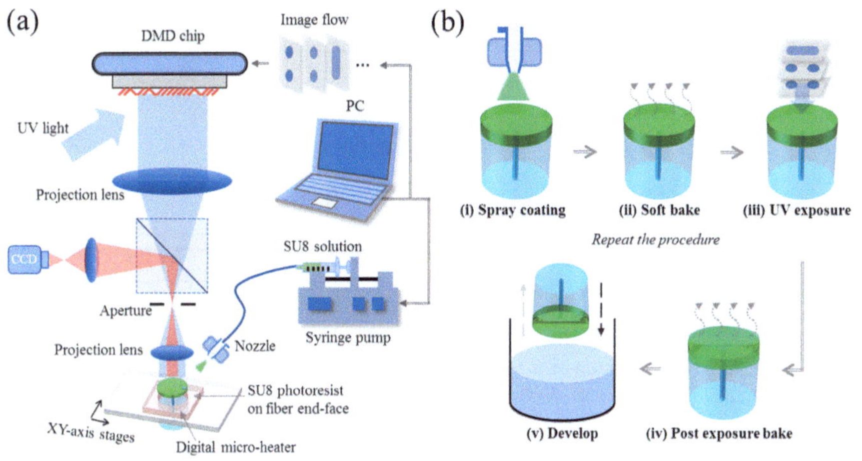

Fig. 2 **a** Schematic diagram for 3-D micro-printing. **b** Printing process on the tip of optical fiber [25]

machine vision metrology, the sample was transferred to the pre-aligned location for optical exposure following soft baking. Next, using self-developed add-on software to slice the image data from the CAD model of the 3D microstructure, optical patterns were created to irradiate the SU-8 film on the optical fiber end-face. The sample was post-baked in situ following exposure. After that, the film was soft baked in place to eliminate the solvent. The film thickness and SU-8 solution concentration were used to optimize the soft-bake duration. With the aid of the digital camera-based machine vision metrology, the sample was transferred to the pre-aligned location for optical exposure following soft baking. However, biosensing applications on optical fibers through plasmonic, surface and fluorescent spectroscopy are possible through high resolution patterning scheme known as nanopatterning. Nanotechnology syntheses are performed using various technologies self-assembly of nano particles [27–29] and scaling down of bulk materials on nanoparticles [30].

2.2 *Light Matter Interactions*

The basic idea behind interferometric biosensors is to identify changes in a sample's refractive index brought on by the conjugate layers' bispecific interaction. A comparison of two equal light routes is made, with one serving as a reference and the other identifying the interaction that modifies the optical path length by altering the refractive index. In interferometric biosensors interference of optical signals with path difference are investigated. These interferometric structures can be developed using Michelson interferometer (MI), Mach–Zehnder interferometer (MZI), Sagnac loop and Fabry–Perot interferometer (FPI) [31] shown in Fig. 3. The designs of MI and

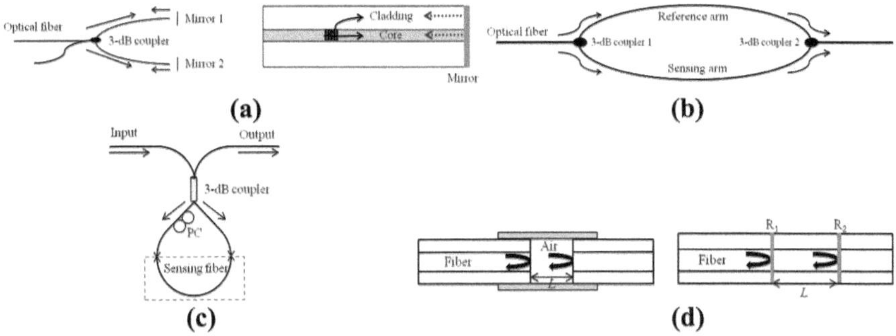

Fig. 3 Schematic representation of **a** Michelson interferometer. **b** Mach–Zehnder interferometer. **c** Sagnac loop and **d** Fabry–Perot interferometer [31]

MZI buildings are nearly identical. The sole distinction is that in MZI, a fiber coupler combines the light split by the fiber splitter, whereas in MI, a mirror reflects the two beams, which are then combined by the same fiber splitter, shown in Fig. 3a, b. MZI structures can be formed with change in geometrical shape of single mode fiber (SMF), the most reported structure is tapering of SMF. As the fiber core diameter reduces, the fundamental mode of the light is modulated for the excitation of higher order modes due to the tapering of the fiber. The medium around the taper has a significant impact on the evanescent field linked to the cladding mode, which results in a phase shift. By applying a certain layer to the tapered area of SMF, a biosensor based on MZI has been created via the tapered fiber.

FPI structure can be easily formed by the selection of materials with distinct refractive indices which leads to reflecting mirrors. Such structure is useful in detection of bovine serum antigen [32], liposome [33], hydrogen sensor [34], microfluidic sensing [35], and many more with high accuracy and sensitivity. As shown in Fig. 3d, FPI structure can be categorized as extrinsic sensors where cavity between reflecting surface is created inside fiber (left in Fig. 3d) and intrinsic sensors where cavity is formed at the end surface of the fiber (right in Fig. 3d).

Recently, Iele et al. have reported on a biosensor based on fiber Bragg grating (FBG) for the early detection of prostate cancer screening [36] and the developed system is shown in Fig. 4. In this configuration, fiber probe is placed between lumen metallic gauge where probe is in direct contact with the membrane to be analyzed during the indentation procedure. The development of fiber probe is based on incorporation of FBG inside fiber forming FPI biosensor. To precisely regulate the indentation depth, the optical probe was moved using a robotic arm (KUKA LBR Med 7) during the indentation experiments across silicone membranes and phantom tissue. Using a 3D printing technique, a customized holder for attaching the optical probe to the robot end effector was created. FBG can be imprinted in specialty fibers like microstructure optical fiber or photonic crystal fiber. Due to unique guiding characteristics for light propagation in these fibers, FBG printing in specialty fibers has potential biochemical sensing applications. Numerous numbers of chemical and biosensing applications are reported in recent articles based on FBG [37, 38]. Some

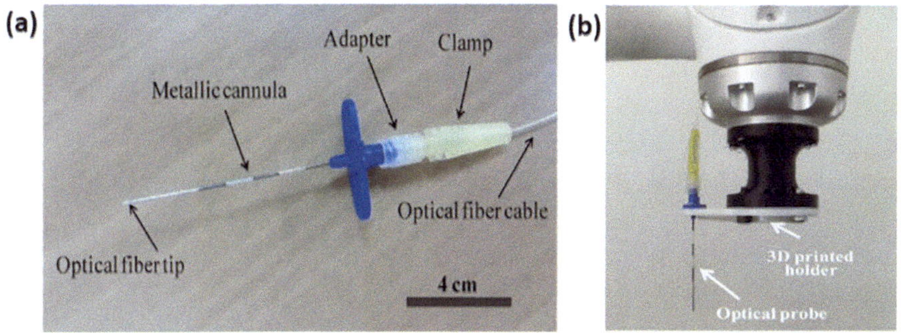

Fig. 4 **a** Schematic diagram of optical fiber probe inserted inside the lumen [36]. **b** Optical fiber probe installation [36]

of the biosensing applications include Nucleic acid based biosensing, antibody based biosensing, aptamer based biosensing, etc.

Further, molecular imprinting on optical fibers is applicable for the sensing of antibiotics like bacteria, fungi, actinomycetes, etc. in plants, animals, or environmental water through fluorescence sensing. It can either reversibly bind the analyte molecule by charge transfer, energy transfer, or excimer formation, or it can experience fluorescence quenching upon analyte binding [39]. In [39], molecular imprinted optical fiber sensors are developed by combining Y-shaped fiber sensors with molecular imprinted nanoparticles (MIN) composite polyethyleneglycol diacrylate hydrogel with replaceable MIN contents. The schematic diagram of complete setup and modification process in MIN is shown in Fig. 5a, b, respectively. As shown in Fig. 5a, optical source is used to provide optical signal of 400 nm for the excitation of fluorescence through Y-shaped optical fiber. The tail end of the fiber is connected through a replaceable MIN packet and other end response is observed at fiber optic spectrometer through optical filter. The filter's function is to lessen laser light source interference. In order to obtain the whole inspection apparatus, the filter is lastly linked to the spectrometer by a plastic cladding optical fiber, and the computer is connected to the computer via a USB interface.

The Y-type optical fiber carries the fluorescence emission from CIP bound in the MINs stimulated by laser source light (400 nm) to the spectrometer. MIN detector packet shown in Fig. 5b is processed through synthesis of molecular imprinted nanoparticles, preparation of PEGDA, and modification of detector.

Characterization process of MIN detector is depicted in Fig. 6 [39]. The resulting hydrogel was clear and firmly attached to the optical fiber in Fig. 6a. The CIP molecule inside the MINs, which is seen in Fig. 6b, exhibited blue fluorescence (468 nm) throughout the hydrogel when it was exposed to a laser light source during the detection procedure. The fluorescence intensities of imprinted nanoparticles bound before and after were recorded, respectively, in accordance with the concept and expectations of the suggested system in order to examine the potential for use in the detection of antibiotics in real-world water samples. When the modified MINs detector was submerged in 1 mL of CIP aqueous solution (0.5 mM) and sonicated

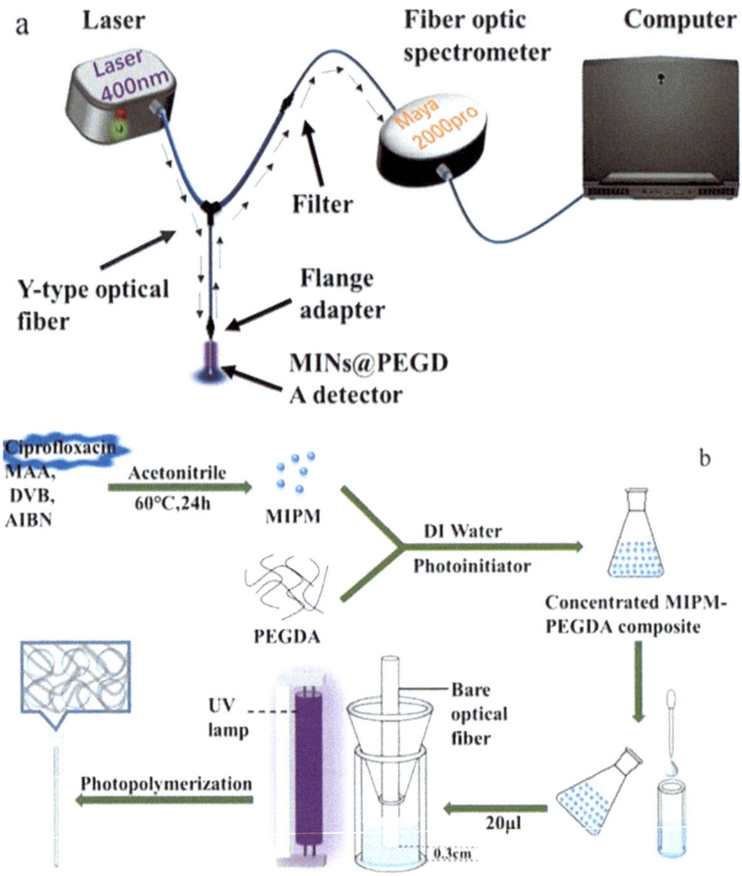

Fig. 5 **a** Schematic diagram of the complete experimental setup for antibiotic sensing through molecular imprinting technique and **b** Modification process for the content of MIN. *MINs@PEGD* molecularly imprinted nanoparticles composite polyethyleneglycol diacrylate hydrogel; *MAA* methylacrylic acid; *DVB* divinylbenzene; *AIBN* 2-methylpropionitrile; *MIPM* molecularly imprinted polymers; *PEGDA* polyethyleneglycol diacrylate [39]

for 20 min, the fluorescence intensity at 468 nm was greatly increased by about five times, as seen in Fig. 6c. This is explained by fluorescence in the modified MINs gel that is stimulated by laser light at 400 nm and released at 468 nm. The produced MINs detector clearly demonstrated an excellent signal response to the fluorescence intensity from CIP in the MINs gel, as seen by this result.

Further, Surface enhanced Raman spectroscopy (SERS) explored in plasmonic nanoparticles is useful in material characterization for vibration and structural monitoring [40]. It takes advantage of inelastic scattering vibrating molecules. In [41], SERS based polystyrene nanosphere optical sensor is developed through nanosphere lithography. A nanosphere lithography technique, which has already been shown to be capable of creating highly reproducible patterns on the fiber tip, was used to

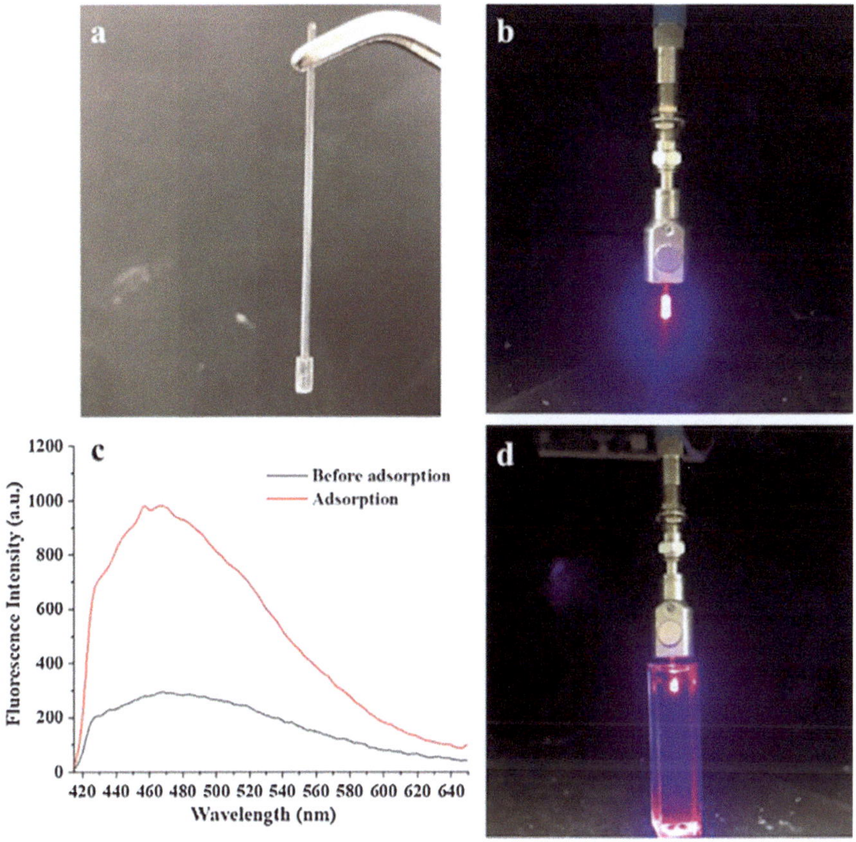

Fig. 6 Characterization of MIN detector [39]

construct the SERS surfaces. As schematically seen in Fig. 7 [41], the SERS active surfaces are made up of hexagonally CPA nanospheres coated in thin gold coatings. In essence, the suggested method depends on polystyrene (PS) nanospheres at the air–water interface self-assembling. The PS nanospheres come in aqueous solution for commercial use. Using a silicon conduit plate and a microsyringe, each nanosphere suspension is discharged at the air–water contact. The desired substrate is dipped into the water bath and raised towards the floating MCC island to transfer the monolayer colloidal crystal (MCC), which has self-assembled at the water surface, onto the planar substrate. Thermal evaporation is used to gradually install the gold layer on the MCC.

Fabrication process for the nanosphere assembly process is discussed in [42] as shown in Fig. 8. The monolayer colloidal crystal-decorated fibers are modified to create various structural motifs and investigate several options for achieving periodic patterns on the optical fiber tips, as seen in Fig. 8. Following the collection of MCC on the fiber, other treatments including sonication, plasma etching, and gold coating

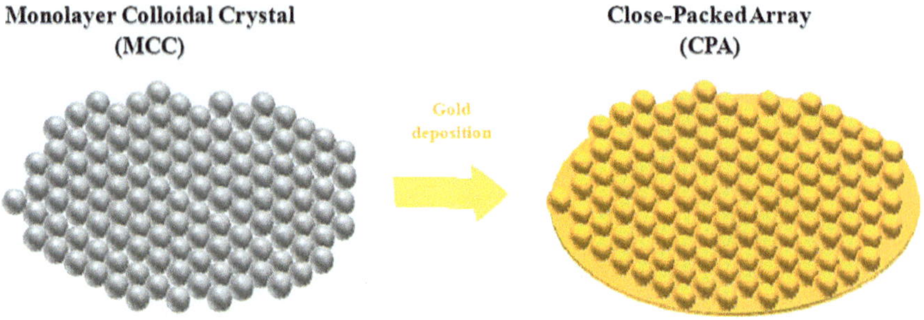

Fig. 7 Surface construction process of surface enhanced Raman spectroscopy [41]

were used to create structures with various characteristics. In particular, a Metallo-dielectric crystal is produced by thermally evaporating a small metallic layer—that is, gold—onto the fiber/MCC assembly, which led to the most straightforward instance of CPA. Following this treatment, as seen in the digital image in Fig. 8, the resultant optical fiber tip exhibits a vivid iridescence brought on by the Bragg reflection of ambient light via the optical fiber tip patterns. The process involves submerging the optical fibre tip in water and then applying ultrasonic therapy to the water's surface after it has been dissolved in a non-polar solvent like hexane or dichloromethane. The nanospheres are formed on water using a suspension of water and alcohol. The addition of ethanol improves the dispersion of nanospheres on the water by lowering the surface tension of the aqueous colloidal solution.

2.3 *Sensing Phenomena*

Plasmonic has attracted a lot of attention in recent decades among the different sensing techniques used on optical fibers because of its amazing benefits, such as the ability to measure surface interactions kinetically in real time with high sensitivity and specificity using a properly designed recognition layer [43–45]. Fiber optic sensors based on surface plasmon resonance have emerged as a possible rival to several sensing methods in recent years due to its ease of use, compact size, affordability, lack of electrical contact between the sensor probe and analyte sample, quick measurement, and real-time sensing. Combination of surface plasmon resonance (SPR) and localized SPR produce hybrid fiber optic sensors which have intensive sensing applications through resonant electron oscillation of metallic nanoparticles on the surface of fiber. There is various prism based SPR hybrid structure reported based on matter-plasmon interactions in single layer and multilayer interactions [46–48]. In order to improve the plasmon–matter interaction, several research projects utilize 2D material/metal film hybrid architectures to SPR setups. The thin gold film in the traditional Kretschmann arrangement, for example, has been replaced with hybrid film-like topologies such as graphene/gold [46], graphene oxide/gold [47],

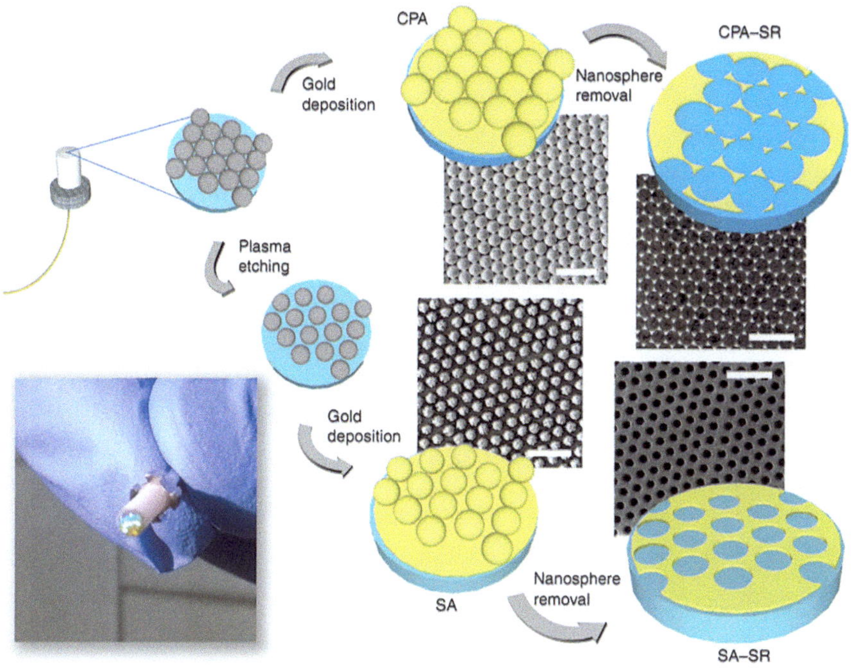

Fig. 8 Schematic of fabrication process of nanoprobe fiber tips. *SA* sparse array; *CPA* close-packed array; *CPA-SR* close-packed array—six-point star [42]

graphene-MoS2/gold [48], etc. In [48], Zeng et al. suggested and created a novel configuration of improved SPR biosensors based on graphene–MoS2 hybrid structures, as illustrated in Fig. 9, based on the special properties of MoS2 nanosheets. The Au thin film used to create the SPR sensing substrate is covered with monolayer graphene and MoS2 nanosheets. A greater electric field augmentation at the sensor interface as a result of this technique will increase the sensitivity to the target analytes.

In Fig. 9, monolayer graphene serves as a bio-recognition component, trapping the target biomolecules via pi-stacking force, while MoS2 layers are utilized to enhance light absorption in order to supply sufficient excitation energy for efficient charge transfer.

Molecular imprinted polymer makes a nanostructure sensing layer which has efficient sensing capability. Chemical and biosensors can greatly benefit from the concept of molecular recognition, which is taken from natural systems. In [49], authors have presented a molecularly imprinted nanocomposite consisting of GNP and SnO_2 NPs (abbreviated as GS) as the sensing layer in a non-enzymatic fiber optic SPR-based biosensor for dopamine detection. An optical fiber's silver-coated, unclad core is covered with a coating of (GS-DA) molecularly imprinted polymer to create the sensor probe. Wavelength interrogation is used to describe the sensor. For optimal sensor performance, the probe's manufacturing parameters are optimized.

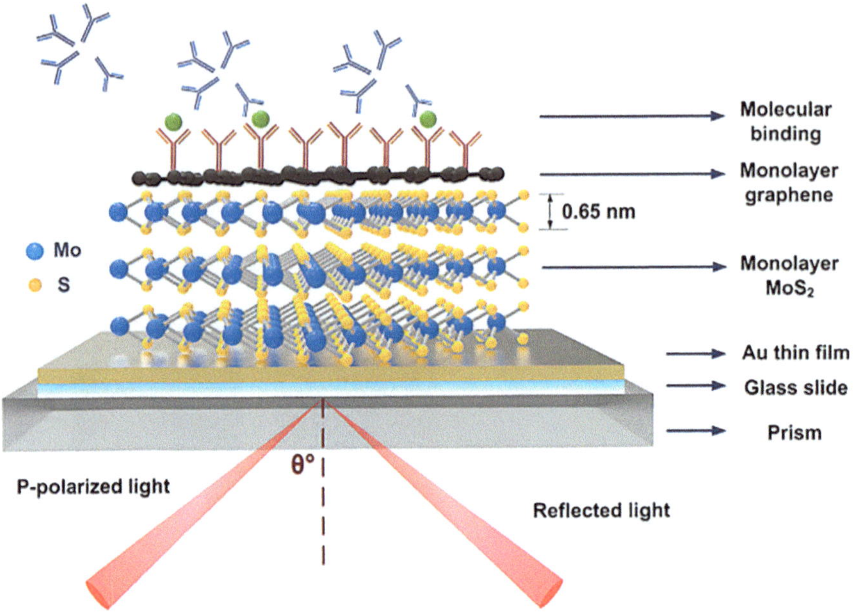

Fig. 9 Multilayer Prism configured graphene–MoS2 hybrid SPR structures [48]

Other potential sensor designs contrast with the optimized sensor. Additionally, the sensor's selectivity is assessed when DA and interferants found in the human body are present.

3 Sensing Recognition

In today's environment, optical sensors and sensing technologies are becoming increasingly significant. Optics may be employed in many different ways to create small, low-cost, stand-off sensing with extraordinary sensitivity and selectivity, from micro-probes to massive devices used in such varied domains as medical diagnostics, defense, and industrial and environmental condition monitoring. In this section, sensing applications have been categorized into three sections based on optical fiber-based sensing, laser based sensing, and biomedical optical sensing. Optical fiber-based sensing covers structure and application of various optical fibers, specialty fibers, macro and nano structures. Laser based sensing includes mid infrared and Terahertz sensing. Biomedical optical sensing covers diagnostic assays, nano plasmonic biological imaging, and spectral histopathology.

3.1 Optical Fiber Based Sensing

Sensing devices based on optical fibers that use optical fibers to track physical quantities and provide an electrical output. Like other optical sensors, optical fiber sensors offer superior performance, resistance to electromagnetic interference, and the capacity to operate in challenging conditions. As discussed earlier, here sensing may be distributed and quasi-distributed for continuous and independent information regarding environmental condition or point types for the discrete point information. Some of the more common physical parameters which can be sensed in distributed sensing are temperature, vibration, strain and multiplexing of signals. In addition, the category of point sensors is quite varied and includes sensors that rely on different kinds of interferometers, like hydrophones and gyroscopes, as well as sensors that depend on the complicated refractive index in bio and chemical sensors or on the change of the polarization state, like current sensors.

It is logical to anticipate more effects in sensing given the growing interest in multimode and multicore fibers for telecom applications [14] for distributed sensing [50]. Because of the long lifespan, high accuracy, small size, quick reaction time, and most importantly, multiplexing, FBGs have continuously drawn a lot of interest. They have found important uses in the power sector, the oil and gas market, and structural health monitoring for civil and aerospace engineering. The relatively high cost of each individual grating, which renders the entire system uncompetitive for applications that frequently replace transductors, and the detection system's cost, which is much higher than that of competing electronic sensors, are the two main issues that FBGs will continue to face over the next ten years. From using simple core/cladding structures and simple materials to introducing wavelength-scale microstructure features to, most recently, integrating multi-material, multi-structure, and multi-functionality in the fiberized platform, specialty fibers [51] have historically undergone a spiral rising evolution. Because of the interactions between the external fields and the propagating optical modes throughout the fiber core, the fiber output signals contain information in the spatial, temporal, frequency, polarization, or phase domains. Because of their specially designed fiber material or materials and structures that improve the interaction between the fiber's internal optical modes and external fields, specialty fibers have an advantage over regular optical fibers. Various applications of specialty fibers in sensing applications include the area of emerging technologies like energy, medical diagnosis, navigation, autonomous systems and robotics. Researchers have used the distributed acoustic sensing system, which is based on the specialty scattering-enhanced optical fiber, for practical engineering applications in a variety of domains, such as hydroacoustic exploration, seismic profile detection, structural health monitoring, and geological and resource exploration [52] as indicated in Fig. 10. In many industrial applications, fluid vortices are employed to efficiently transfer mass, heat, and momentum. It is a difficult undertaking that calls for advanced computational and experimental methods to estimate the vortex convection velocity and frequency. Recently in [53], photonic crystal fiber interferometric is used for design of optical flowmeter in vortex detection. It has been demonstrated

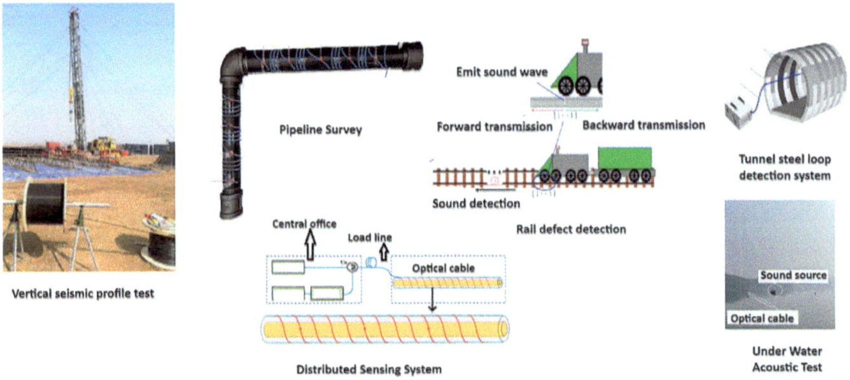

Fig. 10 Sensing area of applications of optical fiber sensors

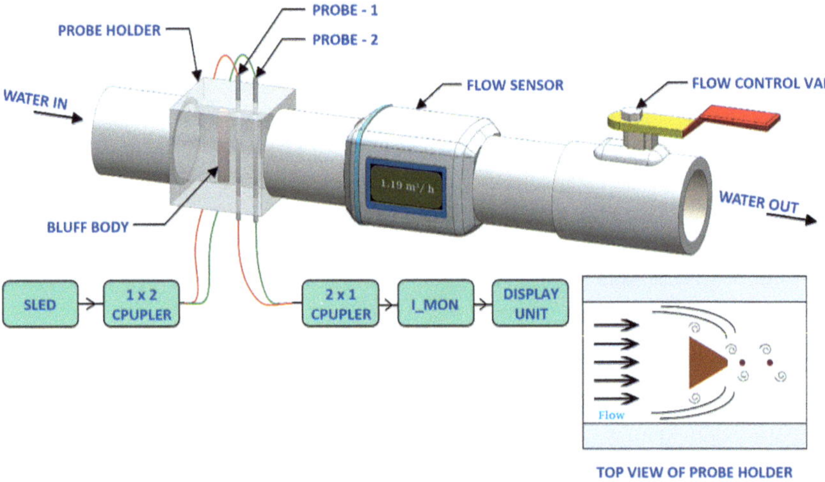

Fig. 11 Schematic diagram for photonic crystal fiber based optical flowmeter

that this method, which is based on the examination of the temporal correlation of the vortex shedding frequency, yields precise and trustworthy measurements of vortex convection velocity under various flow circumstances. Schematic diagram for the detection vortices is shown in Fig. 11. As shown in Fig. 11, water inlet is first passed through probe holder containing trapezoidal bluff body. The vortex generated through bluff body is sense by parallel combination of photonic crystal fiber probes and using correlation techniques convection velocity of fluid is estimated.

There is a growing need for high performance sensors with faster response, smaller footprints, higher sensitivity, and lower power consumption to investigate the boundaries of force detection or molecular interactions in order to comprehend the foundations of physics, biology, and medical science. This has led to significant efforts for micro and nano-engineered optical sensors as a result of the quick advancements in nanotechnology and flexible opto-electronics. In this regard micro/nano fiber (MNF) are widely investigated in the area of fluid parameters like optofluidic absorbance detection [54], multifunctional sensing in microfluidic chips [55], and phases of materials in liquid, gas, or chemicals [56]. In [54], by measuring the absorbance of Fecl3, the idea of using the coiled optical nanofiber for durable and extremely sensitive optofluidic absorbance detection is explored. The integrated MNF, on the other hand, is a multipurpose detector for tracking the microflow state in real time, which is crucial for an optofluidic system's feedback control [55].

3.2 *Laser and Optical Remote Sensing*

As we go into the present era, lasers and optical distant sensing are closely related and probably always will be in order to support further developments in laser technology for applications using Lightwave sensing. Ellipsometers, imaging, spectroscopy, vibrometry, interferometry, frequency comb, lidar, and quantum are among the uses of laser-based sensing. Laser design is constantly at the forefront for advancements in sensitivity, resolution, responsivity, signal to noise ratio, and accuracy, regardless of whether it is stationary or on a moving platform, on the ground or in space, or requires a lot of power or only a few photons [57].

Environmental monitoring and medical diagnosis depend on the real-time detection of biological, chemical, and physical data. To satisfy the demands of the information age, high performance sensors with several functions, even multifunction—must be developed immediately. Laser based sensing is widely used in the area of environmental monitoring, structural monitoring, or biosensing as depicted in Fig. 12 [58].

Mid infra-red sensing explores remote sensing of gases in infra-red wavelength regions like greenhouse gases carbon dioxide and methane in environmental monitoring, hazardous gas detection of combustible gases propane and methane, gases involved in industrial processing, and detection of low concentration gases. By using molecular vibrations triggered by infrared light, it detects and measures gases in the mid-IR region using their distinct absorption spectra. Hazardous gas detection and environmental monitoring are just a few uses for this technology [59–61]. The presence of surface plasmon polaritons in micro-scale structures of doped silicon has created novel and inventive opportunities for photonics, imaging, and sensing. A doped Si plasmonic sensor that works with CMOS is suggested and studied. By adjusting the carrier density and dopant concentration, the plasmon resonance may be adjusted. For the detection of methane and ethane gases, a Si platform-based gas sensor based on a coupled-ring resonator is suggested that is compatible with CMOS technology [60]. The suggested device uses doped silicon materials to guide the light in a plasmonic slot arrangement. The suggested sensor is made of a doped Si wafer on a silicon dioxide substrate, and the wafer is etched to create two suspended non-concentric ring resonators. The device construction is seen in Fig. 13. The radius of

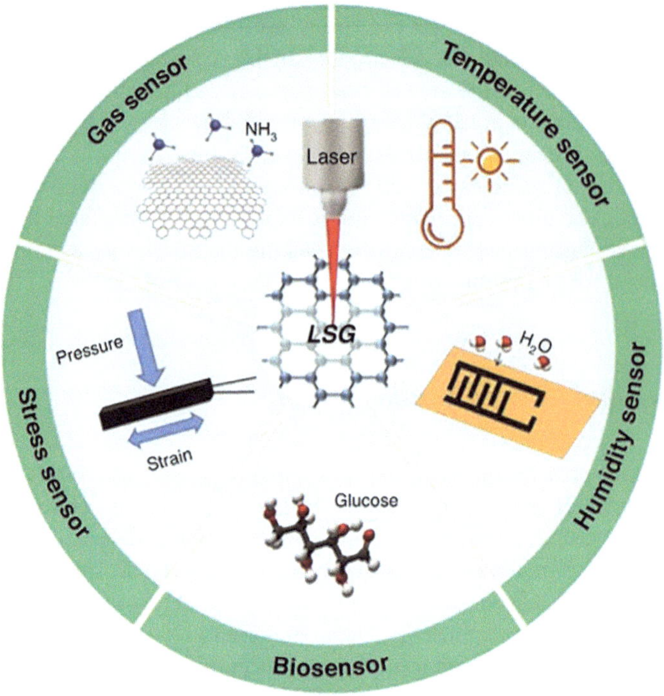

Fig. 12 Laser based broad sensing areas [58]

the smaller ring is R2, while the radius of the bigger ring is R1. Sharp resonance peaks are produced by the splitting of the resonance frequency in a linked resonator design, increasing the system's sensitivity and selectivity.

A smart, inexpensive, multi-gas sensing system that consists of a machine learning algorithm and a mid-infrared micro-spectrometer is demonstrated in [61]. The schematic diagram of meta surface mid infra-red sensing system is depicted in Fig. 14 [61]. Here, the infrared emitter and mid-infra-red meta-surface micro-spectrometer are positioned on each side of a specially designed gas cell. A thermoelectric temperature stabilizer supports the mid-infra-red meta-surface micro-spectrometer. The gas to be detected is poured into the gas cell. The machine learning system is taught to anticipate the gases present by analyzing the micro-spectrometer data. The device has 100% accuracy in detecting the greenhouse gases methane and carbon dioxide at concentrations between 10 and 100%. Additionally, it has a 98.4% accuracy rate in detecting dangerous gases at low concentrations. A concentration of 100 parts per million of ammonia is detectable. Furthermore, 200 ppm, the permitted exposure limit for methyl-ethyl-ketone, may be discovered; this quantity is regarded as low and nonhazardous.

Terahertz sensing in the area of imaging and spectroscopy gives novel technology without any destruction, ionization, and labelling. THz frequency range includes

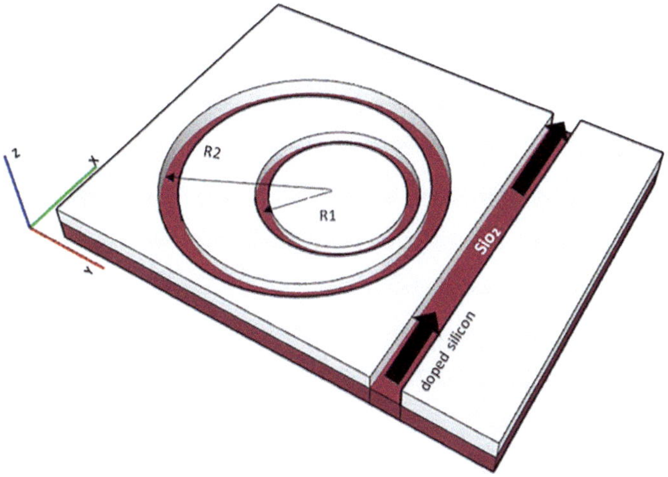

Fig. 13 Mid infra-red sensor device based on non-concentric ring resonators for ethane and methane gas detection [60]

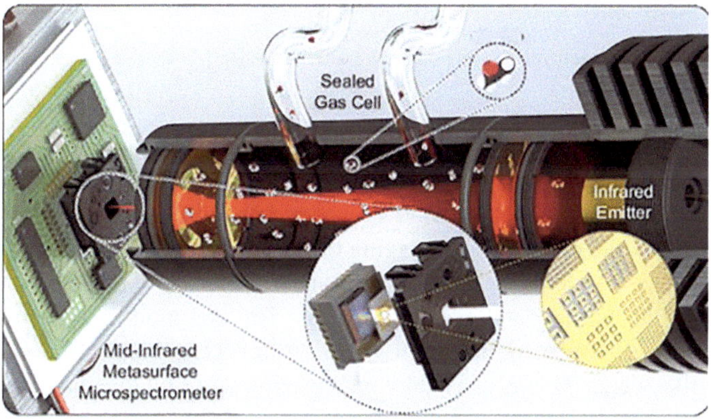

Fig. 14 Mid infra-red multi-gas sensing system [61]

some hundreds of GHz to lower end of mid infra-red region. THz sensing applications are very useful in scientific tools and now a days entering into the commercial use like telecommunication, astronomy, security and defense [62], healthcare industries [63], and non-destructive testing [62, 64]. Frequency spectrum of Terahertz and their sensing applications have been depicted in Fig. 15 [63]. The low photon energy of the radiation, which ranges from 0.4 to 41 meV, makes THz wave-based technology non-invasive and non-ionizing, which makes it appealing for biological applications. Together with spectral fingerprinting, these special features of THz technology allow for accurate and exact label-free features, real-time image-guided cancer tissue removal, and early, more accurate detection.

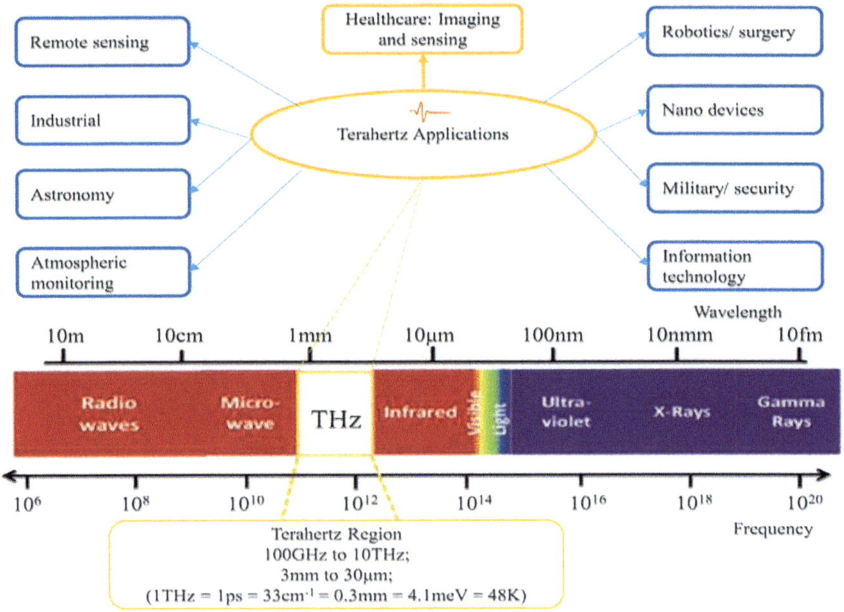

Fig. 15 THz sensing spectrum and its potential applications [63]

Being non-ionizing means that there is no appreciable tissue damage, and that the radiation does not ionize the biomolecules in the target tissue. Regular screening, patient monitoring, and real-time in vivo cancer detection and diagnosis may all be made possible by the non-invasive imaging technique [65, 66]. Compared to X-rays, THz waves' phase sensitivity to polar substances, namely, water and bodily fluids offer stronger absorption and greater contrast. The common endogenous marker for tissue contrast is a markedly high sensitivity of THz radiation to polar liquid molecules like water. This is caused by increased vascularity in cancer and tumors, which raises the local concentration of water molecules and increases the blood supply to the infected tissue [67]. Features associated with THz sensing give the following sensing recognition and research opportunities:

- Open-source THz projects, femtosecond lasers, endoscopic probes, and small and portable sources and detectors
- Beam shaping, nanostructures, machine learning methods, and near field THz imaging and sensing
- Creation of reliable algorithms and THz spectral and imaging databases
- Focused research on potential THz radiation thermal effects
- Building of THz contrast agents, THz endoscopy, and adaptable THz devices that can adapt to surfaces automatically
- High power THz sources, metamaterials, waveguides and surface plasmons, penetration enhancement agents.

3.3 Biomedical Optical Sensing

Clinicians around the world are concentrating on the use of advanced biomedical instrumentation and sensors as essential and efficient tools for patient diagnosis, monitoring, treatment, and overall care due to the growing global population's need for healthcare and the demand for ever-more-sophisticated diagnostic tools. Optical Coherence Tomography probes, force-and shape-sensing catheters in robotic surgery, intra-aortic pressure probes, and temperature monitoring for thermal-based therapies for localized tumors are just a few examples of the wide range of invasive and non-invasive applications in the life sciences, clinical research, medical monitoring, and diagnostics that are best suited for fiber optic-based sensors [68]. Coherence in Optics Using light waves, tomography is a non-invasive imaging method that produces high-resolution, cross-sectional pictures of the retina and other eye components. By creating a cross-sectional map of the retina using the interferometry concept, ophthalmologists may see the many layers and structures in great detail.

Biomedical sensing can be observed through four major areas: biological, physical, chemical, and imaging. Biological sensing involves qualitative and quantitative recognition reactions of biochemical molecules like antigen or ligand and enzyme in the body. Physical sensing finds physiological parameters like heart rate, blood pressure, muscles position, respiration and blood flow rate. Sensing of metabolic parameters and chemical compounds come under the category of chemical sensing through spectroscopic and indicator methods. Internal monitoring of body like endoscopic and photoacoustic observation are covered by imaging sensing. Biomedical optical sensors application in respiratory rate monitoring is shown on Fig. 16 for physical sensing [69]. It is found that Fabry Perot based interferometry or fiber Bragg grating based sensors are more involved in physical monitoring. Such research shows that optical fiber-based sensor reflectance techniques are as accurate as capnometry and may be appropriate for certain oxygen delivery systems, such as portable, continuous respiratory rate monitoring [70]. Temperature sensors or thermometry in physical monitoring helps to investigate many medical processes like photothermal spectroscopy, hot-wire anemometry, intravascular flow measurement, and radiology [71–73]. Chemical or biological parameters monitoring are popularly implemented through fluorescent or surface plasmon based sensors. Direct interaction with the measurand is necessary for optical-fiber biochemical parameter sensing. Traditionally, tapered fibers, diffraction gratings at the fiber's end-face, fiber Bragg gratings at the fiber's end and/or inside the tapered area, and cladding mode excitation have all been employed. Then, surface plasmon resonance effects, made possible by metal coatings like Au or Ag on the fibers in various shapes (such U-shape, tapered, on or at the fiber's end-face), achieved an order of magnitude improvement in the sensitivity of several parameter measurements. Further information on the application of plasmonic in sensing can be found in [74]. Some of the most common biochemical sensing parameters are monitoring pH level, oxygen dissolved, and minerals in the body. Diseases or malfunctioning bodily components can be the cause or result of pH variations. As a result, pH is a crucial quantity for biological systems' sensing.

Among other things, it is preferable to determine pH from ex vivo sources (like urine and saliva) and vivo sources (like intracellular and extracellular pH). Dyes or fluorophores placed in U-shaped [74], on the fiber tip [75], or tapered optical fibers [76] are the most widely used techniques for measuring pH using optical fibers in monitoring fluorescence, absorption, and lifespan. By tapering the fibers, these sensors usually combine the principles of fluorescence with a modification of the evanescent field [75]. In addition to their typical advantages (miniaturization and tolerance to electromagnetic fields), optical fiber sensors are very sensitive. Fiber optic-based optical probes have been used in biological applications for a long time. A versatile way to connect the spectroscope and the sample is by fiber optics. The application often measures a turbid media (such as tissue) in this situation. using Raman, fluorescence, amplitude, and polarized reflectance methods. In addition to fiber-optical probes, improved probes based on gratings have been made possible by developments in micro- and nanofabrication. The fluorescence intensity and spectrum shape of chromophores, which are known to exist in several human tissues, are highly influenced by the properties of the excitation light source. Several clinical investigations have also demonstrated that fluorescence spectroscopy may be used to identify abnormalities in human tissues, including those in the brain, head, breast, skin, and gastrointestinal system.

4 Technology Advancement and Challenges

The services and industrial sectors are undergoing radical change as a result of Industry 4.0 and its primary supporting information and communication technologies. This is particularly true for the health sector, where Big Data, Cloud and fog computing, and the Internet of Things are transforming eHealth and its whole ecosystem, bringing it closer to healthcare 4.0 [77]. Healthcare is under increasing strain due to the world's population expansion, growing demands for better treatments, and an overall improvement in quality of life. As a result, one of the biggest social and economic issues facing the globe today is healthcare, which calls for innovative and cutting-edge scientific and technological solutions.

In addition to the technology at play, development plans also cover regulatory frameworks, enterprise management and work organization, training, and distribution [78, 79]. From a purely technical perspective, this so-called fourth industrial revolution is primarily founded on the idea of Cyber-Physical Systems, or CPS (integration of computing, communication, and control). It also heavily depends on three categories of technologies, or pillars: big data analytics, cloud computing, and the Internet of Things (IoT) paradigm. Optical fiber sensors have significant contributions in technology advancement of healthcare 4.0 as indicated in Fig. 17. Key emerging technologies include smart sensors for wearable technologies embedded in daily use devices like hand watch, head worn, smartphone, etc. Artificial intelligence (AI) and optical sensors are being combined more and more to provide sophisticated

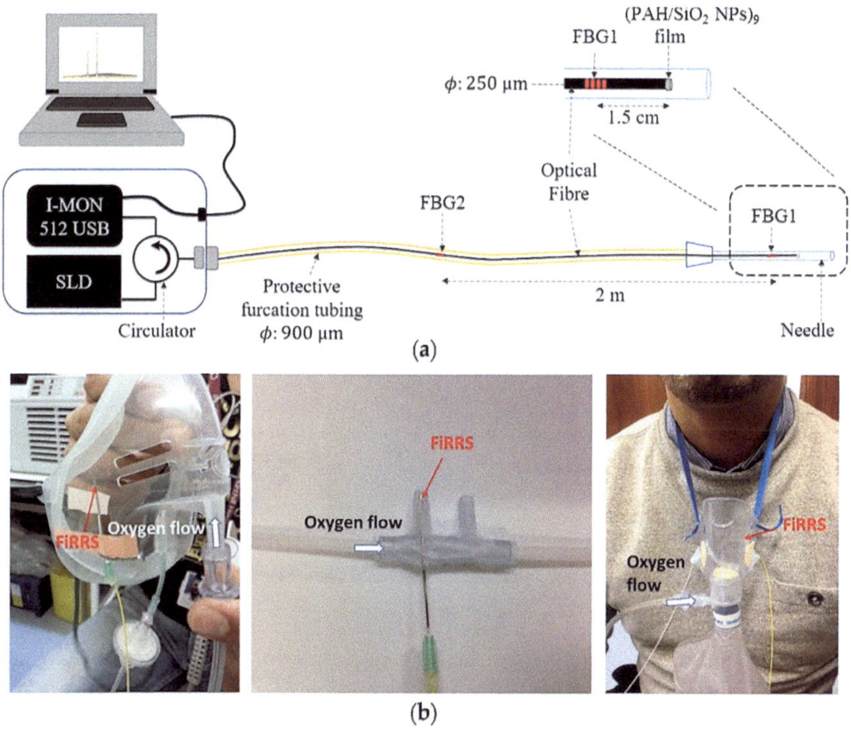

Fig. 16 Biomedical optical sensor application for respiratory rate monitoring. **a** Optical sensor system. **b** Integration of sensing probes at various places of human body [69]

sensing systems. By enabling them to analyse vast volumes of data, identify irregularities, and make predictions more quickly and accurately than with conventional techniques, artificial intelligence (AI) improves optical sensor systems. The robots' flawless functioning is ensured by sensors such as ultrasonic rangefinders and LIDAR (Light Detection and Ranging), which aid in obstacle avoidance and navigation. By making it possible to identify a variety of structural variations and other genetic abnormalities more quickly, accurately, and affordably than with more conventional techniques like sequencing or cytogenetics, optical sensors—in particular, optical genome mapping, or OGM—are transforming the field of genomics. By mapping DNA at precise sites using fluorescent markers, OGM makes it possible to identify large-scale structural variations such as deletions, insertions, duplications, and inversions.

In cloud computing, optical sensors are essential for data transfer and application monitoring. For instance, optical fibers are necessary for fast data movement between data centers, which makes cloud services and storage more effective. To monitor physical characteristics like temperature, distance, or even human behavior for applications like smart homes, optical sensors may also be included into smart

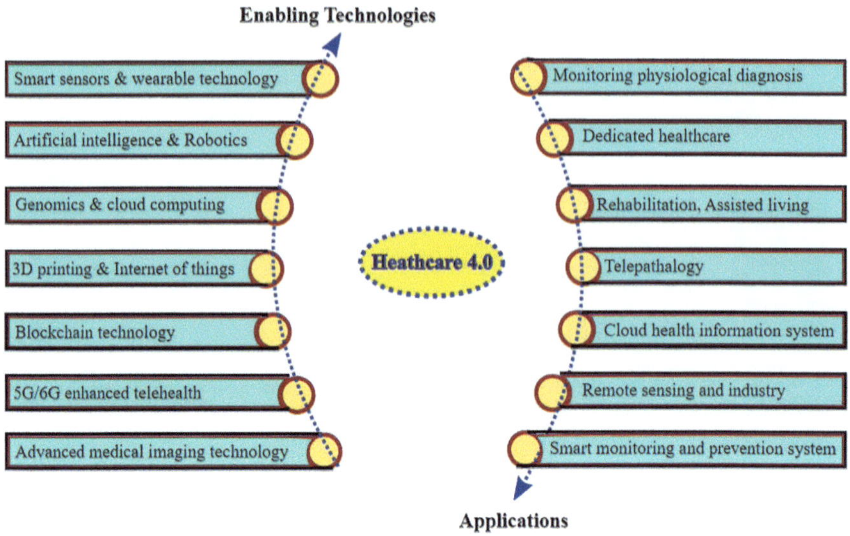

Fig. 17 Enabling technologies and applications for healthcare 4.0

sensor networks. Stereolithography, melting, and sintering are the three main categories of 3D printing techniques. Sintering is a process that produces high-resolution objects by heating a material without melting it. Because they make it possible to detect and analyze light, including its intensity, wavelength, and frequency, optical sensors are essential to the Internet of Things (IoT). These sensors then translate this data into digital signals that IoT devices can interpret and respond to. From recognizing things without touch to monitoring ambient light for automated brightness adjustment, these sensors are employed in many different applications. Blockchain technology with optical sensors may be used to build more transparent and safe systems for a range of uses. Data may be gathered by optical sensors and then stored in a safe, unchangeable ledger using blockchain technology. Applications such as supply chain management, healthcare, and smart cities can benefit greatly from this combination. Healthcare facilities are becoming smart hospitals as a result of the proliferation of devices and technology that provide better, more effective services. 5G makes it possible to combine many incompatible IoT radio networks into a single network and offers seamless connection, doing away with the need to switch between mobile and in-building Wi-Fi networks. When the technology is fully utilized, patients will have increased access to experts who would not otherwise be available and receive treatment sooner. Without having to be physically there, 5G connectivity can let specialists reach these underprivileged patients [80–82]. With benefits including non-invasive examination, high sensitivity, and miniaturization, optical sensors are becoming more and more important in sophisticated medical imaging. These sensors, especially those that use near-infrared (NIR) light, allow for cell tracking, real-time tissue change monitoring, and precise interior structure visualization. To improve diagnostic capabilities, they can also be used in conjunction with other imaging modalities, such as MRI.

5 Summary

Compared to conventional optical sensors, micro- and nanostructured optical sensors perform better. They offer integration across photonic integrated circuits and sophisticated optical sensing methods. Based on the selection of materials and their physical characteristics, several optical cavity shapes have been developed in recent decades to enhance the sensing capabilities of the sensors. Sensor applications in healthcare diagnostics, defense, remote sensing, and industrial settings are defined by the selection of materials, sensing characteristics, size, loss, robustness, selectivity, and sensitivity. In this chapter, types of sensors based on their implementations are discussed with the suitable choices of fabrication platforms. Various kinds of fiber properties are discussed with the corresponding compatible fabrication techniques. Further kinds of sensing in all possible areas are discussed along with the development and challenges in the next generation of optical sensor devices, and their applications are discussed.

References

1. T.M. Daley, B.M. Freifeld, J. Ajo-Franklin, S. Dou, R. Pevzner, V. Shulakova, S. Kashikar, D.E. Miller, J. Goetz, J. Henninges, S. Lueth, Field testing of fiber-optic distributed acoustic sensing (DAS) for subsurface seismic monitoring. Lead. Edge **32**(6), 699–706 (2013)
2. A. Mateeva, J. Lopez, H. Potters, J. Mestayer, B. Cox, D. Kiyashchenko, P. Wills, S. Grandi, K. Hornman, B. Kuvshinov, W. Berlang, Distributed acoustic sensing for reservoir monitoring with vertical seismic profiling. Geophys. Prospect. **62**(4-Vertical Seismic Profiling and Microseismicity Frontiers) (2014), pp. 679–692
3. F. Tanimola, D. Hill, Distributed fibre optic sensors for pipeline protection. J. Nat. Gas Sci. Eng. **1**(4–5), 134–143 (2009)
4. G.A. Wellbrock, T.J. Xia, M.F. Huang, Y. Chen, M. Salemi, Y.K. Huang, P. Ji, E. Ip, T. Wang, March. First field trial of sensing vehicle speed, density, and road conditions by using fiber carrying high speed data, in *2019 Optical Fiber Communications Conference and Exhibition (OFC)* (IEEE, 2019), pp. 1–3
5. N.J. Lindsey, E.R. Martin, D.S. Dreger, B. Freifeld, S. Cole, S.R. James, B.L. Biondi, J.B. Ajo-Franklin, Fiber-optic network observations of earthquake wavefields. Geophys. Res. Lett. **44**(23), 11–792 (2017)
6. P. Jousset, T. Reinsch, T. Ryberg, H. Blanck, A. Clarke, R. Aghayev, G.P. Hersir, J. Henninges, M. Weber, C.M. Krawczyk, Dynamic strain determination using fibre-optic cables allows imaging of seismological and structural features. Nat. Commun. **9**(1), 2509 (2018)
7. J.A. van Herwaarden, M.M. Jansen, E.J.P. Vonken, T. Bloemert-Tuin, R.W. Bullens, G.J. de Borst, C.E. Hazenberg, First in human clinical feasibility study of endovascular navigation with Fiber Optic RealShape (FORS) technology. Eur. J. Vasc. Endovasc. Surg. **61**(2), 317–325 (2021)

8. R. Ahmad, P.S. Westbrook, W. Ko, K.S. Feder, Probing micron-scale distributed contortions via a twisted multicore optical fiber. APL Photonics **4**(6) (2019)
9. P.S. Westbrook, T. Kremp, B. Zhu, W. Ko, Z. Shi, K.S. Feder, Enhanced backscatter fibers for sensing in telecom networks. J. Lightwave Technol. **41**(3), 1010–1016 (2023)
10. G.A. Wellbrock, T.J. Xia, M.F. Huang, J. Fang, Y. Chen, C. Narisetty, D. Peterson, J.M. Moore, A. Scarpaci, P. Westbrook, J. Li, Perimeter intrusion detection with backscattering enhanced fiber using telecom cables as sensing backhaul, in *Optical Fiber Communication Conference* (Optica Publishing Group, 2022), pp. M2F-5
11. A.H. Hartog, *An Introduction to Distributed Optical Fibre Sensors* (CRC Press, 2017)
12. H. Bi, P. Cw, C. Le, Preliminary report on a long fiberscope for examination of stomach and duodenum. Med. Bull. (Ann Arbor, Mich.) **23**(5), 178–180 (1957)
13. Z.Y. Zhao, M. Tang, C. Lu, Distributed multicore fiber sensors. Opto-Electron. Adv. **3**(2), 190024–190031 (2020)
14. X. Sun, Z. Yang, X. Hong, S. Zaslawski, S. Wang, M.A. Soto, X. Gao, J. Wu, L. Thévenaz, Genetic-optimised aperiodic code for distributed optical fibre sensors. Nat. Commun. **11**(1), 5774 (2020)
15. A. Donko, R. Sandoghchi, A. Masoudi, M. Beresna, G. Brambilla, Low-loss micro-machined fiber with Rayleigh backscattering enhanced by two orders of magnitude, in *Optical Fiber Sensors* (Optica Publishing Group, 2018), p. WF75
16. V. Chaudhary, V. Khanna, H.T. Ahmed Awan, K. Singh, M. Khalid, Y.K. Mishra, S. Bhansali, C.Z. Li, A. Kaushik, Towards hospital-on-chip supported by 2D MXenes-based 5th generation intelligent biosensors. Biosens. Bioelectron. **220** (2023)
17. M.L. Torres-Mapa, M. Singh, O. Simon, J.L. Mapa, M. Machida, A. Günther, B. Roth, D. Heinemann, M. Terakawa, A. Heisterkamp, Fabrication of a monolithic lab-on-a-chip platform with integrated hydrogel waveguides for chemical sensing. Sens. (Switz.) **19** (2019)
18. G. Antonelli, A. Mencattini, M. Massimiani, V. Lacconi, J. Filippi, M. Losardo, M. D'Orazio, P. Casti, M. Bragaglia, G. Curci, F. Nanni, Development of integrated optical biosensors based on low-cost stereolithography fabrication and multispectral signature for Lab-On-Chip applications. Sens. Actuators, B Chem. **401**, 135009 (2024)
19. M. Badiola-Mateos, D. Di Giuseppe, R. Paoli, M.J. Lopez-Martinez, A. Mencattini, J. Samitier, E. Martinelli, A novel multi-frequency trans-endothelial electrical resistance (MTEER) sensor array to monitor blood-brain barrier integrity. Sens. Actuators, B Chem. **334**, 129599 (2021)
20. L. Ferrari, L. Rovati, P. Fabbri, F. Pilati, Disposable fluorescence optical pH sensor for near neutral solutions. Sensors **13**(1), 484–499 (2012)
21. N. Salehifar, R.E. Gerald, J. Huang, Fabrication and analysis of a gas sensor utilizing a fabry-perot interferometer with varied thicknesses of ZIF-8 coating on end-faced single-mode optical fiber. IEEE Sens. J. (2024)
22. B.D. Gupta, A. Pathak, A.M. Shrivastav, Optical biomedical diagnostics using lab-on-fiber technology: a review, in *Photonics*, vol. 9, No. 2 (MDPI, 2022), p. 86
23. A. Ricciardi, A. Crescitelli, P. Vaiano, G. Quero, M. Consales, M. Pisco, E. Esposito, A. Cusano, Lab-on-fiber technology: a new vision for chemical and biological sensing. Analyst **140**(24), 8068–8079 (2015)
24. P. Vaiano, B. Carotenuto, M. Pisco, A. Ricciardi, G. Quero, M. Consales, A. Crescitelli, E. Esposito, A. Cusano, Lab on fiber technology for biological sensing applications. Laser Photon. Rev. **10**(6), 922–961 (2016)
25. M. Yao, X. Ouyang, J. Wu, A.P. Zhang, H.Y. Tam, P.K.A. Wai, Optical fiber-tip sensors based on in-situ μ-printed polymer suspended-microbeams. Sensors **18**(6), 1825 (2018)
26. J. Wu, M.J. Yin, K. Seefeldt, A. Dani, R. Guterman, J. Yuan, A.P. Zhang, H.Y. Tam, In situ μ-printed optical fiber-tip CO_2 sensor using a photocross linkable poly (ionic liquid). Sens. Actuators B Chem. **259**, 833–839 (2018)
27. M. Smietana, W.J. Bock, P. Mikulic, J. Chen, Pressure sensing in high-refractive-index liquids using long-period gratings nanocoated with silicon nitride. Sensors **10**(12), 11301–11310 (2010)

28. D. Monzón-Hernández, D. Luna-Moreno, D. Martínez-Escobar, Fast response fiber optic hydrogen sensor based on palladium and gold nano-layers. Sens. Actuators, B Chem. **136**(2), 562–566 (2009)
29. K. Vanmol, T. Baghdasaryan, N. Vermeulen, K. Saurav, J. Watté, H. Thienpont, J. Van Erps, 3D direct laser writing of microstructured optical fiber tapers on single-mode fibers for mode-field conversion. Opt. Express **28**(24), 36147–36158 (2020)
30. J.B. Kim, K.H. Jeong, Batch fabrication of functional optical elements on a fiber facet using DMD based maskless lithography. Opt. Express **25**(14), 16854–16859 (2017)
31. B.H. Lee, Y.H. Kim, K.S. Park, J.B. Eom, M.J. Kim, B.S. Rho, H.Y. Choi, Interferometric fiber optic sensors. Sensors **12**(3), 2467–2486 (2012)
32. L.H. Chen, X.M. Ang, C.C. Chan, M. Shaillender, B. Neu, W.C. Wong, P. Zu, K.C. Leong, Layer-by-layer (chitosan/polystyrene sulfonate) membrane-based Fabry-Perot interferometric fiber optic biosensor. IEEE J. Sel. Top. Quantum Electron. **18**(4), 1457–1464 (2012)
33. J. Wallner, G. Lhota, D. Jeschek, A. Mader, K. Vorauer-Uhl, Application of Bio-Layer Interferometry for the analysis of protein/liposome interactions. J. Pharm. Biomed. Anal. **72**, 150–154 (2013)
34. X. Zhou, F. Ma, H. Ling, B. Yu, W. Peng, Q. Yu, A compact hydrogen sensor based on the fiber-optic Fabry-Perot interferometer. Opt. Laser Technol. **124**, 105995 (2020)
35. S. Wu, G. Yan, B. Zhou, E.H. Lee, S. He, Open-cavity Fabry-Perot interferometer based on etched side-hole fiber for microfluidic sensing. IEEE Photon. Technol. Lett. **27**(17), 1813–1816 (2015)
36. A. Iele, A. Ricciardi, C. Pecorella, A. Cirillo, F. Ficuciello, B. Siciliano, R. La Rocca, V. Mirone, M. Consales, A. Cusano, Miniaturized optical fiber probe for prostate cancer screening. Biomed. Opt. Express **12**(9), 5691–5703 (2021)
37. F. Baldini, M. Brenci, F. Chiavaioli, A. Giannetti, C. Trono, Optical fibre gratings as tools for chemical and biochemical sensing. Anal. Bioanal. Chem. **402**, 109–116 (2012)
38. F. Chiavaioli, F. Baldini, S. Tombelli, C. Trono, A. Giannetti, Biosensing with optical fiber gratings. Nanophotonics **6**(4), 663–679 (2017)
39. Q.D. Huang, C.H. Lv, X.L. Yuan, M. He, J.P. Lai, H. Sun, A novel fluorescent optical fiber sensor for highly selective detection of antibiotic ciprofloxacin based on replaceable molecularly imprinted nanoparticles composite hydrogel detector. Sens. Actuators, B Chem. **328**, 129000 (2021)
40. X. Ma, H. Huo, W. Wang, Y. Tian, N. Wu, C. Guthy, M. Shen, X. Wang, Surface-enhanced Raman scattering sensor on an optical fiber probe fabricated with a femtosecond laser. Sensors **10**(12), 11064–11071 (2010)
41. G. Quero, G. Zito, S. Managò, F. Galeotti, M. Pisco, A.C. De Luca, A. Cusano, Nanosphere lithography on fiber: towards engineered lab-on-fiber SERS optrodes. Sensors **18**(3), 680 (2018)
42. M. Pisco, F. Galeotti, G. Quero, G. Grisci, A. Micco, L.V. Mercaldo, P.D. Veneri, A. Cutolo, A. Cusano, Nanosphere lithography for optical fiber tip nanoprobes. Light Sci. Appl. **6**(5), e16229–e16229 (2017)
43. B.D. Gupta, R. Verma, S.K. Srivastava, *Fiber optic sensors based on plasmonics* (World Scientific, 2015)
44. M. Qi, N.M.Y. Zhang, K. Li, S.C. Tjin, L. Wei, Hybrid plasmonic fiber-optic sensors. Sensors **20**(11), 3266 (2020)
45. Y. Zhao, R.J. Tong, F. Xia, Y. Peng, Current status of optical fiber biosensor based on surface plasmon resonance. Biosens. Bioelectron. **142**, 111505 (2019)
46. O. Salihoglu, S. Balci, C. Kocabas, Plasmon-polaritons on graphene-metal surface and their use in biosensors. Appl. Phys. Lett. **100**(21) (2012)
47. Y.V. Stebunov, O.A. Aftenieva, A.V. Arsenin, V.S. Volkov, Highly sensitive and selective sensor chips with graphene-oxide linking layer. ACS Appl. Mater. Interfaces. **7**(39), 21727–21734 (2015)
48. S. Zeng, S. Hu, J. Xia, T. Anderson, X.Q. Dinh, X.M. Meng, P. Coquet, K.T. Yong, Graphene–MoS2 hybrid nanostructures enhanced surface plasmon resonance biosensors. Sens. Actuators, B Chem. **207**, 801–810 (2015)

49. S. Sharma, B.D. Gupta, Surface plasmon resonance based highly selective fiber optic dopamine sensor fabricated using molecular imprinted GNP/SnO_2 nanocomposite. J. Lightwave Technol. **36**(24), 5956–5962 (2018)
50. Y. Dang, Z. Zhao, X. Wang, R. Liao, C. Lu, Simultaneous distributed vibration and temperature sensing using multicore fiber. IEEE Access **7**, 151818–151826 (2019)
51. P. Russell, Photonic crystal fibers. Science **299**(5605), 358–362 (2003)
52. Y. Sun, H. Li, C. Fan, B. Yan, J. Chen, Z. Yan, Q. Sun, Review of a specialty fiber for distributed acoustic sensing technology, in Photonics, vol. 9, No. 5 (MDPI, 2022), p. 277
53. A.K. Maurya, R. Kumar, V. Arumuru, R. Jha, An all-optical system for transit time estimation in fluids using single source and detector. IEEE Trans. Instrum. Meas. **72**, 1–8 (2023)
54. H. Mei, J. Pan, Z. Zhang, L. Zhang, L. Tong, Coiled optical nanofiber for optofluidic absorbance detection. Acs Sensors **4**(9), 2267–2271 (2019)
55. Z. Zhang, J. Pan, Y. Tang, Y. Xu, L. Zhang, Y. Gong, L. Tong, Optical micro/nanofibre embedded soft film enables multifunctional flow sensing in microfluidic chips. Lab Chip **20**(14), 2572–2579 (2020)
56. X. Jiang, A.J. Qavi, S.H. Huang, L. Yang, Whispering-gallery sensors. Matter **3**(2), 371–392 (2020)
57. M. Livrozet, B. Gronloh, H. Faidel, J. Luttmann, D. Hoffmann, Optical and optomechanical design of the MERLIN laser optical bench, in *International Conference on Space Optics—ICSO 2020*, vol. 11852 (SPIE, 2021), pp. 1176–1189
58. K.Y. Lau, J. Qiu, Broad applications of sensors based on laser-scribed graphene. Light Sci. Appl. **12**(1), 168 (2023)
59. A. Lambrecht, K. Schmitt, Mid-infrared gas-sensing systems and applications, in *Mid-infrared Optoelectronics* (Woodhead Publishing, 2020), pp. 661–715
60. S. Shafaay, S. Mohamed, M. Swillam, Mid-Infrared gas sensing based on electromagnetically induced transparency in coupled plasmonic resonators. Sensors **23**(22), 9220 (2023)
61. J. Meng, S. Balendhran, Y. Sabri, S.K. Bhargava, K.B. Crozier, Smart mid-infrared metasurface microspectrometer gas sensing system. Microsyst. Nanoeng. **10**(1), 74 (2024)
62. M. Strąg, W. Świderski, Non-destructive inspection of military-designated composite materials with the use of Terahertz imaging. Compos. Struct. **306**, 116588 (2023)
63. M. Gezimati, G. Singh, Terahertz imaging and sensing for healthcare: Current status and future perspectives. IEEE Access **11**, 18590–18619 (2023)
64. F. Ellrich, M. Bauer, N. Schreiner, A. Keil, T. Pfeiffer, J. Klier, S. Weber, J. Jonuscheit, F. Friederich, D. Molter, Terahertz quality inspection for automotive and aviation industries. J. Infrared Millimeter Terahertz Waves **41**, 470–489 (2020)
65. X. Yang, X. Zhao, K. Yang, Y. Liu, Y. Liu, W. Fu, Y. Luo, Biomedical applications of terahertz spectroscopy and imaging. Trends Biotechnol. **34**(10), 810–824 (2016)
66. D.M. Mittleman, Twenty years of terahertz imaging. Opt. Express **26**(8), 9417–9431 (2018)
67. A. Banerjee, S. Vajandar, S. and Basu, T., 2020. Prospects in medical applications of terahertz waves. In *Terahertz Biomedical and Healthcare Technologies* (pp. 225–239). Elsevier.
68. L. Bianchi, R. Mooney, Y. Cornejo, C. Hyde, E. Schena, J.M. Berlin, K. Aboody, P. Saccomandi, Fiber Bragg grating sensors-based thermometry of gold nanorod-enhanced photothermal therapy in tumor model. IEEE Sens. J. **22**(12), 11297–11306 (2021)
69. R. Sinha, F.U. Hernandez, C. He, S. Korposh, R. Correia, A.M. Norris, C. Liu, B.R. Hayes-Gill, S.P. Morgan, Development and validation of a novel fibre-optic respiratory rate sensor (FiRRS) integrated in oxygen delivery devices. J. Phys. D Appl. Phys. **54**(12), 124002 (2021)
70. M. Ochoa, J.F. Algorri, P. Roldán-Varona, L. Rodríguez-Cobo, J.M. López-Higuera, Recent advances in biomedical photonic sensors: a focus on optical-fibre-based sensing. Sensors **21**(19), 6469 (2021)
71. L. Bianchi, S. Korganbayev, A. Orrico, M. De Landro, P. Saccomandi, Quasi-distributed fiber optic sensor-based control system for interstitial laser ablation of tissue: theoretical and experimental investigations. Biomed. Opt. Express **12**(5), 2841–2858 (2021)
72. A. Mohammadi, L. Bianchi, S. Asadi, P. Saccomandi, Measurement of ex vivo liver, brain and pancreas thermal properties as function of temperature. Sensors **21**(12), 4236 (2021)

73. Z. Ashikbayeva, A. Aitkulov, M. Jelbuldina, A. Issatayeva, A. Beisenova, C. Molardi, P. Saccomandi, W. Blanc, V.J. Inglezakis, D. Tosi, Distributed 2D temperature sensing during nanoparticles assisted laser ablation by means of high-scattering fiber sensors. Sci. Rep. **10**(1), 12593 (2020)
74. Z. Tang, D. Gomez, C. He, S. Korposh, S.P. Morgan, R. Correia, B. Hayes-Gill, K. Setchfield, L. Liu, A U-shape fibre-optic pH sensor based on hydrogen bonding of ethyl cellulose with a sol-gel matrix. J. Lightwave Technol. **39**(5), 1557–1564 (2021)
75. M. Pospíšilová, G. Kuncová, J. Trögl, Fiber-optic chemical sensors and fiber-optic bio-sensors. Sensors **15**(10), 25208–25259 (2015)
76. O. Podrazký, J. Mrázek, J. Proboštová, S. Vytykáčová, I. Kašík, Š Pitrová, A.A. Jasim, Ex-vivo measurement of the pH in aqueous humor samples by a tapered fiber-optic sensor. Sensors **21**(15), 5075 (2021)
77. G. Aceto, V. Persico, A. Pescapé, Industry 4.0 and health: Internet of things, big data, and cloud computing for healthcare 4.0. J. Ind. Inf. Integr. **18**, 100129 (2020)
78. J. Chanchaichujit, A. Tan, F. Meng, S. Eaimkhong, J. Chanchaichujit, A. Tan, F. Meng, S. Eaimkhong, An introduction to healthcare 4.0, *Healthcare 4.0: Next Generation Processes with the Latest Technologies* (2019), pp.1–15
79. V.V. Estrela, A.C.B. Monteiro, R.P. França, Y. Iano, A. Khelassi, N. Razmjooy, Health 4.0: applications, management, technologies and review. Med. Technol. J. **2**(4), 262–276 (2018)
80. K. Zhan, RETRACTED: Sports and health big data system based on 5G network and Internet of Things system (2021)
81. E. Skondras, A. Michalas, N. Tsolis, D.D. Vergados, A VHO scheme for supporting healthcare services in 5G vehicular cloud computing systems, in *2018 Wireless Telecommunications Symposium (WTS)* (IEEE, 2018), pp. 1–6
82. K. Khujamatov, E. Reypnazarov, N. Akhmedov, D. Khasanov, Blockchain for 5G Healthcare architecture, in *2020 International Conference on Information Science and Communications Technologies (ICISCT)* (IEEE, 2020), pp. 1–5

Advanced 2D Nanomaterials for Remediation of Environmental Pollutants

Smita Kumari, Ragini Singh, Dharmendra Kumar, and Santosh Kumar

Abstract Water scarcity and rapid air pollution have become a huge challenge with the rising population worldwide. Owing to global industrialization, transportation, and radiation, numerous varieties of pollutants are introduced into the different spheres of the environment. These pollutants cause a severe threat to human health and the environment. Considering these issues as an emergency, many improved remediation approaches have been developed. In recent years, 2D materials have gained popularity among researchers in the sustainable remediation of pollutants that emerged in environments. The excellent properties of 2D materials, such as thin structures, strength, larger surface area, and pore size, make them significant and acceptable candidates in remediation and environmental applications. Presently, the chapter emphasizes the ability of 2D materials in the remediation of water and air pollutants. However, the main aim of the chapter is to discuss the properties and application of materials in the removal of pollutants, for example, heavy metals and dyes, from wastewater for the decontamination and purification, removal of seepage oil, and seawater desalination. Evolving 2D materials such as graphene, boron nitride (BN) nanosheets, molybdenum disulfide (MoS_2) and black phosphorus (BP) are discussed in this chapter for remediation purposes. 2D materials introduced to remove particulate matter (PM) for the purification of indoor and environmental air are highlighted. Overview of the latest progress in air filtration. The advantages of effective advanced technologies over conventional methods in improving the quality of air and water

S. Kumari (✉)
Department of Basic and Applied Sciences, School of Engineering and Sciences, G D Goenka University, Gurugram, Haryana, India
e-mail: smitakumari14@gmail.com

R. Singh
Department of Biotechnology, Koneru Lakshmaiah Education Foundation, Vaddeswaram, Andhra Pradesh, India

D. Kumar
Department of Electronics and Communication Engineering, Guru Ghasidas Vishwavidyalaya Bilaspur, Bilaspur, Chhattisgarh, India

S. Kumar
Centre of Excellence for Nanotechnology, Department of Electronics and Communication, Engineering, Koneru Lakshmaiah Education Foundation, Vaddeswaram, Andhra Pradesh, India

B. Ramachandran et al. (eds.), *Advanced Optical Sensors*, Progress in Optical Science and Photonics 38, https://doi.org/10.1007/978-981-95-1687-2_11

and their limitations are reviewed. Further, we have presented the challenges of 2D materials in environmental applications.

Keywords 2D materials · Nanomaterials · Graphene · Boron nitride (BN) nanosheets · Molybdenum disulfide (MoS_2) · Black phosphorus (BP) · Remediation

1 Introduction of 2D Materials for Remediation Purposes

Over the last few decades, excessive consumption of natural resources, the proliferation of manufacturing industries, and unregulated energy consumption and production have resulted in widespread environmental pollution. These changes pose a threat to flora and fauna, as well as to humans, and as a result, environmental protection is a top priority. Chemical exposures in aqueous and gaseous forms are also a concern [1]. As a result, even a trace amount of these contaminants can cause significant environmental damage and cause human toxicity. Therefore, it is necessary to investigate modern remediation techniques utilizing two-dimensional materials.

Worldwide, water pollution poses serious health risks due to the release of large quantities of toxic heavy metals and dyes that enter into the environment, either accidentally or intentionally. The textile industries are the main cause of discharging a large volume of wastewater containing dyes such as indigoid, xanthene, azo derivatives, phthalocyanine derivatives, and anthraquinone in water sources [2]. These pollutants enter the water sources like rivers, lakes, and seas and consequently destroy the normal ecological conditions, which affect the diverse flora and fauna as well as threaten human life. Additionally, industrial processes such as plating, leather tanning, stainless steel fabrication, and pigment fabrication led to the discharge of heavy metals like arsenic (As), chromium (Cr), lead (Pb), aluminum (Al), etc. in water bodies. Among these heavy metals, few remain in the water in a different state, such as Cr as trivalent (Cr(III)) and hexavalent (Cr(VI)), owing to toxicity in all types of living organisms [3]. These metals exhibit a variety of chemical, physical, and toxicological properties and induce carcinogenesis and mutagenesis in biological cells. Anthropogenic and natural sources like industrial discharge from chimneys, volcanic eruptions, construction sites, and emissions from vehicles emit various gaseous pollutants and specific heavy metals such as Hg, Pb, etc., which lead to contaminating the air. This toxic air inhaled by human beings results in serious lung disease and increased mortality rate, therefore, researchers work towards developing advanced filtration-based green technology such as HEPA filters, activated carbon filters, photocatalytic oxidation-based technology, etc. for the purification of polluted air [4, 5].

Nowadays, oil spillage or leakage that occurs during transportation is a major cause of water pollution, and because of the unique chemical properties of both oil and water, it is very challenging to separate them. The traditional techniques used to date are through a mechanical process and chemicals like detergent and sorbents, which are not effective or eco-friendly, as results completely impact the performance of separation. But researchers have found solutions that are very effective and eco-friendly. Here, we will elaborate on the different applications of potential 2D-based materials such as mesh-based, fabric-based, and fiber-based materials for the separation of oil from oil/water mixtures. Fabric-based materials have provided a very promising approach for application in oil/water separation; for example, cotton fabrics using the dip-coating method consist of ZnO nanoparticles and bare fabric modified with TiO_2 [6]. Later, several combinations and modifications like Cu mesh via oxidation modification method, stainless-steel modification using acid and coating with 2D materials like TiO_2 and graphene are reported to be effective up to 99% in the application of oil/water separation [7–9]. Desalination is another challenge in water purification that causes insufficient availability of clean drinking water. Most of the population are living in the vicinity of the coastline, and they face a major problem of drinking water scarcity. This problem could be solved by desalination of seawater. Monovalent and divalent ions are the composition in seawater that is not easily removed. Several industries used membranes for blocking ions, but the drawbacks are that these cannot block monovalent ions owing to their smaller size than the pores of membrane-based materials used [10]. Moreover, filtration techniques established on thin-film membranes consisting of a polyamide layer along with subnanometer pores are accountable for the elimination or blocking of monovalent ions, and thus the issue gets solved and the technique is commercialized [11]. But due to degradation, low permeability, low resistance to chemicals, and membrane fouling, researchers must invent other durable materials, high-water permeability, outstanding chemical resistance, low cost, and less time-consuming techniques [12, 13].

Further, alternative 2D materials have emerged and showed outstanding results, which blocked ions, either monovalent or divalent, through plane pores or stacked nanosheets. Thus, these 2D nanosheets can be economically served and produced at a huge range owing to their capable electrostatic repulsion properties, pores, and nanochannels, which blocked the ion, especially monovalent ions, for desalination [14]. Graphene-based materials and molybdenum disulfide (MoS_2) received more improvement in the process of desalination, which has been discussed in detail later [15]. In recent years, research interest in effective aqueous pollution remediation has exploded. In addition, graphitic carbon nitride (g-C_3N_4) is a promising candidate for applications such as liquid splintering and protecting the environment due to its low cost as well as its environmentally friendly nature. A variety of other materials, such as MoS_2, are also being used in conjunction with g-C_3N_4 by researchers. MoS_2 is an excellent material for a variety of applications, such as field-effect transistors (FETs), electroluminescent sensors, and high-efficiency light energy conversion [16]. A photocatalyst with light-excited conduction band electrons has been widely used in recent years to convert toxic metals to less harmful, which is a viable method for environmental remediation.

2 Types and Properties of 2D Materials

2.1 Graphene

Carbon atoms in hybrid sp^2 orbitals form a two-dimensional hexagonal honeycomb structure in graphene. Schafhautl described the expansion behavior of graphite in the presence of concentrated sulfuric acid for the first time in 1840. Later, after a century of intensive research, a graphene foundation has been established. Although pure graphene requires extensive preparation and has limited applications due to insurmountable defects, it is a promising material. Several methods, like chemical, thermal, and electrochemical methods, are applied for the preparation of graphene. Among these, the electrochemical method is preferred by most researchers because it is a straightforward method for the preparation of graphene and its derivatives with high quality, and the materials used in the process are non-hazardous and eco-friendly [17]. Environmental pollution and energy scarcity are primary concerns to maintain sustainable development. This requires the introduction of novel methodologies and materials. By functionally compounding various materials, the material properties can be enhanced. Graphene is a carbon-based nanomaterial that has recently attracted considerable attention due to its unique properties and variety of applications in photonics and nanoelectronics. Graphene has high mechanical strength (approximately 200 times that of steel), high carrier mobility (up to 200,000 $cm^2\ V^{-1}\ s^{-1}$), a high thermal conductivity (approximately 3080–5150 $Wm\ K^1$), a high degree of flexibility, excellent optical properties (transparency of approximately 97.3%), and a high specific surface area (2675 $m^2\ g^1$, twice that of carbon nanotubes) [18]. Previously, electrochemical sensing was considered an effective tool because it allowed for the modification of electronic properties by varying the Fermi level of the electrode material surface via an external power supply.

Pure graphene has limited commercial applications due to its lack of catalytic properties and zero band gap. Potential applications of 2D graphene can be enhanced by tuning its physical and chemical properties. Additionally, it can be converted to a three-dimensional structure (foam-like materials), a one-dimensional structure (nanoribbons), or a zero-dimensional structure (quantum dots) [19]. Doping another element into graphene can also alter its electronic and physicochemical properties, resulting in the loss or addition of free electrons. The dopant and graphene Fermi levels' density of states (DoS) positions aid in determining charge transfer.

Nowadays, graphene-based composite materials are also extremely popular due to their ultrahigh specific area, superior thermal and mechanical stability, controllable surface properties, sufficient active sites, unique porous structure, light transmittance, and outstanding carrier transport. Graphene has a broad range of environmental applications [20]. These extraordinary properties of graphene-based composites enable the development of highly efficient adsorbents for sample pre-treatment, which is critical for the development of novel extraction methods for environmental analysis.

2.2 Black Phosphorus

There is no bandgap between the valence and conduction bands, which limits graphene's use as a semiconductor in nano- and optoelectronics. The band gap is critical in a variety of electronic applications because it controls the flow of electrons. Additionally, applications in optoelectronic devices are limited due to the incompatibility of silicon and graphene. Numerous studies have been conducted to introduce the band gap property into graphene but have not yielded satisfactory results. Following that, other graphene-like materials, such as boron nitride (BN), have been introduced, each of which exhibits superior mechanical, thermodynamic, and optoelectronic properties [21]. Similarly, black phosphorus (BP), another layered structure, has been introduced [22]. In terms of stability, BP is the most stable of the phosphorus allotropes and has a high density. Because it is chemically inert, most organic solvents do not dissolve it [23]. It is a metallic conductor with an orthorhombic metallic luster. BP crystallizes in four different ways: orthogonal, simple cubic, amorphous, and diamond. BP has a lamellar structure that is like that of graphene. At a layer distance of 5.3, the van der Waals forces between the interlayer atoms become active. A honeycomb-like structure is formed even though not all the atoms within a single layer are on one plane. 3*p* hybrid orbitals connect P and its surrounding neighbors, forming a folded, layered structure that's extremely stable. BP is a direct band gap semiconductor with a variable band gap that can be operated by the number of layers used and the level of stress applied. As the number of BP layers increases, the size of the band gap shrinks exponentially.

2.3 Boron Nitride

Boron nitrides (BN) are synthesized with equal amounts of boron (B) and nitrogen (N) [24]. It has a similar crystal structure to carbon and is found in a variety of crystal forms. Balmain first synthesized it in 1842 using potassium cyanide (KCN) and boric acid (H_3BO_3). As carbon analogs, boron nitride-based materials have similar properties to diamond, making them the second hardest material. White graphene is another name for hexagonal boron nitride (h-BN). Physical properties and chemical and thermal stability are superior in BN structures, such as high-temperature inert environment h-BN. Sensory and bioconjugation applications benefit from the electrical, mechanical, and optical properties of BN materials. Microwave transparent, BN materials are nontoxic and chemically inert. BN allotropes have a strong effect due to the ionic nature of the B-N bond. It can form sp^2 or sp^3 bond structures. Structures based on h-BN have low electrical conductivity and high thermal stability, making them ideal for scientific research and industrial use. They cannot be used as thermoelectric bulk materials, however, because of their low electrical conductivity and high thermal conductivity. An additional set of parameters determines thermal conductivity, including atomic mass and acoustic mode frequency, anisotropy, and bond strength. They have a higher thermal conductivity than hexagonal allotropes due to the cubic structure being more symmetric and isotropic than hexagonal.

2.4 Molybdenum Disulfide (MoS_2)

MoS_2 is a popular material for electrocatalysis and energy storage systems because of its unique layered structure and electronic properties. The van der Waals forces allow MoS_2 monolayers to be stacked on top of one another. Diverse coordination modes of Mo and S atoms lead to the formation of several distinct phases, including 2H, 3R, and 1T [25]. It is also possible to alter the intrinsic properties of MoS_2 by doping heteroatoms into its structure. Also, energy conversion and storage may be improved due to the large surface area of the material and its high density of active sites. The physicochemical properties of MoS_2 can be improved by rationally designing heterostructures composed of MoS_2. Chemical vapor deposition (CVD), intercalation, hydrothermal, and solid-phase pyrolysis can all be used to structure MoS_2 and MoS_2-based nanomaterials [25]. With these methods, MoS_2-based nanomaterials with a variety of morphologies can be produced, including nanospheres, nanoflowers, mesoporous MoS_2, and porous structures with a hierarchical hierarchy. As a result, energy conversion and storage performance are enhanced [26]. MoS_2's narrow band gap (1.29 eV) makes it ideal for photocatalytic applications that require visible-light generation. It is also possible to incorporate alkaline metal ions into MoS_2's interlayers because of its unique layered structure, which is beneficial for energy storage applications. In addition to its electrocatalytic activity, it has a unique electronic structure. MoS_2 catalysts can also be used to remove organic contaminants from water using photocatalytic degradation technology [23].

3 Cause of Heavy Metals and Dyes in the Environment

Global industrialization and urbanization have introduced enormous toxic pollutants into the environment. These hazardous pollutants have a detrimental effect on the quality of the natural resources such as water and air utilized for the betterment of the human race, consequently threatening human health. Hazardous, toxic heavy metals and dyes enter the atmosphere through natural and anthropogenic activities.

3.1 Source of Heavy Metals in the Environment

Since the decade, the term "heavy metals" was defined either due to the high density or high atomic weight. But nowadays the definition of heavy metals has been designated as metalloids and metallic elements. These metallic chemical elements are highly toxic for environmental and health concerns. Whereas the toxicity is not dependent upon the metalloid's weight. Sometimes the heavy metal known as a gold element is nontoxic; however, lighter metals such as aluminum and selenium are toxic for both humans and the environment [27, 28]. Generally, pollutants, when present in the environment above the defined concentration, have become life-threatening. But the presence of certain heavy metals such as cadmium, arsenic, nickel, lead, mercury, zinc, chromium, and cobalt are highly toxic even in lower concentrations than other pollutants. However, the primary source of heavy metal occurrence on the terrestrial as well as in aquatic environments owing to the surplus activities has been led by a human. This is because of vast mining of metals and leaching of metals, such as from effluent discharge of fertilizer industries and dye industries, which are adding a huge quantity of heavy metals into the water bodies. The waste released from roadworks, automobile exhaust, smelting, waste dumps, and burning of fossil fuels distributes the heavy metals in different forms into the environment [28–30].

The second major source of metal pollution is an advanced practice in agricultural activities. Nowadays, the excessive use of insecticides and pesticides for the progress of cultivation has led to the release of unwanted heavy metals, consequently exploiting the quality of the agricultural field and water resources at a careless rate. Heavy metals do not enter the water sources directly; they persist in the soil for a long time, and gradual leaching of metals deteriorates the quality of groundwater and surface water. The natural sources of heavy metals are volcanic eruptions, forest fires, weathering from rocks, chemical weathering of minerals, and soil leaching. After soil leaching and rock weathering, metals reach downstream and severely contaminate the water ecosystem [31].

3.2 Source of Dyes in the Water Ecosystem

Furthermore, dyes are another major cause of environmental water pollution. These dyes are introduced into the environment through tanneries, paper and pulp industries, distilleries, textile industries, and dyestuff with large-volume effluents as colored wastewater. Among these, textile industries are the largest producer of dyes as contaminants and are highly concerned for the environment. Textile industries employ a large volume of water during the process of fabric dyeing; as a result, they discharge watery waste in huge quantities in the form of wastewater and dye waste into the water resources. These discharges of textile industries have high biological oxygen demand (BOD) and chemical oxygen demand (COD) and consist of persistent color, which threatens the aquatic environment [32, 33] by blocking the entry of light sources in the water bodies [34], resulting in the reduced photosynthetic process [35]. These water contaminants are difficult to treat in one go. Various methods and technologies have been applied frequently for treatment and to decontaminate the environmental resources from pollutants. Some chemical dyes that are frequently applied in textile and other industries are azo derivatives, phthalocyanine derivatives, indigoid, xanthene, anthraquinone, and arylmethane. To remove these pollutants from water sources, various effective technologies have been developed that will be discussed later to achieve clean natural resources.

4 Potential of 2D Materials in Wastewater Decontamination

4.1 Adsorption-Based 2D Materials for Removal of Heavy Metals and Dyes

Graphene and its derivatives as absorbents

The invention of 2D graphene must upsurge as an important material owing to its remarkable physico-chemical properties and has gained great interest among researchers. Graphene and its derivatives, graphene oxide (GO) and reduced graphene oxide (rGO), have been proven to be potent and promising adsorption-based 2D materials for the removal of inorganic (heavy metals) and organic (dyes) wastes from wastewater and air environments. This promising adsorbent is effectively used for environmental remediation purposes and wastewater treatment because of its tremendous surface area and chemical properties [36, 37]. One of the outstanding characteristics of pristine GO is its functional groups (carboxylic, hydroxyl, and epoxide) on the surface, which showed suitability for environmental remediation purposes, specifically for wastewater treatment. The best inexpensive and effective technique for wastewater treatment is adsorption-based, which occurs with the formation of a molecular layer on solid materials or adsorbents on the interaction with liquid solute or adsorbate [38, 39]. This interaction may be both a physical and chemical reaction reliant on the interaction forces at the interface of adsorbate and adsorbent, such as an ionic bond, covalent bond, and van der Waals forces.

Adsorption of heavy metals

Heavy metals from industries and agricultural runoff contaminated the water resources, hence increasing the lack of safe and pure drinking water. Towards this, modified graphene, GO, and rGO materials have been developed for the removal efficiency of dyes and heavy metals from wastewater that depends upon the pH, dose, ionic strength, temperature, contact time, and concentration. The key mechanism of ion exchange that takes place in the adsorption process is electrostatic forces between the positively charged heavy metal ion and the negatively charged surface of rGO and GO. GO rapidly aggregates irreversibly in water due to inert surface chemistry and strong π-π forces [40]. This impedes the capability of GO for environmental remediation purposes. So, few chemical modifications assist in solving the problem of GO aggregation and enhancing the dispersibility, hence, the surface area increased. The remediation of heavy metals from water sources is enhanced by attaching functional groups or soluble polymers with GO; for instance, trioctylamine-impregnated GO

(TOA-EGO) and ethylenediamine-reduced graphene oxides (ED-RGO) [41]. The interaction mechanisms between heavy metals and TOA-EGO are lone pair-π and electrostatic interactions. All through the green method, GO associated with dimethyl has successfully changed the functional groups to remove heavy metals from wastewater by enhancing adsorption capacity, selectivity, and elevated adsorption kinetics [42, 43].

Several factors are responsible for influencing the adsorption or removal of heavy metals from the environment. One of the crucial factors is pH, which is specific to the environment and affects the adsorption capability of metals; hence, it differs the concentration of contaminants to several extents. If we are applying GO-based material for metal adsorption, then we should emphasize the pH of the solution because it strongly affects the adsorption efficiency of heavy metals. While different metals have different adsorption approaches on different kinds of GO-based material due to variations in pH [44]. We will discuss here the different mechanisms of metal adsorption at different pH. The deprotonation and protonation reaction of functional groups present on GO-based materials at different pH largely affect the binding efficiency of metals. For example, at high pH, GO is easily protonated and creates a large surface area for the adsorption of heavy metals [45]. Moreover, another reason for enhanced metal adsorption at high pH is due to electrostatic interaction among the negatively charged GO surface and positively charged heavy metal ions [46]. Additionally, when the pH of the solution is low, H^+ ion concentration is eventually increased, creating a competitive inhibition in the solution and reducing the adsorption capacity of metal ions with the surface of GO-based materials [47]. However, to understand the leaching of metals, we can perform a pH–stat leaching test in which at lower pH cation release is enhanced, whereas towards higher pH anion increases. In addition, pH_{stat} test data will be presented as "U," "V," and" "L"-shaped curves for heavy metal adsorption from the environment [48]. Further, we will discuss a few studies related to the pH of the solution.

The lower pH (2.0–3.0) [49] favors the removal of heavy metals chromium, Cr (VI); cobalt, Co (II); iron, Fe (II); zinc, Zn (II); and copper, Cu (II), etc. But GO nanosheets have been studied for good adsorption ability for Co (II) and Cd (II) ions at pH 6.0. Because of the richness of lead, Pb (II), in wastewater, graphene with Fe_3O_4 has been widely used owing to its magnetic property. However, layered GO nanosheets have increased their efficiency to adsorb Pb (II) ions from wastewater as the pH increases from 1.0 to 8.0 [50]. The adsorption of nickel (II) is achieved best at pH 2.0–5.5 due to electrostatic force between Ni (II) and GO [51]. Besides pure graphene and composite adsorbents, GO adsorbent is chemically modified with 2-pyridinecarboxaldehyde thiosemicarbazone (2-PTSC) on the surface for the removal of mercury, Hg (II) [52].

Adsorption of dyes

Dyes into the aquatic ecosystem are mainly hazardous to aquatic organisms, so the eradication of the pigments is urgent for the survival and safety of aquatic animals. To adsorb organic dyes from wastewater, Fe_3O_4/porous graphene nanocomposites play a useful adsorbent. The magnetic property of the adsorbent enhances the adsorption efficacy of graphene and the separation ease of the magnetic stuff. The adsorption process with any 2D material is mainly affected by contact time, dye concentration, and pH differences. The adsorption of dyes increases at high pH levels, and mostly all dyes are adsorbed at pH 7.0. The kinetics of dye adsorption on the Fe_3O_4/porous graphene nanocomposite is analyzed using Lagergren first-order and pseudo-second-order models [53].

The process of removal of heavy metals and dyes chiefly obeys the Freundlich model, Langmuir model, Lagergren first-order, and pseudo-second-order kinetics. Besides, the adsorption of heavy metals is mainly attributed to external ion exchange, complexation, and electrostatic fascination. The adsorption kinetics can be calculated by a given formula. The first-order rate can be given as in Eq. (1).

$$\log(qe - q) = \log(qe) - k_1 t/2.303 \tag{1}$$

Pseudo-second-order kinetic model can be represented as in Eq. (2).

$$^{t}/_{q} = {}^{1}/_{k_2 q_e^2} + {}^{1}/_{qe} \tag{2}$$

The mathematical representation of the Langmuir model is expressed as

$$\frac{C_e}{Q_e} = \frac{1}{Q_o b} + \frac{C_e}{Q_o} \tag{3}$$

where Q_e (mg/g) is the equilibrium concentration of metal ion on the adsorbent, C_e (mg/L)

is the equilibrium concentration of metal ion, Q_0 (mg/L) is the monolayer sorption ability, and b is the Langmuir constant.

The mathematical representation of the Freundlich model can be represented as in Eq. (4)

$$\log(Q_e) = \log(k_f) + \frac{\log C_e}{\eta} \tag{4}$$

where C_e (mg/L) is the equilibrium of metal ion concentration, K_f ($mg1^{-1}/nL^{1}/n\,g^{-1}$) is adsorption capacity, and 1/n is the empirical parameter linked to adsorption intensity.

However, GO plays a significant role in the adsorption of several dyes, such as methylene blue (MB), rhodamine B, and methyl violet, which also follow pseudo-second-order reaction and Langmuir models. This adsorption is dependent on pH and due to excellent electrostatics between the contact of compounds and GO [54]. But GO shows more adsorption efficacy of cationic dyes compared to anionic dyes owing to robust electrostatic repulsion [37].

4.2 MoS_2 Nanosheets and Layer-Stacked Membranes for Heavy Metal Remediation

Adsorption efficiency of MoS_2 nanosheets for heavy metals

MoS_2 is an emerging 2D material widely studied as a transition metal dichalcogenide (TMD) having an exceptional structure (Fig. 11.1) that makes it unique from other 2D materials. MoS_2 provides a promising adsorbent technology owing to the presence of intrinsic sulfur atoms having a high affinity to different heavy metal ions, hence enhancing the removal efficiency from wastewater. Synthesis approaches of MoS_2 affect its performance in the removal of heavy metals because of distinct morphologies, crystal configurations, physicochemical properties, active sites, and thicknesses in different preparation methods [55]. In each MoS_2 monolayer, both sulfurs are linked with molybdenum via covalent bonds as a tri-layer and explore the sulfur outward, which highly interacts with heavy metal ions. In multiple layers, sulfur is linked via weak van der Waals forces to form a bulk stack. This bulk structure restricts the free surface area to interact with ions due to small interlayer spacing (0.63 nm); hence, it is hardly used for ion removal [56]. Generally, all heavy metals can be successfully adsorbed by MoS_2 nanosheets [57, 58] except Hg^{2+}, which can be adsorbed by multilayer MoS_2 surfaces [59].

MoS_2 monolayers have potential in heavy metal adsorption. Further, MoS_2 monolayers are used to synthesize layer stack membranes for the removal of heavy metals from water. The need for synthesizing layer stack membranes is because in MoS_2 monolayer nanosheets, the retention time of toxic metals is very short. Therefore,

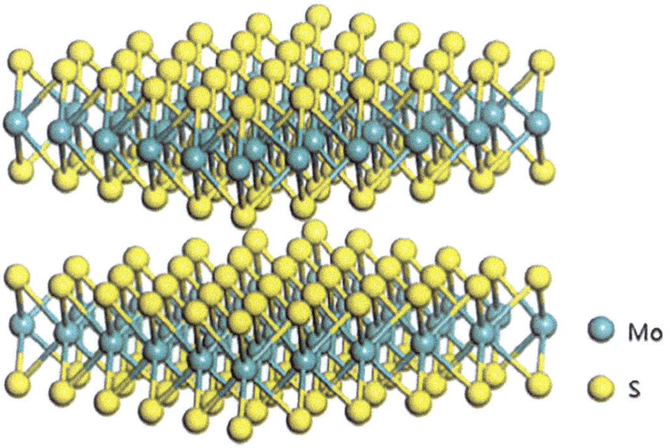

Fig. 11.1 Crystal structure of 2D material molybdenum disulfide. Reprinted with permission from Jia et al. [59] Copyright 2017, Royal Society of Chemistry

to increase the retention time, the thickness of MoS_2 increased to enhance the treatment capacity of the membrane. As a result, the layer stack membrane for wastewater treatment capability is enhanced compared to monolayer.

Adsorption mechanism of MoS_2 nanosheets

To achieve the maximum adsorption capacity of MoS_2 it is extremely essential to understand the mechanism behind it. The adsorption mechanism of metal ions and MoS_2 nanosheets has been studied by various research groups, and still, progress is going on. Three explored mechanisms—specifically electrostatic interaction, redox reaction, and complexation—are the key to the maximum uptake efficiency of MoS_2 for heavy metal ions. Complexation is believed to be the primary adsorption mechanism [60]. The establishment of a complex between different heavy metals and S atoms owing to a different affinity for ions exhibits the variation in the adsorption capacity of MoS_2 for heavy metals. Because of the negative charge, MoS_2 interacts with heavy metal ions via electrostatic interaction. However, the major interaction is the complexation mechanism discussed above, but still being a minor adsorption mechanism, electrostatic interaction is dominant for Cu^{2+} and Co^{2+} which cannot establish a complexation mechanism with MoS_2 [57]. Few heavy metals show multilayer adsorption. For example, Hg^{2+} establishes a complex with sulfur present on MoS_2 to assign an inner layer, but the outer layer is created from electrostatic interaction between the MoS_2 negative charges and cation Hg^{2+} [59].

Another mechanism is known as a redox-reaction mechanism, which exhibits particularity on heavy metals and relies only on redox potential. The specific circumstance in this mechanism is that it shows adsorption on MoS_2 when the redox potential of any pair of heavy metals is greater than SO_4^{2-}/MoS_2 pair and MoO_4^{2-} then only it is categorized as adsorptive degradation. In this case, 1T-MoS_2 is used for metal removal owing to its metastable property, and electrons are released to reduce metals in aqueous solution [61, 62].

4.3 Catalyst Based 2D Materials for Removal of Heavy Metals and Dyes

2D materials such as rGO nanocomposites, transition metal dichalcogenides (TMDCs), titanium dioxide (TiO_2), MoS_2, and black phosphorus (BP) are employed as photocatalysts for the removal of inorganic and organic pollutants for wastewater treatment.

Photocatalysts in environmental applications

Photocatalysts are employed in the degradation of various organic contaminants such as dyes, volatile organic compounds, and pesticides in wastewater by an advanced oxidation process. During the photocatalytic process, different reactive species are produced, such as oxygen radicals ($^{\cdot}O^{-}_2$), hydroxyl radicals ($^{\cdot}OH$) which gain more visibility for their potential application in the eradication of organic contaminants

[63]. The basic principle of photocatalysis demonstrates that when semiconductor materials are treated with light, UV (ultraviolet) or visible light creates holes in the valence band (VB) and excites the electron (conduction band) to generate various potential reactive species, which is the chief agent for contaminant degradation (Fig. 11.2). The same mechanism works for the wastewater treatment process; the holes (h^+) generated during the process are surrounded by a water molecule (H_2O) and form a potent oxidizing agent ($^{\cdot}OH$ radicals). These oxidizing agents react with heavy metals and dyes as a potent contaminant in wastewater and oxidize it into simple products, gases, and H_2O.

Titanium dioxide (TiO_2)

The promising photocatalyst TiO_2 is known to be a widely applied catalyst for environmental applications. The significant properties like chemical stability, reactivity, and notably wide band gap energy (3.2 eV) categories TiO_2 as a potent photocatalyst for the degradation of heavy metals and dyes from contaminated water sources. Due to the relatively wider band gap, this catalyst is active only under UV radiation. Further, without showing dependency and restriction only on UV radiation, other new approaches must develop to achieve the visible light photocatalyst. Consequently, on doping TiO_2 with S, C non-metallic elements, reduced band gap energy may perhaps tune to visible light. The band gap of TiO_2 has been reduced, and

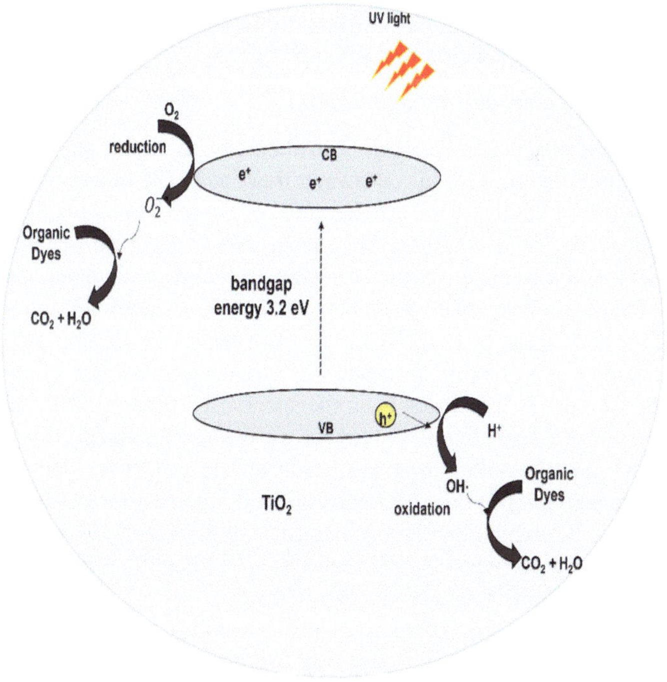

Fig. 11.2 General principle of photocatalytic process in semiconductor TiO_2

synthesized $MnFe_2O_4$/g-C_3N_4/TiO_2 nanocomposite has proven an effective photocatalyst under visible light. The photocatalyst $MnFe_2O_4$/g-C_3N_4/TiO_2 was exhibited for methylene oxide (MO) degradation [64]. TiO_2 is very supportive in the degradation of mixed dye due to its photocatalytic properties and overcomes the limitation of only one type (cationic and anionic) of adsorption removal by other 2D material [65].

Reduced graphene oxide

2D heterostructures are attracting great attention from researchers in several organic contaminant remediations. For the degradation of a dye such as MB, 2D nanocomposites have been studied a lot. As in earlier discussion, rGO nanocomposites have great potential in the removal of heavy metals. rGO nanocomposites with $Co_xZn_{1-x}Fe_2O_4$ also act as an efficient photocatalyst for MB degradation under visible light [66–68].

Black phosphorus

Another novel 2D material is BP nanosheets, which exhibit efficient photocatalytic activity. It is a highly promising candidate for photocatalyst owing to its tunable band gap (0.3–2.0 eV), wide-range UV absorption, high carrier mobility, and enhanced reduction–oxidation capability [69, 70]. To enhance the photocatalytic activity of BP, various stable non-metals, such as graphene oxide and C_{60}, are used to construct heterojunctions [71, 72]. One such eco-friendly nanocomposite photocatalyst, which consists of BP nanosheets and ZnAl-layered double hydroxides (LDH), showed enhanced MB (99%) degradation in visible light [73].

Electrocatalysts in environmental application

Besides photocatalysis, electrocatalysis is an alternative effective technique used for the degradation of organic dyes in wastewater treatment. The fundamental principle of electrocatalysis is based on the acceleration of the rate of electrochemical reactions on the surface of the electrode. The mechanism of the electrolysis process is, first, the reactant molecule adsorbs on the surface of the electrocatalyst; second, after adsorption, charge transfer takes place between reactant molecules and electrocatalyst. And finally, export of charge carriers from the electrode and desorption of product molecules. And finally, desorption of product and transfer of charge from the electrode take place. The focus is on the specific 2D catalyst used for the electrocatalytic oxidative degradation of dyes for wastewater treatment. However, there are several 2D materials like hexagonal boron nitride (hBN) and BP that display electrocatalytic activity. But due to its bad conductivity and poor catalytic function, pristine hBN and BP exhibit hardly any electrocatalytic activity. However, the electrical properties have been developed in few cases by coupling nanocomposites, for example, graphene and BN heterostructures [74].

Moreover, graphene-based (rGO, GO) electrocatalysts and their nanocomposites are extensively used for dye removal for decontamination of wastewater. Reduced graphene nanocomposites, rGO/TiO_2/carbon cloth (CC) electrodes, are applied by using 0.9 V potential for the 15% degradation of MB. Electrocatalytic oxidation of rhodamine B (RhB) dye was performed by applying electrocatalyst anthraquinone@rGO nanohybrid cathodes (AQ@rGO-NC). The robust interactions between rGO and anthraquinone molecules ensure the effective cathodic production of reactive oxygen species (ROS), like H_2O_2, and cause RhB degradation [75]. As an electrocatalyst, rGO/TiO_2 is also explored for the electroreduction of acid orange II and RhB dyes [76]. Graphene derivatives along with nanocomposites have effectively achieved improved oxidative degradation of an organic compound for wastewater treatment as an electrocatalyst.

4.4 Fenton-Based Catalyst in Environmental Application

Heterogeneous Fenton catalysts, homogeneous Fenton catalysts, and photo-Fenton reaction processes are actively involved in the oxidation of heavy metals and dye degradation. Homogenous Fenton catalysts function at low pH (2.8) and produce iron sludge in an aqueous medium [77]. While in a heterogeneous Fenton reaction, degradation of dyes for wastewater treatment is performed easily at neutral pH and solves the issue of huge production of iron sludge [78]. Heterogeneous Fenton catalysts, iron-based materials, are considered superior catalysts owing to their low cost, high catalytic efficiency, relatively low toxicity, and ease of performance under visible light.

The basic principle of the Fenton catalyst is based on the advanced oxidation process in water. In this process, H_2O_2 and ferrous ions are applied to produce $OH^{\cdot}$ radicals, which are the major factor to initiate oxidation reaction for degradation of organic compounds from wastewater [79]. The reaction that takes place in a traditional Fenton reaction is given as [80]:

$$Fe^{2+} + H_2O_2 \overset{k_1=40-80L/(mol.sec)}{\longrightarrow} Fe^{3+} + OH^- + HO \tag{5}$$

$$Fe^{3+} + H_2O_2 \overset{k_2=9.1\times10^{-7}L/(mol.sec)}{\longrightarrow} Fe^{2+} + .O_2H + H^+ \tag{6}$$

$$Fe^{3+} + +.O_2H \overset{k_3=(0.33-21)\times10^{6}L/(mol.sec)}{\longrightarrow} Fe^{2+} + O_2 + H^+ \tag{7}$$

$$Fe(II) + H_2O_2 \rightarrow Fe(III) + ^{\cdot}OH + OH^- \; k = 40 - 80\,M^{-1}\,s^{-1} \tag{8}$$

$$Fe(III) + H_2O_2 \rightarrow Fe(II) + HO_2^{\cdot} + H^+ \; k = 0.001 - 0.01\,M^{-1}\,s^{-1} \tag{9}$$

$$\textit{Organic matter} + ^{\cdot}OH \rightarrow \textit{Degradation products} \tag{10}$$

In Eqs. 5, 6, and 7, k1, k2, and k3 denote kinetic constants. However, the Fenton reaction activity is limited in the traditional Fenton reaction due to the low value of k_2 and limits the redox cycle process of Fe^{2+}, thus significantly affecting the production of OH· radicals and the overall reaction. This disadvantage has been overcome by different reducing agents to enhance the Fe^{2+} redox cycle, which is directly responsible for producing ROS. But owing to certain drawbacks of using reducing agents, different approaches have been discovered to gain the visibility for regeneration of Fe^{2+}. Further, to enhance the activity of the Fenton reaction, we emphasized various 2D materials like transition metal dichalcogenides, MoS_2, TiO_2, and GO.

To understand the importance of the Fenton reaction, we will find the function of a few 2D materials from some examples. Among different 2D materials, MoS_2 is drawing attention among researchers due to its exceptional photocatalytic properties [81, 82]. The key reason to opt for MoS_2 in comparison to other 2D materials is its narrow band gap, which enhances the possibility of absorbing visible light. MoS_2 alone could be exploited as a co-catalyst in a homogenous Fenton reaction for quick regeneration of Fe_2^+ [83]. However, iron-based 2D molybdenite (2DM-Fe_2^+) functions as a heterogeneous Fenton catalyst. For example, the degradation efficacy of MB on activation of H_2O_2 (produced during the Fenton reaction) by the 2DM-Fe_2^+ approach is very low in the absence of light, but as the light illuminated, 100% degradation was achieved. Therefore, participation of both H_2O_2 (HO· generated from H_2O_2 breakdown) and light irradiation is important for the Fenton process [84]. Similarly, as a photo-Fenton catalyst, the graphene surface is anchored with abundantly present Fe_3O_4, and iron oxyhydroxide (FeOOH), which act as an active site on the graphene surface and show catalytic activity in MB degradation [81, 82, 85]. Also, we have discussed in the adsorption section that the properties of graphene and its derivatives also act as an efficient candidate for the removal of metals and dyes from the water surface.

5 Potential of 2D Materials in Air Purification

5.1 Basic Concept and Source of Particulate Matter

Particulate matter (PMs) is an atmospheric aerosol that enters the environment through various anthropogenic and natural origins. PM has become a global challenge, mostly in developing countries. Industrial discharge, combustion of fossil fuels, and emissions from vehicles in the form of black carbon, nitrates, sulfates, organic carbon, gaseous pollutants, and specific elements such as Hg, Pb, etc., are the major compositions of anthropogenic sources. The composition of pollutants varies depending on the process that occurs. A major natural source is soil dust caused by wind acts, and other biological PMs are pollens, plant waxes, and spores [86].

PM has a severe impact on the environment and health. The size of the PM differs and is majorly categorized as $PM_{0.3}$, $PM_{2.5}$, and PM_{10}. When the diameter of PM is less than 10 μm, it is designated as PM_{10}. $PM2_{.5}$ is known as "fine PM" with a diameter less than 2.5 μm. Owing to its smaller size, it can be moved long distances. Such a smaller fraction of toxic pollutants when entering inside the body through inhalation causes acute effects, and its toxicity level is greater than others. Several conventional methods exhibit insignificant results of removal of such toxic PM pollutants from the environment. Therefore, due to the weak surface affinity of conventional technologies to PMs, 2D material-based advanced technologies are being developed for the complete treatment of indoor and outdoor air quality.

Advanced technologies in air filtration

For purifying polluted air from the environment, various advanced technologies have emerged. Techniques used for the purification of PM and other existing toxic pollutants in the air are filtration technologies. Different materials such as glass fiber air filter material, activated carbon fiber (ACF), super fine glass fiber, and TiO_2 photocatalyst-based material are being used in filtration technology. Fibrous filters (HEPA), glass fiber filters, and wire mesh filters (2–10 μm) are types of filters that are used for indoor air applications [87, 88]. An air filter can be of many types, but a high-efficiency particulate air (HEPA) filter has a strong ability to filter 99.97% of PM of size 0.3 μm [5]. Another filter is an adsorption-based granular activated carbon filter (GAC), mainly used for filtration of gas-phase air pollutants [89]. Electrostatic air filter (ES) is another type of filter applied in the removal of pollen, smoke, and dust particles that are less than 1 μm.

Emerging 2D materials provide great opportunity in various multidisciplinary research. Photocatalytic oxidation (PCO) technology can be a promising technique to remove indoor contaminants, mainly VOCs. TiO_2 photocatalyst is an emerging photocatalytic technology to purify indoor air that is based on the photocatalytic mechanism (discussed earlier) [4]. This developed photocatalyst effectively removed distinct air pollutants from the environment like NH_3, SO_2, NO_x, CO, VOC, etc.

5.2 2D Material in Oil–Water Separation

Mesh-based Materials

Functionalization of meshes with super wettability has drawn the attention of researchers for their application in the separation of oil from water/oil mixtures. Because of their high separation efficiency and rate, simplicity, and low cost, meshes have been considered as perfect candidates for wide application in real-world industrial applications. Due to high mechanical strength, copper and stainless steel meshes are extensively used in the oil/water separation technique. The most commonly used metal meshes are copper oxides (CuO or Cu $(OH)_2$) which can be prepared by etching Cu meshes with HCl, NaOH, HNO_3, and ammonium persulfate (APS) to obtain a mesh surrounded by CuO or Cu $(OH)_2$ nano/microwires [90, 91]. Song et al. demonstrated the hydrophobic stainless steel mesh for oil separation with 94% efficiency and filter oil with the purity of 99 wt% [92]. Hydrophobic Cu mesh with a rough structure was prepared by a chemical etching process. Further, the use of polymer for the functionalization of metal meshes provides the wetting properties important for oil/water separation. Superhydrophilic meshes can be prepared by using cationic and anionic polyelectrolytes. Functionalization with polyelectrolytes leads to the preparation of meshes with superhydrophilic and underwater superoleophobic properties for oil/water separation. Polyacrylamide (PAM) hydrogels exhibit superhydrophilic and underwater superoleophobic properties and can be used in mesh preparation [93]. PAM can efficiently separate the vegetable oil, gasoline, crude oil, and diesel from the oil/water mixture.

Further, the wettability of Cu mesh can be enhanced by the fabrication of Cu_2OHPO_4 over Cu mesh via in situ growth [67, 68]. APS and sodium hydrogen phosphate are responsible for the flower-like nanoflakes rough structure over the Cu mesh surface. The hydrophilicity of Cu mesh has been governed by the solution pH, and at 9.7 pH, Cu mesh exhibits superhydrophilicity with an oil sliding angle of 1.4°. Additionally, Cu mesh has excellent chemical and separation stability and mechanical properties. It can also be well applied in the photodegradation of aqueous pollutants. Recently, Yuan et al. demonstrated the formation of a rough nano-needle structure over Cu mesh via an oxidation modification method, and its oil/water separation efficiency is reported to be 99.8% [8]. This study also confirms that the hydrophilicity of Cu mesh can be enhanced by the rough surface. It has been reported that the hydrophobicity of Cu mesh can be increased by its modification by Cu_2O, as it is difficult for Cu_2O to form a chemical bond with water. In this context, Lei et al. synthesized the superhydrophobic Cu mesh modified with Cu_2O with oil–water separation efficiency of 95% [94]. Further, Chen et al. also synthesized the superhydrophobic Cu mesh using graphene oxide (GO) with high efficiency of > 99.49% and reusability [95].

Stainless steel has several advantages, like heat and corrosion resistance, high mechanical strength, and low-cost synthesis, and thus found its application in the oil–water separation [9]. The chemical stability of stainless steel is more than that of Cu mesh and thus difficult to modify by using strong acids and bases. Zhang et al. designed the superhydrophilic stainless steel by using polystyrene-co-maleic anhydride and hydrophilic polyethyleneimine and used it in oil/water separation [96]. The coating provides superhydrophilicity and underwater superoleophobic properties, as well as underwater anti-oil-fouling activity. Further, tannic acid and polyvinylpyrrolidone modification provide the stability to special wettability stainless steel [61, 62]. Meng et al. demonstrated the oil/water separation using superhydrophilic stainless steel coated with waterborne polyurethane and TiO_2 [7]. The separation efficiency of synthesized material is reported to be 99%, and it remains as high as 98.5% after 210 repeated tests. Wettability of modified mesh maintained in NaOH (pH 14), HCl (pH 1), hot water (80 °C), and NaCl solution. Further, Du et al. also reported the TiO_2-modified superhydrophobic stainless steel material. Additional modification by polydimethylsiloxane and graphene provides resistance to acidic and alkaline conditions, UV light-exposed environments, salt, and organic solvents. Additionally, mesh also exhibits properties like antifouling, reusability, superhydrophilicity, and photodegradation [97].

Fabric-Based Material

Even though metallic meshes have been extensively used as a promising agent for oil/water separation, they experience several drawbacks like high cost, heavy weight, and easy corrosion. To overcome these limitations, fabric-based materials have been used, as they are corrosion-free and have other advantages like flexibility, low cost, and light weight. For selective oil/water separation, superhydrophobic/superoleophilic and superoleophobic/superhydrophilic fabrics are required [98]. Zhou et al. demonstrated the superhydrophobic and superoleophilic cotton fabric coated with polyaniline and fluorinated alkylsilane via the chemical vapor deposition (CVD) method [6]. Growth of nano/micro inorganic crystal over a fabric surface in a controlled way tunes its surface roughness. Coated fabric separates the oil/water mixture with a high efficiency of 97.8%. Superhydrophobicity and separation efficiency of material remain constant even after repeated use 30 times. Fabric can resist harsh conditions like corrosion, humidity, high temperature, and mechanical forces. As synthesized fabric can be reused several times, it can be effectively used for oil-spill cleanup with extraordinary durability. The dip-coating method provides special wettability to fabrics and thus has been extensively used to prepare fabric with superhydrophobic/superoleophilic and superhydrophilic/superoleophobic characteristics. Zhang et al. demonstrated the preparation of superhydrophobic cotton fabrics using the dip-coating method consisting of ZnO NPs and polystyrene on the surface and its application in the separation of oil/water mixtures [43]. This fabric exhibits an excellent wettability property with a water contact angle (WCA) over 153°. In other studies, the bare fabric was dipped in TiO_2 solution and coated with stearic acid to produce a dual structure on the fabric surface [99]. Further, Li et al. also showed the preparation of superhydrophobic filtration fabric via the dip-coating method using poly(dimethyl siloxanes) (PDMS) and SiO_2 dispersion [100]. Toluene and ethanol mixture the surface roughness and induce microphase separation.

Fiber-based Materials

Porous structure and the surface-to-volume ratio of nanofibrous matrices can be easily tuned and thus can be synthesized by the electrospinning technique. Electrospun fibers can be effectively used in oil/water separation due to their nanoscale pore structure and high specific surface areas. Several studies reported on the oil/water separation property of nanofiber membranes fabricated by the electrospinning method. The carbon-SiO_2 composite nanofibrous membrane has been utilized in oil/water separation applications [101]. Its interpenetrating network improves the properties of the nanofibrous membrane with good flexibility and better mechanical properties. It also exhibits excellent thermal and chemical stability, as its properties are maintained at high temperatures. This silicone oil-coated membrane exhibits both hydrophobic and super oleophilic properties with oil CA and WCA of 0° and 144°, respectively. Further, Ma et al. also demonstrated the oil/water separation using cellulose acetate/polyimide electrospun nanofibrous membranes [102]. The mechanical strength of the membrane was enhanced by a further coating of polybenzoxazine layer containing SiO_2 NPs. The WCA and oil CA of nanofibrous membranes was reported to be 160° and 0°, respectively, with a high tensile strength of 200 MPa. Further, the pH-responsive electrospun fibrous membrane was reported to exhibit opposite behavior at acidic and basic pH, as depicted by the oil-to-water selective property at pH 1 and the opposite property at pH 14 [103]. Although nanofibrous material has a promising effect on oil/water separation, it suffers major challenges for industrial application, i.e., poor mechanical strength and fragile behavior. Further research on the analysis of separation phenomenon and control during the oil/water separation and collection process is necessary.

6 Application of 2D Material in Seawater Desalination

Currently, water scarcity is one of the most critical issues, and polymeric membranes used for desalination experience major drawbacks like less water permeability, high cost, and energy consumption [14]. To overcome these challenges, the use of 2D material is in progress with advantages like enhanced water permeability and chemical resistance. The 2D material-triggered desalination process depends mainly on two principles, i.e., Donnan electrostatic exclusion and size exclusion. Due to the electrostatic Donnan effect, hydrated ions can be blocked, and thus it plays a crucial role in selective transportation [104]. Ions can be blocked either on the nanopore's edges or on line edges and surfaces of nanosheets. Moreover, in the size exclusion mechanism, small-sized water molecules are permeated, and hydrated ions are blocked. Water desalination using 2D material can be achieved either by in-plane nanopores on single monolayers or stacked nanosheets with nanochannels in between [105].

Graphene-based membranes with well-defined nanochannels and high water permeability can be a promising tool for purification and desalination [15]. Pei et al. demonstrated the reduced graphene oxide membrane fabricated with dopamine followed by vacuum filtration [15]. In a comparison of pristine graphene oxide membranes, the synthesized membrane facilitates a high rate of water permeation with a high solute retention rate. These excellent properties may be attributed to the interaction of the graphene oxide functional group with ions or water molecules. Surwade et al. demonstrated the graphene oxide monolayer membrane via CVD on a copper foil catalyst [106]. Generated nanopores enhance the selectivity of the membrane for water molecules in comparison to dissolved ions, i.e., K^+, Na^+, Li^+, and Cl^-. Synthesized membranes exhibit a 100% salt rejection property with high water permeability.

Molybdenum disulfide (MoS_2) 2D material has been reported to gain extra advantages in comparison to other 2D materials used in water desalination. Pores of MoS_2 in the same area of other 2D materials hold more water and facilitate fast transportation. Reports also showed that the energy barrier of water molecules to pass the

MoS_2 membrane is lower than that of other used materials [107, 108]. Heiranian et al. demonstrated the fishbone, hourglass structure of Mo-only pores with hydrophobic centers and hydrophilic edges within 1 nm length facilitates water permeation in comparison to counterparts [108]. Authors reported that single-layer MoS_2 effectively rejects ions and allows water transportation at a higher rate, whereas more than 88% of ions are rejected by a membrane with pore size ranging between 20 to 60 Å. Further, Hirunpinyopas et al. demonstrated that the thin laminar membrane of 5 μm of exfoliated MoS_2 can effectively reject ~ 99% of ions found in seawater and maintain water fluxes higher (~ 5 times) in comparison to graphene oxide [109]. Further, Ma et al. demonstrated the intercalating GO and MoS_2 nanosheets for efficient water desalination [110]. Due to the strong van der Waals interactions between MoS_2 and GO and the weak interaction between MoS_2 and H_2O molecules, the MoS_2/GO filtration membrane interlayer stabilizing force can be highly enhanced. Thus, even in the absence of additional cross-linkers, the synthesized membrane is highly stable in a diverse environment (pH from 3 to 11). In a comparison of only GO membrane, MoS_2/GO hybrid membrane showed enhanced water flux with improved salt removal activity and high stability.

7 Factors Affecting Remediation of Heavy Metals, Dyes, and Oil Separation from Water

Several environmental factors, like humic acids, pH, temperature, and low molecular weight organic acids, affect the transportation, transformation, bioavailability of heavy metals to microorganisms, and valence state of heavy metals. At acidic pH, heavy metals form ionic species, and the presence of more protons can saturate the metal-binding site. The attraction between metal cations and adsorbent reduces due to the presence of hydrogen ions, which leads to increased toxicity [111]. Increased temperature also increases the adsorbate diffusion rate across the external boundary layer. The bioavailability of heavy metals also increases with increasing temperature due to high solubility [112]. However, with the temperature rise, microbial metabolism and enzyme activity also increase, which accelerates the bioremediation process [111].

Different dyes have been used in the textile industry, and thus industrial wastewater varies in composition and pH [113]. Optimization of abiotic factors also offers great advantages to developing bioreactors for the bioremediation process. At acidic or neutral pH, decolorization of dye proceeds via fungi and fungal enzymes, whereas, at alkaline pH, it takes place by bacteria and bacterial enzymes. Temperature is also an important factor for the remediation process; at high temperatures, microorganisms' activity reduces due to deactivation of enzymes and slow growth and reproduction rate. Phytoremediation of textile dyes is one of the economically efficient and environmentally friendly methods that depends on soil properties as well as the plant's absorption and degradation capacity [114, 115].

Azoreductase is an important enzyme used for the discoloration of textile dyes. It induces the reductive cleavage of azo bonds to release aromatic amines, which are further degraded into CO_2 and H_2O. Many peroxidase enzymes, such as lignin peroxidase and Mn peroxidase, are frequently used in the discoloration of dyes [116]. However, these enzymes tend to be deactivated by the inhibitors present in a polluted environment.

Several factors affect the oil/water separation, like mechanical strength and durability, as the material used can be easily damaged and hence can alter the oil/water separation performance. Pores of oil/water separators tend to clog by floating material in oceans, which also affects the performance. Further, due to the small pore size, the membrane suffers lower separation capacity, which limits the separation of oil/water. In ships, oil/water separators are designed properly to work in an ideal condition, but there are several sources to induce contamination in bilge water. The bilge water is a mixture of various grades of oil in water, with suspended solids, rust, detergents, paint chips, cargo dust, chemicals, etc.

8 Advantages of 2D Materials Over Conventional Methods

The remarkable properties of emerging 2D materials that have been discussed are advanced and promising technology over conventional methods in the treatment of wastewater, air purification, and oil separation from water sources. Technologies that are used in the conventional methods are chemical precipitation, the ion exchange method, and electrolytic degradation technology, which have several disadvantages. In chemical precipitation, huge quantities of chemicals are required to eliminate metals. During the process, a surplus amount of sludge is produced; also, the precipitation rate of metal is slow, and sludge disposal in the environment has a severe impact. The ion exchange technique is very sensitive to the pH of the system. Whereas in the electrolytic degradation method, the major restricting factor is the electrode corrosion that is applied in the system. Due to the drawback in the conventional methods, various advanced 2D materials have emerged to overcome the problem raised. However, the key point is that in 2D materials-based technologies, there is no use of extra chemicals during the whole process. Due to its ultra-thin size, enhanced charge density in TiO_2, and efficient directed interface charge separation in MoS_2, it displayed improved photocatalytic activity. The invention of 2D/2D heterostructures is advantageous to enhance the stability of photocatalysts. Adsorption-based 2D materials used in dye degradation and heavy metal reductions are inexpensive, effective, and have short treatment times and are mostly eco-friendly. Even though homogenous photo-Fenton catalysts are very potential in oxidizing organic pollutants of wastewater, there are still a few limitations. For illustration, homogenous photo-Fenton catalysts can function at pH 2.8, but at this pH, a huge amount of iron sludge is generated at the end. This creates another serious environmental issue. So, because of the narrow pH range and production of iron sludge, advanced heterogeneous photo-Fenton catalysts need to be developed that oxidize the contaminant of wastewater at neutral pH efficiently. To overcome the drawbacks, such as constant damage of catalyst material and acidic conditions in basic Fenton catalyst reactions, the application of 2D-based Fenton catalysts, as mentioned in previous segments, has solved the limitations.

2D materials TiO_2, MoS_2, and graphene oxide are efficiently applied for oil separation from the water surface and desalination of water. For the desalination of seawater, MoS_2/GO hybrid has been a highly studied membrane due to its high water permeability and enhanced salt removal. For separation of oil from the water surface, Cu mesh using graphene oxide (GO) and TiO_2-modified superhydrophobic stainless steel materials are highly effective. Owing to material stability against strong acid and base as well as reusability, long-term storage, and high mechanical strength, copper and stainless steel enhance the ability of 2D material in oil separation. Therefore, we believe that the 2D-based approach retains great promise in industries, agriculture, and households to remediate heavy metals and dyes contamination and maintain water quality and indoor/outdoor air for the community.

9 Challenges Exist in 2D Materials

Nowadays, high contamination of Pb in drinking water is of great concern, and 2D materials of high affinity to Pb are still lacking. The development of highly selective adsorbent 2D material for target contaminants is extremely important. Because the most challenging thing is the presence of non-targeted interfering contaminants that hinder effective degradation or removal of organic and inorganic contaminants. Graphene is successfully growing for purification and treatment of wastewater, but the aggregation nature of graphene hampers the interaction of contaminants on its surface, consequently limiting the purification process. To overcome this limitation, various functional groups are attached to the surface of graphene to enhance the surface area. Moreover, the synthesis of graphene and its derivatives at an industrial scale is still challenging. However, the high cost and complex production process of 2D material are still challenging in a commercial application. At a large scale, the complete separation of nanosized adsorbent MoS_2 is a great concern. The sustainable and renewable technologies photocatalysis and electrocatalysis are still challenging for stable catalysts and high efficiency. Although enormous research is going on to enhance its efficiency, the stability of materials, its fabrication, fundamental research, and industrial applications remain a challenge for most 2D materials used for the purification and separation of water, air, oil, and seawater desalination from the environment.

10 Summary

In a recent situation, there is an urgent requirement for advanced technologies that remove all types of pollutants present in water and air. The development of emerging technologies is also the priority for water/oil separation and desalination, which ensures the overcoming of water scarcity. In this respect, 2D materials have appeared to be only an alternative strategy for remediation of environmental pollutants like heavy metals and dyes from the different ecosystems. 2D materials, including graphene, boron nitride, molybdenum disulfide, and black phosphorus, have been discussed to be effectively engaged for remediation purposes. Textiles industries, mining metals, leaching of metals, and paper and pulp industries are the source of several types of heavy metals and dyes. Graphene and its derivatives are adsorption-based 2D materials that enhance the remediation of heavy metals and dyes, which depends upon the pH, ionic strength, temperature, and contact time. Adsorption-based MoS_2 nanosheets and layer-stacked membranes are efficient for heavy metal remediation. The mechanism of adsorption-based MoS_2 nanosheets showed how interaction occurs between materials and pollutants. Further, catalyst-based 2D materials, TiO_2, MoS_2, BP, GO, and rGO perform effectively, resulting in heavy metals and dyes remediation. Photocatalysis, electrocatalysis, and Fenton-based catalysts are different catalyst-based techniques in which 2D materials are employed for environmental remediation purposes. TiO_2 acts as an efficient photocatalyst for the purification of indoor and outdoor air. For the separation of oil/water mixed polluted water, mesh-based, fabric, and fiber-based 2D materials have proven to be highly effective in remediation.

However, several factors like temperature, pH, concentration, other bonded substrates, mechanical strength, etc., limit the remediation of metal dyes present in water and air particles. MoS_2, TiO_2, 2D/2D heterostructures, and homogenous photo-Fenton catalysts have enhanced the photocatalytic activity during remediation; hence, they are advantageous over conventional methods. Still, various challenges exist with 2D materials that need to be solved for a green environment.

References

1. I. Ahmed, S.H. Jhung, Covalent organic framework-based materials: synthesis, modification, and application in environmental remediation. Coord. Chem. Rev. **441**, 213989 (2021)
2. M. Berradi, R. Hsissou, M. Khudhair, M. Assouag, O. Cherkaoui, A. El Bachiri, A. El Harfi, Textile finishing dyes and their impact on aquatic environs. Heliyon **5**(11), e02711 (2019)
3. M. Tumolo, V. Ancona, D. De Paola, D. Losacco, C. Campanale, C. Massarelli, V.F. Uricchio, Chromium pollution in European water, sources, health risk, and remediation strategies: an overview. Int. J. Environ. Res. Public Health **17**(15), 5438 (2020)
4. A. Aghighi, F. Haghighat, Evaluation of nano-titanium dioxide (TiO2) catalysts for ultraviolet photocatalytic oxidation air cleaning devices. J. Environ. Chem. Eng. **3**(3), 1622–1629 (2015)
5. J.P. Brincat, D. Sardella, A. Muscat, S. Decelis, J.N. Grima, V. Valdramidis, R. Gatt, A review of the state-of-the-art in air filtration technologies as may be applied to cold storage warehouses. Trends Food Sci. Technol. **50**, 175–185 (2016)

6. X. Zhou, Z. Zhang, X. Xu, F. Guo, X. Zhu, X. Men, B. Ge, Robust and durable superhydrophobic cotton fabrics for oil/water separation. ACS Appl. Mater. Interfaces **5**, 7208–7214 (2013)
7. W. Meng, P. Li, Y. Lan, X. Shi, S. Peng, Q. Hongqiang, X. Jianzhong, Green fabrication of superhydrophilic and underwater superoleophobic coatings with applications in oil-water separation, photocatalysis and fire-retardance. Sep. Purif. Technol. **233**, 115988 (2020)
8. J. Yuan, C. Cui, B. Qi, J. Wei, M.A. Qaisrani, Study on oil-water separation of selective-wettability meshes with different micro/nano structures. Colloids Surf. A **584**, 124026 (2020)
9. M. Zhu, Y. Liu, M. Chen, X. Zhiheng, L. Li, Y. Zhou, Metal mesh-based special wettability materials for oil-water separation: a review of the recent development. J. Petrol. Sci. Eng. **205**, 108889 (2021)
10. B. Tansel, Significance of thermodynamic and physical characteristics on permeation of ions during membrane separation: hydrated radius, hydration free energy and viscous effects. Sep. Purif. Technol. **86**, 119–126 (2012)
11. W.J. Koros, C. Zhang, Materials for next-generation molecularly selective synthetic membranes. Nat. Mater. **16**(3), 289–297 (2017)
12. P. Marchetti, M.F. Jimenez Solomon, G. Szekely, A.G. Livingston, Molecular separation with organic solvent nanofiltration: a critical review. Chem. Rev. **114**(21), 10735–10806 (2014)
13. D.H. Seo, S. Pineda, Y.C. Woo, M. Xie, A.T. Murdock, E.Y. Ang, Y. Jiao, M.J. Park, S.I. Lim, M. Lawn, F.F. Borghi, Anti-fouling graphene-based membranes for effective water desalination. Nat. Commun. **9**(1), 1–12 (2018)
14. J. Safaei, P. Xiong, G. Wang, Progress and prospects of two-dimensional materials for membrane-based water desalination. Mater. Today Adv. **8**, 100108 (2020)
15. J. Pei, L. Xiantao Zhang, H.J. Huang, H. Xuejiao, Fabrication of reduced graphene oxide membranes for highly efficient water desalination. RSC Adv. **6**, 101948–101952 (2016)
16. E.D. Koutsouroubi, I. Vamvasakis, M.G. Minotaki, I.T. Papadas, C. Drivas, S.A. Choulis, G. Kopidakis, S. Kennou, G.S. Armatas, Ni-doped MoS2 modified graphitic carbon nitride layered hetero-nanostructures as highly efficient photocatalysts for environmental remediation. Appl. Catal. B: Environ. **297**, 120419 (2021)
17. S.K. Tiwari, S. Sahoo, N. Wang, A. Huczko, Graphene research and their outputs: status and prospect. J. Sci. Adv. Mater. Dev. **5**(1), 10–29 (2020)
18. G. Liu, Z. Xiong, L. Yang, H. Shi, D. Fang, M. Wang, P. Shao, X. Luo, Electrochemical approach toward reduced graphene oxide-based electrodes for environmental applications: a review. Sci. Total Environ. 146301 (2021)
19. S.J. Lee, J. Theerthagiri, P. Nithyadharseni, P. Arunachalam, D. Balaji, A.M. Kumar, J. Madhavan, V. Mittal, M.Y. Choi, Heteroatom-doped graphene-based materials for sustainable energy applications: a review. Renew. Sustain. Energy Rev. **143**, 110849 (2021)
20. T. Huang, X. Tang, K. Luo, Y. Wu, X. Hou, S. Tang, An overview of graphene-based nanoadsorbent materials for environmental contaminants detection. TrAC Trends Anal. Chem. 116255 (2021)
21. Q. Weng, X. Wang, X. Wang, Y. Bando, D. Golberg, Functionalized hexagonal boron nitride nanomaterials: emerging properties and applications. Chem. Soc. Rev. **45**(14), 3989–4012 (2016)
22. C. Xing, J. Zhang, J. Jing, J. Li, F. Shi, Preparations, properties and applications of low-dimensional black phosphorus. Chem. Eng. J. **370**, 120–135 (2019)
23. S. Wu, K.S. Hui, K.N. Hui, 2D black phosphorus: from preparation to applications for electrochemical energy storage. Adv. Sci. **5**(5), 1700491 (2018)
24. V. Sharma, H.L. Kagdada, P.K. Jha, P. Śpiewak, K.J. Kurzydłowski, Thermal transport properties of boron nitride based materials: a review. Renew. Sustain. Energ. Rev. **120**, 109622 (2020)
25. C. Liu, Q. Wang, F. Jia, S. Song, Adsorption of heavy metals on molybdenum disulfide in water: a critical review. J. Mol. Liq. **292**, 111390 (2019)
26. J. Huo, R. Ge, Y. Liu, J. Guo, L. Lu, W. Chen, C. Liu, H. Gao, H. Liu, Recent advances of two–dimensional molybdenum disulfide based materials: synthesis, modification and applications in energy conversion and storage. Sustain. Mater. Technol. **24**, e00161 (2020)

27. J. Briffa, E. Sinagra, R. Blundell, Heavy metal pollution in the environment and their toxicological effects on humans. Heliyon **6**(9), e04691 (2020)
28. P.B. Tchounwou, C.G. Yedjou, A.K. Patlolla, D.J. Sutton, Heavy metal toxicity and the environment. *Molecular, Clinical and Environmental Toxicology* (2012), pp. 133–164
29. V. Masindi, K.L. Muedi, Environmental contamination by heavy metals. Heavy Metals **10**, 115–132 (2018)
30. C.H. Walker, R.M. Sibly, D.B. Peakall, *Principles of Ecotoxicology* (CRC Press, 2005)
31. D. Kołodyńska, Chelating agents of a new generation as an alternative to conventional chelators for heavy metal ions removal from different waste waters, in *Expanding Issues in Desalination* (InTech Publishers, 2011), pp. 339–371
32. T. Setiadi, Y. Andriani, M. Erlania, Treatment of textile wastewater by a combination of anaerobic and aerobic processes: a denim processing plant case. Southeast Asian Water Environ. **1** (2006)
33. X. Wang, X. Gu, D. Lin, F. Dong, X. Wan, Treatment of acid rose dye containing wastewater by ozonizing–biological aerated filter. Dyes Pigm. **74**(3), 736–740 (2007)
34. M. Imran, D.E. Crowley, A. Khalid, S. Hussain, M.W. Mumtaz, M. Arshad, Microbial biotechnology for decolorization of textile wastewaters. Rev. Environ. Sci. Bio/Technol. **14**(1), 73–92 (2015)
35. M.M. Hassan, C.M. Carr, A critical review on recent advancements of the removal of reactive dyes from dyehouse effluent by ion-exchange adsorbents. Chemosphere **209**, 201–219 (2018)
36. G. Ersan, O.G. Apul, F. Perreault, T. Karanfil, Adsorption of organic contaminants by graphene nanosheets: a review. Water Res. **126**, 385–398 (2017)
37. Z.B. Khalid,, M. Nasrullah, A. Nayeem, Z.A. Wahid, L. Singh, S. Krishnan, Application of 2D graphene-based nanomaterials for pollutant removal from advanced water and wastewater treatment processes, in *Adapting 2D Nanomaterials for Advanced Applications* (American Chemical Society, 2020), pp. 191–217
38. Z.A. Alothman, A.Y. Badjah, I. Ali, Facile synthesis and characterization of multi walled carbon nanotubes for fast and effective removal of 4-tert-octylphenol endocrine disruptor in water. J. Mol. Liq. **275**, 41–48 (2019)
39. A.A. Basheer, I. Ali, Stereoselective uptake and degradation of (±)-o, p-DDD pesticide stereomers in water-sediment system. Chirality **30**(9), 1088–1095 (2018)
40. H.Y. Mao, Y.H. Lu, J.D. Lin, S. Zhong, A.T.S. Wee, W. Chen, Manipulating the electronic and chemical properties of graphene via molecular functionalization. Prog. Surf. Sci. **88**(2), 132–159 (2013)
41. H.L. Ma, Y. Zhang, Q.H. Hu, D. Yan, Z.Z. Yu, M. Zhai, Chemical reduction and removal of Cr (VI) from acidic aqueous solution by ethylenediamine-reduced graphene oxide. J. Mater. Chem. **22**(13), 5914–5916 (2012)
42. D.R. Dreyer, A.D. Todd, C.W. Bielawski, Harnessing the chemistry of graphene oxide. Chem. Soc. Rev. **43**(15), 5288–5301 (2014)
43. M. Zhang, C. Wang, S. Wang, J. Li, Fabrication of superhydrophobic cotton textiles for water–oil separation based on drop-coating route. Carbohyd. Polym. **97**, 59–64 (2013)
44. J.Y. Lim, N.M. Mubarak, E.C. Abdullah, S. Nizamuddin, M. Khalid, Recent trends in the synthesis of graphene and graphene oxide based nanomaterials for removal of heavy metals—a review. J. Ind. Eng. Chem. **66**, 29–44 (2018)
45. X. Wang, Z. Chen, S. Yang, Application of graphene oxides for the removal of Pb (II) ions from aqueous solutions: experimental and DFT calculation. J. Mol. Liq. **211**, 957–964 (2015)
46. D. Gu, J.B. Fein, Adsorption of metals onto graphene oxide: Surface complexation modeling and linear free energy relationships. Colloids Surf., A **481**, 319–327 (2015)
47. Ö. Gerçel, H.F. Gerçel, Adsorption of lead (II) ions from aqueous solutions by activated carbon prepared from biomass plant material of Euphorbia rigida. Chem. Eng. J. **132**(1–3), 289–297 (2007)
48. A. Król, K. Mizerna, M. Bożym, An assessment of pH-dependent release and mobility of heavy metals from metallurgical slag. J. Hazard. Mater. **384**, 121502 (2020)

49. H. Jabeen, V. Chandra, S. Jung, J.W. Lee, K.S. Kim, S.B. Kim, Enhanced Cr(vi) removal using iron nanoparticle decorated graphene. Nanoscale **3**, 3583–3585 (2011)
50. G. Zhao, X. Ren, X. Gao, X. Tan, J. Li, C. Chen, Y. Huang, X. Wang, Removal of Pb (II) ions from aqueous solutions on few-layered graphene oxide nanosheets. Dalton Trans. **40**(41), 10945–10952 (2011)
51. X. Li, F. Li, L. Fang, Effect of *Paecilomyces cateniannulatus* on the adsorption of nickel onto graphene oxide. Korean J. Chem. Eng. **32**(12), 2449–2455 (2015)
52. A. Tadjarodi, S.M. Ferdowsi, R. Zare-Dorabei, A. Barzin, Highly efficient ultrasonic-assisted removal of Hg (II) ions on graphene oxide modified with 2-pyridinecarboxaldehyde thiosemicarbazone: adsorption isotherms and kinetics studies. Ultrason. Sonochem. **33**, 118–128 (2016)
53. G. Bharath, E. Alhseinat, N. Ponpandian, M.A. Khan, M.R. Siddiqui, F. Ahmed, E.H. Alsharaeh, Development of adsorption and electrosorption techniques for removal of organic and inorganic pollutants from wastewater using novel magnetite/porous graphene-based nanocomposites. Sep. Purif. Technol. **188**, 206–218 (2017)
54. I. Ali, X.Y. Mbianda, A. Burakov, E. Galunin, I. Burakova, E. Mkrtchyan, A. Tkachev, V. Grachev, Graphene based adsorbents for remediation of noxious pollutants from wastewater. Environ. Int. **127**, 160–180 (2019)
55. M. Chhowalla, H.S. Shin, G. Eda, L.J. Li, K.P. Loh, H. Zhang, The chemistry of two-dimensional layered transition metal dichalcogenide nanosheets. Nat. Chem. **5**(4), 263–275 (2013)
56. K. Ai, C. Ruan, M. Shen, L. Lu, MoS_2 nanosheets with widened interlayer spacing for high-efficiency removal of mercury in aquatic systems. Adv. Func. Mater. **26**(30), 5542–5549 (2016)
57. M.J. Aghagoli, M.H. Beyki, F. Shemirani, Application of dahlia-like molybdenum disulfide nanosheets for solid phase extraction of Co (II) in vegetable and water samples. Food Chem. **223**, 8–15 (2017)
58. S. Tong, H. Deng, L. Wang, T. Huang, S. Liu, J. Wang, Multi-functional nanohybrid of ultrathin molybdenum disulfide nanosheets decorated with cerium oxide nanoparticles for preferential uptake of lead (II) ions. Chem. Eng. J. **335**, 22–31 (2018)
59. F. Jia, X. Zhang, S. Song, AFM study on the adsorption of Hg^{2+} on natural molybdenum disulfide in aqueous solutions. Phys. Chem. Chem. Phys. **19**(5), 3837–3844 (2017)
60. A.E. Gash, A.L. Spain, L.M. Dysleski, C.J. Flaschenriem, A. Kalaveshi, P.K. Dorhout, S.H. Strauss, Efficient recovery of elemental mercury from Hg (II)-contaminated aqueous media using a redox-recyclable ion-exchange material. Environ. Sci. Technol. **32**(7), 1007–1012 (1998)
61. C.-F. Wang, W.-N. Wang, S.-W. Kuo, Y.-W. Chiang, J.-H. Hung, K.-J. Lee, Biocompatible meshes with appropriate wettabilities for underwater oil transportation/collection and highly effective oil/water separation. Langmuir **34**, 11442–11448 (2018)
62. Z. Wang, A. Sim, J.J. Urban, B. Mi, Removal and recovery of heavy metal ions by two-dimensional MoS_2 nanosheets: performance and mechanisms. Environ. Sci. Technol. **52**(17), 9741–9748 (2018)
63. Y. Nosaka, A.Y. Nosaka, Generation and detection of reactive oxygen species in photocatalysis. Chem. Rev. **117**(17), 11302–11336 (2017)
64. K. Vignesh, A. Suganthi, B.K. Min, M. Kang, Photocatalytic activity of magnetically recoverable $MnFe_2O_4$/g-C_3N_4/TiO_2 nanocomposite under simulated solar light irradiation. J. Mol. Catal. A: Chem. **395**, 373–383 (2014)
65. G. Chandrabose, A. Dey, S.S. Gaur, S. Pitchaimuthu, H. Jagadeesan, N.S.J. Braithwaite, V. Selvaraj, V. Kumar, S. Krishnamurthy, Removal and degradation of mixed dye pollutants by integrated adsorption-photocatalysis technique using 2-D MoS_2/TiO_2 nanocomposite. Chemosphere **279**, 130467 (2021)
66. S. Nazim, T. Kousar, M. Shahid, M.A. Khan, G. Nasar, M. Sher, M.F. Warsi, New graphene-CoxZn1− xFe_2O_4 nano-heterostructures: magnetically separable visible light photocatalytic materials. Ceram. Int. **42**(6), 7647–7654 (2016)

67. C. Zhang, T. Zhang, J. Huang, T. Yan, C. Li, L. Liu, L. Wang, F. Jiao, Copper hydroxyphosphate nanosheets-covered robust membranes with superhydrophilicity and underwater ultralow adhesive superoleophobicity for oil/water separation and visible light photodegradation. Colloids Surf., A **584**, 124000 (2020)
68. S. Zhang, B. Li, X. Wang, G. Zhao, B. Hu, Z. Lu, T. Wen, J. Chen, X. Wang, Recent developments of two-dimensional graphene-based composites in visible-light photocatalysis for eliminating persistent organic pollutants from wastewater. Chem. Eng. J. **390**, 124642 (2020)
69. P. Nakhanivej, X. Yu, S.K. Park, S. Kim, J.Y. Hong, H.J. Kim, W. Lee, J.Y. Hwang, J.E. Yang, C. Wolverton, J. Kong, Revealing molecular-level surface redox sites of controllably oxidized black phosphorus nanosheets. Nat. Mater. **18**(2), 156–162 (2019)
70. X. Zhu, S. Huang, Q. Yu, Y. She, J. Yang, G. Zhou, Q. Li, X. She, J. Deng, H. Li, H. Xu, In-situ hydroxyl modification of monolayer black phosphorus for stable photocatalytic carbon dioxide conversion. Appl. Catal. B **269**, 118760 (2020)
71. K. Ge, Y. Zhang, D. Wang, Z. Li, J. He, C. Fu, Y. Yang, M. Pan, L. Zhu, Highly stable reduced graphene oxide wrapped black phosphorus heterostructure with superior photocatalytic performance under visible light. ACS Appl. Mater. Interfaces. **12**(17), 20035–20043 (2020)
72. M. Shi, S. Huang, N. Dong, Z. Liu, F. Gan, J. Wang, Y. Chen, Donor–acceptor type blends composed of black phosphorus and C 60 for solid-state optical limiters. Chem. Commun. **54**(4), 366–369 (2018)
73. J. Yang, R. Jing, P. Wang, D. Liang, H. Huang, C. Xia, Q. Zhang, A. Liu, Z. Meng, Y. Liu, Black phosphorus nanosheets and ZnAl-LDH nanocomposite as environmental-friendly photocatalysts for the degradation of Methylene blue under visible light irradiation. Appl. Clay Sci. **200**, 105902 (2021)
74. M.P. Levendorf, C.J. Kim, L. Brown, P.Y. Huang, R.W. Havener, D.A. Muller, J. Park, Graphene and boron nitride lateral heterostructures for atomically thin circuitry. Nature **488**(7413), 627–632 (2012)
75. G. Zhang, Y. Zhou, F. Yang, FeOOH-catalyzed heterogeneous electro-Fenton system upon anthraquinone@ graphene nanohybrid cathode in a divided electrolytic cell: catholyte-regulated catalytic oxidation performance and mechanism. J. Electrochem. Soc. **162**(6), H357 (2015)
76. D. Wang, X. Li, J. Chen, X. Tao, Enhanced photoelectrocatalytic activity of reduced graphene oxide/TiO_2 composite films for dye degradation. Chem. Eng. J. **198**, 547–554 (2012)
77. L. Clarizia, D. Russo, I. Di Somma, R. Marotta, R. Andreozzi, Homogeneous photo-Fenton processes at near neutral pH: a review. Appl. Catal. B **209**, 358–371 (2017)
78. A.N. Soon, B.H. Hameed, Heterogeneous catalytic treatment of synthetic dyes in aqueous media using Fenton and photo-assisted Fenton process. Desalination **269**(1–3), 1–16 (2011)
79. Z. Miao, X. Gu, S. Lu, M.L. Brusseau, N. Yan, Z. Qiu, Q. Sui, Enhancement effects of reducing agents on the degradation of tetrachloroethene in the Fe (II)/Fe (III) catalyzed percarbonate system. J. Hazard. Mater. **300**, 530–537 (2015)
80. C. Ma, S. Feng, J. Zhou, R. Chen, Y. Wei, H. Liu, S. Wang, Enhancement of H_2O_2 decomposition efficiency by the co-catalytic effect of iron phosphide on the Fenton reaction for the degradation of methylene blue. Appl. Catal. B **259**, 118015 (2019)
81. H. Wang, F. Wen, X. Li, X. Gan, Y. Yang, P. Chen, Y. Zhang, Cerium-doped MoS2 nanostructures: Efficient visible photocatalysis for Cr (VI) removal. Sep. Purif. Technol. **170**, 190–198 (2016)
82. P. Wang, L. Wang, Q. Sun, S. Qiu, Y. Liu, X. Zhang, X. Liu, L. Zheng, Preparation and performance of Fe_3O_4@ hydrophilic graphene composites with excellent Photo-Fenton activity for photocatalysis. Mater. Lett. **183**, 61–64 (2016)
83. X. Ji, Z. Han, J. Li, Y. Deng, X. Han, J. Zhao, X. Zhao, C. Chen, MoSx co-catalytic activation of H_2O_2 by heterogeneous hemin catalyst under visible light irradiation. J. Colloid Interface Sci. **557**, 301–310 (2019)
84. C. Liu, B. Yang, J. Chen, F. Jia, S. Song, Synergetic degradation of methylene blue through photocatalysis and Fenton reaction on two-dimensional molybdenite-Fe. J. Environ. Sci. **111**, 11–23 (2022)

85. S. Su, Y. Liu, X. Liu, W. Jin, Y. Zhao, Transformation pathway and degradation mechanism of methylene blue through β-FeOOH@ GO catalyzed photo-Fenton-like system. Chemosphere **218**, 83–92 (2019)
86. V. Després, J.A. Huffman, S.M. Burrows, C. Hoose, A. Safatov, G. Buryak, J. Fröhlich-Nowoisky, W. Elbert, M. Andreae, U. Pöschl, R. Jaenicke, Primary biological aerosol particles in the atmosphere: a review. Tellus B: Chem. Phys. Meteorol. **64**(1), 15598 (2012)
87. G. Liu, M. Xiao, X. Zhang, C. Gal, X. Chen, L. Liu, S. Pan, J. Wu, L. Tang, D. Clements-Croome, A review of air filtration technologies for sustainable and healthy building ventilation. Sustain. Cities Soc. **32**, 375–396 (2017)
88. A.E. Setekleiv, H.F. Svendsen, Operation and dynamic behavior of wire mesh pads. Chem. Eng. Sci. **68**(1), 624–639 (2012)
89. A. Khazraei Vizhemehr, F. Haghighat, C.S. Lee, H. Kholafaei, N. Lakdawala, Evaluation of gas-phase filter performance for a gas mixture. CLEAN–Soil Air Water **43**(4), 469–478 (2015)
90. H.J. Lee, W.S. Choi, 2D and 3D bulk materials for environmental remediation: air filtration and oil/water separation. Materials (Basel) **13** (2020)
91. F. Xia, L. Jiang, Bio-inspired, smart, multiscale interfacial materials. Adv. Mater. **20**, 2842–2858 (2008)
92. J. Song, S. Huang, Y. Lu, X. Bu, J.E. Mates, A. Ghosh, R. Ganguly, C.J. Carmalt, I.P. Parkin, W. Xu, C.M. Megaridis, Self-driven one-step oil removal from oil spill on water via selective-wettability steel mesh. ACS Appl. Mater. Interfaces **6**, 19858–19865 (2014)
93. Z. Xue, S. Wang, L. Lin, L. Chen, M. Liu, L. Feng, L. Jiang, A novel superhydrophilic and underwater superoleophobic hydrogel-coated mesh for oil/water separation. Adv. Mater. **23**, 4270–4273 (2011)
94. S. Lei, X. Fang, F. Wang, M. Xue, O. Junfei, C. Li, W. Li, A facile route to fabricate superhydrophobic Cu_2O surface for efficient oil–water separation. Coatings **9**, 659 (2019)
95. C. Chen, B. Chen, Graphene oxide coated meshes with stable underwater superoleophobicity and anti-oil-fouling property for highly efficient oil/water separation. Sci. Total Environ. **696**, 133777 (2019)
96. G. Zhang, X. Jia, J. Xing, S. Shen, X. Zhou, J. Yang, Y. Guo, R. Bai, A facile and fast approach to coat various substrates with poly (styrene-co-maleic anhydride) and polyethyleneimine for oil/water separation. Ind. Eng. Chem. Res. **58**, 19475–19485 (2019)
97. J. Du, C. Zhou, L. Chen, J. Cheng, P. Pi, J. Zuo, W. Shen, S. Jin, L. Tan, L. Dong, Gate-embedding strategy for pore size manipulation on stainless steel mesh: toward highly efficient water-in-oil nanoemulsions separation. Ind. Eng. Chem. Res. **58**, 15288–15296 (2019)
98. C. Cao, M. Ge, J. Huang, S. Li, S. Deng, S. Zhang, Z. Chen, K. Zhang, S.S. Al-Deyab, Y. Lai, Robust fluorine-free superhydrophobic PDMS–ormosil@fabrics for highly effective self-cleaning and efficient oil–water separation. J. Mater. Chem. A **4**, 12179–12187 (2016)
99. X. Zhang, T. Geng, Y. Guo, Z. Zhang, P. Zhang, Facile fabrication of stable superhydrophobic SiO_2/polystyrene coating and separation of liquids with different surface tension. Chem. Eng. J. **231**, 414–419 (2013)
100. K. Li, X. Zeng, H. Li, X. Lai, H. Xie, Facile fabrication of superhydrophobic filtration fabric with honeycomb structures for the separation of water and oil. Mater. Lett. **120**, 255–258 (2014)
101. M.H. Tai, P. Gao, B.Y. Tan, D.D. Sun, J.O. Leckie, Highly efficient and flexible electrospun carbon-silica nanofibrous membrane for ultrafast gravity-driven oil-water separation. ACS Appl. Mater. Interfaces **6**, 9393–9401 (2014)
102. W. Ma, Q. Zhang, S.K. Samal, F. Wang, B. Gao, H. Pan, X. Haijun, J. Yao, X. Zhan, S.C. De Smedt, C. Huang, Core–sheath structured electrospun nanofibrous membranes for oil–water separation. RSC Adv. **6**, 41861–41870 (2016)
103. C.-T. Liu, Y.-L. Liu, pH-induced switches of the oil- and water selectivity of crosslinked polymeric membranes for gravity-driven oil–water separation. J. Mater. Chem. A **4**, 13543–48 (2016)

104. M. Zhang, K. Guan, Y. Ji, G. Liu, W. Jin, X. Nanping, Controllable ion transport by surface-charged graphene oxide membrane. Nat. Commun. **10**, 1253 (2019)
105. Y. You, V. Sahajwalla, M. Yoshimura, R.K. Joshi, Graphene and graphene oxide for desalination. Nanoscale **8**, 117–119 (2016)
106. S.P. Surwade, S.N. Smirnov, I.V. Vlassiouk, R.R. Unocic, G.M. Veith, S. Dai, S.M. Mahurin, Water desalination using nanoporous single-layer graphene. Nat. Nanotechnol. **10**, 459–464 (2015)
107. Z. Cao, V. Liu, A.B. Farimani, Why is single-layer MoS2 a more energy efficient membrane for water desalination? ACS Energy Lett. **5**, 2217–2222 (2020)
108. M. Heiranian, A.B. Farimani, N.R. Aluru, Water desalination with a single-layer MoS_2 nanopore. Nat. Commun. **6**, 8616 (2015)
109. W. Hirunpinyopas, E. Prestat, S.D. Worrall, S.J. Haigh, R.A.W. Dryfe, M.A. Bissett, Desalination and nanofiltration through functionalized laminar MoS2 membranes. ACS Nano **11**, 11082–11090 (2017)
110. J. Ma, X. Tang, Y. He, Y. Fan, J. Chen, H. Yu, Robust stable MoS2/GO filtration membrane for effective removal of dyes and salts from water with enhanced permeability. Desalination **480**, 114328 (2020)
111. B.E. Igiri, S.I.R. Okoduwa, G.O. Idoko, E.P. Akabuogu, A.O. Adeyi, I.K. Ejiogu, Toxicity and bioremediation of heavy metals contaminated ecosystem from tannery wastewater: a review. J. Toxicol. **2018**, 2568038 (2018)
112. B.A. Bandowe, M.B. Musa, L. Boamah, E. Nyarko, F.K. Saalia, W. Wilcke, Polycyclic aromatic compounds (PAHs and oxygenated PAHs) and trace metals in fish species from Ghana (West Africa): bioaccumulation and health risk assessment. Environ. Int. **65**, 135–146 (2014)
113. L. Pereira, M. Alves. Dyes—environmental impact and remediation, in, *Environmental Protection Strategies for Sustainable Development* (Springer, 2012)
114. N.R. Rane, V.V. Chandanshive, A.D. Watharkar, R.V. Khandare, T.S. Patil, P.K. Pawar, S.P. Govindwar, Phytoremediation of sulfonated Remazol Red dye and textile effluents by Alternanthera philoxeroides: an anatomical, enzymatic and pilot scale study. Water Res. **83**, 271–281 (2015)
115. U. Tahir, S. Sohail, U.H. Khan, Concurrent uptake and metabolism of dyestuffs through bio-assisted phytoremediation: a symbiotic approach. Environ. Sci. Pollut. Res. **24**, 22914–22931 (2017)
116. B. Lellis, C.Z. Fávaro-Polonio, J.A. Pamphile, J.C. Polonio, Effects of textile dyes on health and the environment and bioremediation potential of living organisms. Biotech. Res. Innovat. **3**, 275–290 (2019)

MIX
Papier aus verantwortungsvollen Quellen
Paper from responsible sources
FSC® C105338

If you have any concerns about our products,
you can contact us on
ProductSafety@springernature.com

In case Publisher is established outside the EU,
the EU authorized representative is:
Springer Nature Customer Service Center GmbH
Europaplatz 3, 69115 Heidelberg, Germany

Printed by Libri Plureos GmbH
in Hamburg, Germany